Advances in improving animal feed safety

It is widely recognised that agriculture is a significant contributor to global warming and climate change. Agriculture needs to reduce its environmental impact and adapt to current climate change whilst still feeding a growing population, i.e. become more 'climate-smart'. Burleigh Dodds Science Publishing is playing its part in achieving this by bringing together key research on making the production of the world's most important crops and livestock products more sustainable.

Based on extensive research, our publications specifically target the challenge of climate-smart agriculture. In this way we are using 'smart publishing' to help achieve climate-smart agriculture.

Burleigh Dodds Science Publishing is an independent and innovative publisher delivering high quality customer-focused agricultural science content in both print and online formats for the academic and research communities. Our aim is to build a foundation of knowledge on which researchers can build to meet the challenge of climate-smart agriculture.

For more information about Burleigh Dodds Science Publishing simply call us on +44 (0) 1223 839365, email info@bdspublishing.com or alternatively please visit our website at www.bdspublishing.com.

Related titles:

Advances in postharvest management of cereals and grains
Print (ISBN 978-1-78676-352-5); Online (ISBN 978-1-78676-354-9, 978-1-78676-355-6)

Developing alternative animal feed ingredients
Print (ISBN 978-1-83545-210-3); Online (ISBN 978-1-83545-211-0, 978-1-83545-212-7)

Developing animal feed products
Print (ISBN 978-1-78676-463-8); Online (ISBN 978-1-78676-465-2, 978-1-78676-466-9)

Chapters are available individually from our online bookshop:
https://www.bdspublishing.com/webshop/

climate•SMART•publishing
IN AGRICULTURAL SCIENCE

BURLEIGH DODDS SERIES IN AGRICULTURAL SCIENCE
NUMBER 185

Advances in improving animal feed safety

Edited by Dr Regiane R. Santos, Schothorst Feed Research, The Netherlands

burleigh dodds
SCIENCE PUBLISHING

Published by Burleigh Dodds Science Publishing Limited
82 High Street, Sawston, Cambridge CB22 3HJ, UK
www.bdspublishing.com

Burleigh Dodds Science Publishing, 1518 Walnut Street, Suite 900, Philadelphia, PA 19102-3406, USA

First published 2026 by Burleigh Dodds Science Publishing Limited

Library of Congress Control Number: 2026942316

British Library Cataloguing in Publication Data
A catalogue record for this book is available from the British Library

ISBN 978-1-83545-311-7 (print)
ISBN 978-1-83545-312-4 (ePub)
ISBN 978-1-83545-313-1 (PDF)
ISSN 2059-6936 (print)
ISSN 2059-6944 (online)

DOI: 10.19103/AS.2026.0172

Typeset by Deanta Global Publishing Services, Dublin, Ireland

Contents

Series list

Title Series number

Achieving sustainable cultivation of maize - Vol 1 001
From improved varieties to local applications
Edited by: Dr Dave Watson, CGIAR Maize Research Program Manager, CIMMYT, Mexico

Achieving sustainable cultivation of maize - Vol 2 002
Cultivation techniques, pest and disease control
Edited by: Dr Dave Watson, CGIAR Maize Research Program Manager, CIMMYT, Mexico

Achieving sustainable cultivation of rice - Vol 1 003
Breeding for higher yield and quality
Edited by: Prof. Takuji Sasaki, Tokyo University of Agriculture, Japan

Achieving sustainable cultivation of rice - Vol 2 004
Cultivation, pest and disease management
Edited by: Prof. Takuji Sasaki, Tokyo University of Agriculture, Japan

Achieving sustainable cultivation of wheat - Vol 1 005
Breeding, quality traits, pests and diseases
Edited by: Prof. Peter Langridge, University of Adelaide, Australia

Achieving sustainable cultivation of wheat - Vol 2 006
Cultivation techniques
Edited by: Prof. Peter Langridge, University of Adelaide, Australia

Achieving sustainable cultivation of tomatoes 007
Edited by: Dr Autar Mattoo, USDA-ARS, USA and Prof. Avtar Handa, Purdue University, USA

Achieving sustainable production of milk - Vol 1 008
Milk composition, genetics and breeding
Edited by: Dr Nico van Belzen, International Dairy Federation (IDF), Belgium

Achieving sustainable production of milk - Vol 2 009
Safety, quality and sustainability
Edited by: Dr Nico van Belzen, International Dairy Federation (IDF), Belgium

Achieving sustainable production of milk - Vol 3 010
Dairy herd management and welfare
Edited by: Emeritus Prof. John Webster, University of Bristol, UK

Ensuring safety and quality in the production of beef - Vol 1 011
Safety
Edited by: Prof. Gary Acuff, Texas A&M University, USA and Prof. James Dickson, Iowa State University, USA

Ensuring safety and quality in the production of beef - Vol 2 012
Quality
Edited by: Emeritus Prof. Michael Dikeman, Kansas State University, USA

Achieving sustainable production of poultry meat - Vol 1 013
Safety, quality and sustainability
Edited by: Prof. Steven C. Ricke, University of Arkansas, USA

Achieving sustainable production of poultry meat - Vol 2 014
Breeding and nutrition
Edited by: Prof. Todd Applegate, University of Georgia, USA

Introduction

Animal feed products must be safe for animal consumption and need to be free from pathogens, chemicals and toxins that would be harmful if consumed by the animal. In order to address these challenges, the animal feed sector must find new and innovative ways to improve animal feed product safety in order to detect and control contamination of feed products.

This collection provides an overview of animal feed supply chains and highlights key microbiological risks affecting animal feed, such as bacterial and viral pathogens and other contaminant risks such as mycotoxins and environmental contaminants. The book also covers technologies to improve batch control, specifically techniques for detecting and analysing key contaminant risks and also micro-scaling technologies for more precise feed control.

Part 1 Microbiological risks

Part 1 begins by examining the microbiological risks of animal feed products and their effects, specifically bacterial and viral contamination.

The first chapter focuses on foodborne bacterial pathogen contamination of animal feed products and their effects. Chapter 1 firstly provides an overview of some of the general concepts of foodborne pathogens in feed. It then goes on to provide specific examples of key foodborne pathogens, such as *Listeria*, *Campylobacter*, *Clostridia* and other spore-formers. A section on pathogenic *Escherichia coli* is also provided, which is then followed by an overview of the pathogen most closely associated with animal feed contamination: *Salmonella*.

Chapter 2 explores current and emerging strategies for mitigating bacterial contamination in animal feed, emphasising practical, regulatory, and technological considerations relevant to feed safety and animal health. It firstly summarises current physical and chemical intervention strategies for pathogen control in animal feeds. Physical interventions include thermal processing through rendering and pelleting, and chemical interventions include organic acids and aldehydes, with recent interest in plant-derived phytochemicals as natural antimicrobial alternatives. It then highlights future directions for advancing feed safety and biosecurity. Effective feed hygiene depends on strong biosecurity practices and hazard analysis within feed mills to identify and manage critical control points. The chapter looks ahead to future trends in research, including multifaceted solutions where physical, chemical, and procedural controls are integrated into a cohesive management plan to improve animal feed safety.

The next chapter provides a comprehensive overview of viral pathogen contamination of animal feed products, their effects and control management. Chapter 3 begins by reviewing the literature on the risk of feed and feed ingredients for the transport and transmission of viral pathogens. It then provides information on means currently in place to mitigate this risk, and summarises how this new risk factor has changed behaviour throughout global agriculture. Finally, as much of the published literature on feed risk has centred on viral diseases of pigs, the chapter will primarily focus on swine and the risk of African swine fever virus transmission through feed, but will also reference recent work in poultry and the transmission of avian influenza virus (AIV). The goal of the chapter is to provide a comprehensive summary of the literature on the risk of feed for viral transmission to animals and to educate readers on current actions taken by the global swine industry to mitigate it.

Part 2 Other feed contaminants risks

Part 2 examines other feed contaminant risks, including mycotoxin and environmental contaminant risk assessment and control.

Chapter 4 aims to elucidate the actual threat posed by the most prevalent mycotoxins in agricultural commodities, animal feed, and related products with respect to the occurrence of foodborne illnesses and diseases. The underestimated hazard of joint mycotoxin uptake in low contamination levels is discussed with regards to synergistic or additive interaction between key mycotoxins, and possible lower limit values and control measures are suggested. Various critical aspects of risk assessment, hygiene control and regulation of mycotoxins are discussed. The effectiveness of current mycotoxin regulations and control measures is evaluated with regards to human and animal health, and the potential hazard of masked mycotoxins in feedstuffs is reviewed. Appropriate control and preventive measures throughout the food/feed supply chain are proposed to prevent key foodborne diseases. In addition, potential veterinary hygiene controls and risk assessment strategies are suggested in practical cases of foodborne diseases to avoid mycotoxin contamination and associated health issues.

The next chapter reviews techniques for controlling mycotoxin contamination of animal feed products. Chapter 5 firstly compares the differences in risk between emerging and masked mycotoxins and their toxicological effects. Pre-harvest interventions are examined - these interventions are essential for ensuring host plants are healthy and protected from pathogens. These strategies aim to reduce host-pathogen interactions, limit fungal growth and spread, and create environments conducive to healthy plant growth. The chapter goes on to review post-harvest interventions such as physical, chemical, and biological techniques that are used to decontaminate feed and make it safer for farm animals. Post-harvest strategies deter mycotoxin development and accumulation during storage, processing, and distribution. This chapter argues that to establish an effective mycotoxin management program, there is the need for a holistic approach to mitigate mycotoxin contamination in animal feed. Finally, future trends in research are explored, including using predictive technologies such as biomarker technologies and machine learning through predictive modelling.

Chapter 6 provides an overview of the risk assessment related to chemical contaminants in animal feed, the transfer and uptake of environmental contaminants into animal-derived food products, and the methods available for decontamination and detoxification of feeds. It provides an overview of environmental contaminants, including persistent organic pollutants (POPs), heavy metals and other toxins. It emphasises the importance of feed safety for both animal and human health, outlines legislative frameworks, and discusses species-specific contaminant issues and mitigation strategies. These strategies include approaches to decontamination and detoxification of animal feeds.

Part 3 Technologies to improve batch control

Part 3 identifies technologies that are used to improve batch control. This includes techniques in analysing bacterial and viral pathogens, mycotoxins, and micro-scaling technologies for more precise feed.

Chapter 7 begins by reviewing molecular techniques used in identifying microorganisms in animal feed. This review discusses previous attempts to characterise the feed microbial ecology by cultural methods followed by an examination of the earlier molecular approaches. This will lead to a discussion on 16S rRNA gene sequencing as a tool to enhance understanding of the microbial ecology of animal feeds, highlighting recent advancements across various feed types. Developments in molecular sequencing, utilising the variable region of the 16S rRNA gene, can be used to generate microbiome taxonomic identification of resident microbial populations on feeds, thereby characterising overall microbial ecology responses in feeds and feed ingredients.

Chapter 8 summarises techniques for detecting viral pathogens in animal feed products and manufacturing surfaces. It begins by providing a brief history of viral pathogens in animal feed ingredients. The chapter moves on to provide an overview of current applications of diagnostic sampling and analysis in animal feed production. This is followed by a review of the challenges in detecting viral pathogens. A section on the different categories of detection methods is also provided, followed by an overview of the practices than can be used for sample collection and analysis. The chapter also describes best practices for sample collection, analysis and interpretation, then provides examples of utility to improve biosecurity programs.

Chapter 9 examines techniques for detecting mycotoxins in animal feed products. Mycotoxins, toxic metabolites of fungi (e.g., Aspergillus, Fusarium) in optimal temperature and humidity, contaminate a large range of global crops and animal feed. They cause multi-toxicity (e.g., carcinogenicity), harm animal health, enter the human food chain via animal products (e.g., milk, meat), and damage the agricultural economy, making accurate detection critical. This chapter reviews mycotoxin detection in animal feed, covering sample collection, preservation, and pretreatment. It also reviews detection methods (chromatographic, rapid, emerging technologies), regional contamination, and masked mycotoxin detection. Future directions focus on the potential of multiplex detection, eco-friendly methods, and data standardisation to safeguard feed safety and the food chain.

Finally, the primary focus for Chapter 10 is improved batching and micro-scaling systems for more precise feed control. The advancement of animal nutrition and feed safety assurance increasingly rely on refining feed production systems. Modern feed manufacturing is a precision-focused process enabled by automation, real-time sensing, and data-informed decision-making. The chapter presents an overview of advanced batch control techniques in animal feed production. It examines how programmable logic controllers (PLCs), human-machine interfaces (HMIs), and industrial control algorithms accurately coordinate ingredient dosing, mixing, and delivery. Alongside these controls, micro-scaling tools such as Near-Infrared Spectroscopy (NIRS) and Hyperspectral Imaging provide detailed insights into nutrient composition, enabling tailored rationing to meet specific dietary needs. Integration of Internet of Things (IoT) architecture with Artificial Intelligence (AI) and Machine Learning (ML) brings a new dimension to feed manufacturing - one that supports responsive formulation adjustments, condition-based Maintenance, and system-wide autonomy.

Part 1

Microbiological risks

Chapter 1

Foodborne bacterial pathogen contamination of animal feed products and their effects

Steven C. Ricke, Elena G. Olson and Haley A. Tarcin, University of Wisconsin-Madison, USA

1 Introduction

Bacterial foodborne pathogens continue to persist in animal production systems. There are many reasons for this. First, many pathogens are zoonotic and can infect a wide array of hosts, including humans (Chlebicz and Śliżewska, 2018). Consequently, the presence of such pathogens in food animals represents a central concern in foodborne disease transmission and a public health issue. Foodborne pathogens can inhabit the animal's gastrointestinal tract (GIT) and become members of the local microbial community (Dunkley et al., 2009; Horrocks et al., 2009; Davis et al., 2019). Once established, foodborne pathogens can coexist with the indigenous GIT microbial population without clinical symptoms (Chlebicz and Śliżewska, 2018; Sargeant et al., 2021). This carrier state can pose a significant challenge for identifying and tracking pathogens in animal populations and is further complicated by their widespread dissemination within the environment, via air, dust, and water, and

http://dx.doi.org/10.19103/AS.2026.0172.06

across geographical locations (Pillai and Ricke, 2002; Miller et al., 2008; Park et al., 2008; Arnold et al., 2023). Coupled with the ability of many pathogens to survive harsh environmental conditions, continual exposure of animals to these organisms can also increase the risk of infection (Beales, 2004; Horn and Bhunia, 2018).

Depending on the circumstances, foodborne pathogens can also become pathogenic to animals, invading internal tissues and causing disease. Physiological status of the host, such as age and health, can play an important role in pathogen clearance. Additionally, subclinical conditions can facilitate pathogen invasion and lead to animal mortality. One such obvious example is the delay in microbial development and maturation in young animals due to management or environmental stresses. A lack of a mature microbiota can promote pathogen colonization and/or the incursion of GIT dysbiosis in older animals (Ricke, 2017; Welch et al., 2021; Costello et al., 2025; Swanson et al., 2025). Regardless of the animal's age or health status, feed remains a consistent reservoir of pathogens. This review will describe some of the foodborne pathogens of concern that may be associated with animal feed as well as methodologies for identifying and tracking pathogens in feed during processing and manufacturing.

2 Foodborne pathogens in feed: general concepts

2.1 Pathogen exposure pathways in livestock production systems

During their production lifespan, livestock can encounter a myriad of foodborne pathogens daily (Table 1). Exposure can occur in pasture settings, such as when ruminants graze or when poultry flocks are raised on pasture (Looper et al., 2006; Rothrock et al., 2017; Sibanda et al., 2018; Nazareth et al., 2021; Ricke et al., 2023). Over the past several decades, animal production systems have become increasingly intensive, and opportunities for animal-to-animal contact have increased. This is particularly evident in densely populated poultry housing or increased dairy herd sizes, and large cattle feedlots, where animals from many sources are intermingled and held in close proximity to one another (Barham et al., 2002; Adhikari et al., 2009; Wigley, 2024). Livestock housing reflects several potential environmental factors that may influence pathogen exposure. Animals and poultry can be exposed to pathogens from aerosols/dusts generated in houses, presence of rodents, insects, and other potential disease-carrying vectors, as well as litter quality, and the type of bedding material (Henzler and Opitz, 1992; Miller et al., 2008; Olafson et al., 2016; Amin et al., 2022; Liegsalz et al., 2025). Livestock management certainly plays a key role in minimizing the impact of some of these factors. For example,

Table 1 Foodborne pathogens associated with animal feeds

Bacterial pathogen	Primary animal species affected	Economic effects in animal production	Human health effects/ public health impact
Salmonella enterica	Poultry, cattle, swine, ruminants, pets, other livestock	Asymptomatic carriage complicates detection; contamination of feed and feed mills leads to downstream contamination of meat and eggs; extensive costs associated with control measures, surveillance, and interventions	Major cause of foodborne illness; >1 million cases/year in the U.S., hospitalizations, and deaths
Listeria monocytogenes	Ruminants (cattle, sheep, goats), poultry	Causes encephalitis, septicemia, abortion, and mortality in ruminants; silage-associated infections result in production losses	Severe listeriosis with high mortality (20–30%), especially in pregnant women, elderly, immunocompromised individuals
Campylobacter jejuni	Poultry (primary reservoir)	Ingestion of toxin from silage can cause paralysis and death, leading to acute herd losses	Neurotoxin can cause paralysis and death in humans
Clostridium botulinum	Ruminants, wild birds	Ingestion of toxin from silage can cause paralysis and death, leading to acute herd losses	Neurotoxin can cause paralysis and death in humans
Clostridium perfingens	Poultry	Causes necrotic enteritis ranging from subclinical disease to mortality; reduced performance and increased losses, especially following removal of antibiotic growth promoters	Food poisoning in humans
Pathogenic *Escherichia coli* (e.g. STEC, APEC)	Cattle (STEC reservoirs), poultry (APEC), calves	Calf scours; poultry extraintestinal infections leading to significant economic losses	Shiga toxin-producing strains cause severe foodborne diseases; cattle are a major source of human outbreaks

strict biosecurity measures can be implemented to prevent transmission of foodborne pathogens, such as *Campylobacter*, within poultry flocks (Sibanda et al., 2018).

2.2 Feed ingredients and manufacturing as sources of contamination

An often-overlooked contributing factor is the hygienic status of the animal feed being delivered to the livestock farm. Animal feed production involves several steps, starting with the introduction of the primary components of the feed, followed by the inclusion of individual feed ingredients that provide small amounts of nutrients, including vitamins, minerals, and other supplements. Primary raw materials can be derived from cereal grains, oilseeds, or animal-based proteins and fats (Ricke, 2018; Ge et al., 2025). Additional supplements can help meet specific dietary requirements to maximize performance. They include feed-grade amino acids and protein supplements (Lee et al., 2023). Ultimately, the combination of ingredients from various sources increases the risk of contamination associated with the final product.

Feed processing at the feed mill is a complex process as outlined by Jones and Ricke (1994). Initially, raw ingredients are transported to the feed mill, unloaded, and temporarily stored until processing. Grinding and mixing are conducted to achieve uniformity, followed by the production of mash and transfer to a finished storage location. Feeds are further conditioned with steam, then extruded and pelletized. The resulting feeds are bagged or stored in bulk until they are delivered to the livestock operation. Because feed production involves numerous processing steps and repeated environmental exposures, the potential for contamination with foodborne pathogens is high. These opportunities range from contaminated raw ingredients entering the feed mill to equipment, storage environments, and transportation conditions. Vehicles and workers involved in feed handling represent additional potential contamination sources (Ricke, 2018). Furthermore, despite thermal processing during pelleting, recontamination can occur post-pelleting via condensation (Jones and Richardson, 2004; Jones, 2011). Although it is possible to identify locations in the feed mill that pose a risk of contamination and include in the hazard analysis of critical control points (HACCP) strategic response, biosecurity measures across locations, such as rendering plants, can improve preventative measures (Jones and Ricke, 1994; Wierup and Kristoffersen, 2014; Pandey et al., 2020).

Several pathogens have been associated with feed contamination (Olson et al., 2023). Some pathogens, such as *Salmonella*, have a long history of being identified in animal feeds (Williams, 1981a, 1981b, 1981c; Maciorowski et al., 2004; Wales et al., 2010; Jones, 2011; Wierup and Widell, 2014; Tomičić et al.,

2019; Sargeant et al., 2021). Other foodborne pathogens are less commonly attributed to feeds and may only have coincidental associations. There may be several reasons for this. First, *Salmonella* has been a consistent target for identification. Given its importance in foodborne illnesses and the ongoing outbreaks, considerable effort has been made to identify sources and drive a groundswell of technological advancements in detection and quantification (Ricke et al., 2018). The sought-out identification made it easier to track *Salmonella* across different ecosystems, including feeds. The ecology and environmental dissemination of *Salmonella* provide numerous opportunities for feed contamination at various stages of production. Other foodborne pathogens have certainly drawn public attention as contaminants of human food. Still, animal feed has been much less considered as a potential source, leading to less information on pathogen occurrence. This does not necessarily preclude their potential importance as feed contaminants; rather, it reflects the limited research attention given to their association with feeds and feed ingredients. The following sections will discuss major foodborne pathogens, summarize documented occurrences reported in previous research, and examine their relationship with animal feeds.

3 *Listeria:* feed and silage-based systems

Listeria species are facultatively anaerobic Gram-positive bacteria, which are intracellular foodborne pathogens (Bierne et al., 2018). *Listeria monocytogenes* causes foodborne listeriosis and can lead to relatively high mortality rates of 20–30% (Milillo et al., 2012a; Bodie et al., 2023). Listeriosis most commonly affects immunocompromised individuals, including the elderly, pregnant women, newborns, and infants (Milillo et al., 2012a; Lakicevic et al., 2022). The primary contaminated products include ready-to-eat refrigerated foods, deli meats, dairy products, smoked salmon, fresh vegetables, and salads, among others (Hanning et al., 2008; Milillo et al., 2012a). *Listeria monocytogenes* is a psychrotroph that can withstand multiple environmental stresses, including heat, low pH, high osmolarity, and elevated temperatures (Doyle et al., 2001; Bucur et al., 2018; Arcari et al., 2020; Lakicevic et al., 2022). Exposure to these stresses during food preservation has been proposed to increase virulence expression and infectivity (Horn and Bhunia, 2018). In addition, the pathogen can be problematic in food processing due to its ability to form biofilms on equipment, thereby increasing its resistance to sanitizers (Kalmokoff et al., 2001; Shi and Zhu, 2009).

Animals also suffer from listeriosis. Listeriosis in ruminants can lead to encephalitis, septicemia, and abortion (Rocha et al., 2013). Numerous *Listeria* species have been isolated in cattle feedlots, sheep, goats, dairy, and cattle farms (Skovgaard and Morgen, 1988; Nightingale et al., 2004; Callaway et al.,

2006). Some studies identified feed as a contamination source (Nightingale et al., 2004; Callaway et al., 2006). Silage can also be contaminated with *Listeria* (Ryser et al., 1997; Vongkamjan et al., 2012; Driehuis et al., 2018). Forage type, final product quality, and pH contribute to the frequency of *Listeria* occurrence (Ryser et al., 1997). The ability of *Listeria* to grow in low-oxygen environments may help explain survival during silage fermentation (Lungu et al., 2009; Davis et al., 2019; Roberts et al., 2020). While *Listeria* may not always be isolated directly from silage, *L. monocytogenes*-specific bacteriophages have been detected, suggesting prior bacterial presence (Vongkamjan et al., 2012).

Listeria contamination has been reported throughout poultry production, ranging from broiler flocks to meat processing (Rothrock, Jr. et al., 2017, 2019; Ricke et al., 2023). Environmental exposure, particularly in pasture or free-range systems, may increase the risk of contact with *Listeria* species, which have been isolated from soil, feces, and carcass rinses (Milillo et al., 2012b; Locatelli et al., 2017; Golden et al., 2019a, 2019b). Additional sources, such as wild birds, may also contribute (Cao et al., 2018; Huff et al., 2008,2009). Whether poultry feed serves as a primary source is unclear, though it represents a potential route for *Listeria* contamination. Survival of the pathogen on raw peanuts and pecan kernels (Brar et al., 2015) indicates the potential for contamination of dry cereal grains and animal feeds. The persistence of *Listeria* within the feed mill highlights its ability to survive under processing and storage conditions, allowing for recontamination after thermal processing (Blank et al., 1996; Whyte et al., 2003; Olson et al., 2023). This behavior underscores the importance of monitoring critical control points within the mill and implementing effective post-processing interventions. While thermal and radiation strategies may be substantial in reducing *L. monocytogenes* in feed (Farag et al., 1990), preventing moisture accumulation, maintaining equipment sanitation, and reducing airborne transmission are equally essential to limit feedborne contamination and downstream exposure.

4 *Campylobacter:* ecology and association with feed

Campylobacter species are among the most prominent foodborne pathogens worldwide (Awad et al., 2018). In poultry, *C. jejuni* readily colonizes the GIT (Horrocks et al., 2009; Sahin et al., 2015; Indikova et al., 2015; Awad et al., 2018). Colonization dynamics appear to be linked to the microbiome and nutrient availability, as *Campylobacter* relies heavily on amino acids and peptides rather than carbohydrates for energy and growth (Awad et al., 2016; Hofreuter, 2014; Indikova et al., 2015; Awad et al., 2018). Consequently, dietary protein composition and feed-associated microbial activity may influence *Campylobacter* persistence (Olson et al., 2024a, 2025). Although confirmation of direct transmission of *Campylobacter* via feed has remained elusive, the organism's ability to exist in a viable nonculturable state suggests that feed could serve as a potential vehicle for *Campylobacter* exposure (Ge et al., 2013; Silva, 2011; Ricke, 2005, 2018). For example, studies suggest that *C. jejuni* can survive and grow in feed after inoculation (Alves et al., 2017; Olson et al., 2023).

Incorporating molecular approaches into feed microbiological testing has improved detection sensitivity, particularly for fastidious organisms such as *Campylobacter*, which may persist undetected by conventional methods. When bacterial colonies from commercial poultry feed were isolated on tryptic soy agar and identified via 16S rRNA gene sequencing, eight of ten isolates contained approximately 1% reads assigned to the *Campylobacter* genus (Olson et al., 2021). Follow-up studies analyzing DNA directly extracted from various poultry feed formulations confirmed the presence of *Campylobacter* in chick starter diets as a part of the core microbiome but not in other feed types (Olson et al., 2024b,c). Collectively, these studies suggest that *Campylobacter* may be present at low levels in feed matrices, and its presence may be influenced by feed formulation.

5 *Clostridia* and other spore formers as indicators of feed hygiene

5.1 Clostridium botulinum: *silage, spores and neurotoxin risk*

Clostridia are anaerobic spore-forming bacteria that can be isolated from soil, sewage, and marine sediments and inhabit the GIT of animals and humans (Hatheway, 1990). Several clostridial species are widely distributed among animals, humans, and the environment. Some *Clostridium* species are toxigenic and can cause disease in animals and humans. The two primary disease-causing species to consider for feed management are *C. botulinum* and *C. perfringens*. The neurotoxin produced by *C. botulinum* can result in paralysis and death (Lindström et al., 2010). Wild birds inhabiting water environments where *C. botulinum* spores are frequently present, such as wetlands, are prone to botulism (Anniballi et al., 2013; Rasetti-Escargueil et al., 2020). In feed production, elevated moisture levels and anaerobic conditions favorable to clostridial proliferation underscore the importance of feed management in preventing clostridial disease in animals. In ruminants, *C. botulinum* has been associated with silage, where ingestion of the toxin can cause mortality (Driehuis et al., 2018). Furthermore, specific management practices, including plastic wrapping of silages, may promote the growth of *C. botulinum* and its toxin production (Notermans et al., 1981; Lindström et al., 2010). Nevertheless, although *C. botulinum* can be present across various types of silage, its multiplication is inhibited by rapid acidification during the final stages of silage fermentation (Driehuis et al., 2018).

5.2 Clostridium perfringens: *necrotic enteritis, diet and feed contamination*

Clostridium perfringens is a notorious enteric pathogen affecting both humans and animals (Hatheway, 1990; Van Immerseel et al., 2004; Wilson et al., 2005; Broom, 2017; Santos et al., 2022). In poultry, *C. perfringens* is associated with various stages of necrotic enteritis (NE), ranging from subclinical dysbiosis to mild infections to mortality and outbreaks (Wilson et al., 2005). Increases in this disease have been observed alongside the removal of antibiotic growth promoters in recent years (Van Immerseel et al., 2009; Broom, 2017). NE can manifest as overt disease with intestinal lesions, mucosal damage, dehydration, decreased food intake, diarrhea, and depressed behavior (Wilson et al., 2005; Villagrán-de-la Mora et al., 2020). More often, NE occurs in a subclinical form, in which mild intestinal conditions can affect bird performance without clinical signs (Wilson et al., 2005). Diagnosing NE is challenging because *C. perfringens* naturally occurs in the poultry gut and farm environment; however, many strains are not pathogenic (Moore, 2016; Villegrán-de-la Mora et al., 2020). Proliferation of pathogenic strains depends on factors impacting intestinal health, such as high-protein or nonstarch polysaccharide diets, coccidiosis, mycotoxins, and wet litter (Broom, 2017). *Clostridium perfringens* has been isolated from various poultry feeds, though its relationship to the proliferation of specific pathogenic strains remains to be determined (Xylouri et al., 1997; Tessari et al., 2014; Moore, 2016; Udhayavel et al., 2017; Munoz et al., 2021).

Given this variability, identifying a consistent nonpathogenic sporogenic indicator microorganism may improve monitoring of feed contamination risk. For example, Okelo et al. (2008) proposed using *Bacillus stearothermophilus* spores as a potential surrogate to assess the sterilization efficacy of feed extrusion. Since the presence of nonpathogenic *C. perfringens* is frequent, indicator spore formers can help predict the presence of other potentially pathogenic spore-forming taxa. Maciorowski et al. (2007) have previously suggested that *Clostridium* itself can be used as an indicator of soil contamination and the hygienic status of feed production. However, more needs to be understood about the ecological and physiological relationships among nonclostridial spore formers, nonpathogenic *C. perfringens*, and other clostridial species, as well as with pathogenic *C. perfringens* strains.

6 Pathogenic *Escherichia coli*

6.1 Pathogenesis, reservoirs and food safety implications

Nonpathogenic facultative *E. coli* are considered commensal residents of the intestine and commonly isolated from feces (Mason and Richardson, 1981;

Nataro and Kaper, 1998; Baker et al., 2016a). Pathogenic *E. coli* serotypes infect animals and humans and produce a Shiga-like toxin similar to that produced by *Shigella dysenteriae*. Hence, they are collectively referred to as a Shiga toxin-producing *E. coli* (STEC). These strains also carry the *eae gene*, which encodes intimin, a protein that facilitates attachment and invasion of intestinal epithelial cells (Baker et al., 2016a). Detection methods for STEC are based in part on the presence of toxin-encoding and virulence genes (Baker et al., 2016a,b). Pathogenic *E. coli* isolates cause foodborne diseases and are classified by different pathogenic mechanisms into groups such as enterotoxigenic, enteropathogenic, enterohemorrhagic, enteroaggregative, enteroinvasive, and diffuse-adhering *E. coli* (Nataro and Kaper, 1998; Baker et al., 2016b).

Farm animals and wildlife are major reservoirs of STEC (Baker et al., 2016a; Perez et al., 2025). While pathogenic *E. coli* foodborne disease can originate from poultry products, avian pathogenic *E. coli* (APEC) is primarily associated with poultry extraintestinal infections, such as air sacculitis and septicemia, leading to significant economic losses in chickens and turkeys (Huff et al., 2008, 2009; Dziva and Stevens, 2008; Perez et al., 2025). Pathogenic *E. coli* strains can cause calf scours and can serve as reservoirs for human infection (Kolenda et al., 2015). Cattle are the primary source of foodborne outbreaks linked to contaminated beef products (Arthur et al., 2008; Baker et al., 2016a). Increased scrutiny was implemented in the early 1990s after major restaurant-associated outbreaks had occurred, which, in turn, prompted federal regulatory action and extensive research on *E. coli* O157:H7 to track its dissemination and develop intervention approaches (Barham et al., 2002; Arthur et al., 2008; Baker et al., 2016a).

Pathogenic *E. coli* can live asymptomatically in the GIT and survive on cattle hides, soil, feces, irrigation water, and other environmental niches, and can be transmitted via insect vectors, along with dust and aerosols (Miller et al., 2008; Callaway et al., 2014; Baker, 2016a). Fecal shedding of pathogenic *E. coli* in cattle, particularly "super shedders," facilitates transmission within feedlots, while diet composition and feeding conditions can further influence shedding levels (Callaway et al., 2014; Munns et al., 2015; Baker et al., 2016a). These factors have direct implications for beef processing, as *E. coli* O157:H7 can transfer from cattle hides, contaminated during fasting or lairage, to carcasses during processing, increasing the risk of beef contamination (Arthur et al., 2008).

6.2 *Pathogenic* Escherichia coli *in feeds and silages: risk and control strategies*

Although feeds are not necessarily considered a primary source of pathogenic *E. coli* in ruminants, *E. coli* O157:H7 has been isolated from feed samples collected from dairy farms and feedlots (Hancock et al., 1998). Moreover, seasonal weather might affect its abundance and persistence. Nazareth et al. (2021) analyzed commercial bagged feed and forage from integrated organic cattle operations and found that *E. coli*-positive samples were more frequent during summer months. Additionally, how forage is processed may dictate the level of exposure. In well-managed silage, rapid and adequate acidification creates an environment with a low pH, unfavorable to *E. coli* proliferation. Conversely, insufficiently fermented or buffered silage can support the persistence of pathogenic *E. coli* (Driehuis et al., 2018). Silages and feeds can be treated with chemicals to limit pathogenic *E. coli*. Tajkarimi et al. (2008) demonstrated that anhydrous ammonia was effective in eliminating *E. coli* O157:H7 from cottonseed. Gosling et al. (2021) observed reductions in an APEC pathogenic avian strain with a commercial phytochemical/carboxylic acid blend, albeit with limitations due to low inoculum recovery. In summary, feed is unlikely to serve as a primary source of transmission for adult ruminants. Still, monogastric animals such as chickens, with fewer GIT microbial barriers, may be more susceptible to feedborne *E. coli*. Feed treatments could help reduce exposure to avian pathogenic *E. coli* in poultry, and similar benefits may extend to young calves, whose underdeveloped rumen offers limited protection against opportunistic pathogens.

7 *Salmonella*

7.1 Pathogenesis and epidemiology of feed-associated salmonellosis

Salmonellosis is a foodborne disease caused by *Salmonella,* a Gram-negative genus within the family *Enterobacteriaceae*. Two species are recognized: *S. bongeri* and *S. enterica*. *Salmonella enterica* consists of six subspecies: *enterica*, *arizonae*, *diarizonae*, *salamae*, *indica*, and *houtenae* (Ferrari et al., 2019; Aljuwayd et al., 2025). Subspecies of *enterica* alone include more than 1,500 serovars and account for the majority of animal and human infections (Ferrari et al., 2019). Generally, *Salmonella* infection occurs via the fecal-oral route. Following survival and passage through the acidic stomach environment, the bacterium can attach and invade intestinal epithelial cells, further proliferating inside intestinal cells and macrophages (Foley et al., 2013). Foodborne salmonellosis manifests in symptoms such as abdominal cramps, vomiting, diarrhea, fever, and dehydration (Aljuwayd et al., 2025). While generally self-limiting in healthy adults, the disease can be fatal in immunocompromised individuals (Aljuwayd et al., 2025).

While food products such as raw vegetables can be contaminated with *Salmonella*, animals and animal-based products are a critical source of foodborne *Salmonella* (Hanning et al., 2009; Ferrari et al., 2019). Certain serovars, such as the poultry-specific S. Gallinarum and S. Pullorum, are confined to one host and therefore are not zoonotic. However, foodborne serovars such as S. Typhimurium and S. Enteritidis can infect humans and multiple animal hosts. Animals, including ruminants, swine, poultry, and pets, can serve as reservoirs for *Salmonella* (Foley et al., 2008; Stevens et al., 2009; Rukambile et al., 2019; Abrego et al., 2025). Once colonized, animals can remain asymptomatic *Salmonella* carriers and continue undetected. This becomes an issue when animal products, such as meat or eggs, enter the food chain, exposing consumers to the risk of infection. In the U.S., foodborne salmonellosis was reported to cause over one million illnesses, more than 20,000 hospitalizations, and approximately 450 deaths annually (Scallan et al., 2011; Ferrari et al., 2019; O'Bryan et al., 2022). Consequently, numerous interventions have been implemented across all stages of production and processing.

7.2 Salmonella *in poultry production: perception, risk and preharvest control*

While many animal species have been implicated in foodborne salmonellosis, poultry has received the most attention. Although other pathogens, such as *Campylobacter*, are also strongly linked to poultry and are equally important from a public health standpoint, there is an inherent public perception that associates poultry with *Salmonella* (Wigley, 2024). Poultry may account for up to 18% of all foodborne salmonellosis, and recent trends suggest that specific serovars, such as S. Enteritidis, may be increasing in prevalence within eggs (Painter et al., 2013; O'Bryan et al., 2022; Gast et al., 2024). For years, poultry has remained a public concern. One of the several potential reasons for this is undoubtedly the rapid development of vertical integration in the U.S. poultry industry's live production and processing. This vertical integration has opened the door for more opportunities to increase the dissemination potential across greater numbers of birds. On a global scale, increases in S. Enteritidis-contaminated eggs have been speculated to derive from increased centralization of sources and international trade in breeding stock (Gast et al., 2024). In the U.S., movement toward aviary-style operations for laying hens has been suggested as a potential contributing factor (Holt, 2021; Gast et al., 2024). Concurrently, improved detection and whole genome sequencing (WGS) have enabled *Salmonella* tracking and differentiation at the strain level (Gast et al., 2024).

Preventing *Salmonella* has traditionally focused on processing, including the use of antimicrobials and HACCP protocols in the plant (O'Bryan et al., 2022; Aljuwayd et al., 2025; Dittoe et al., 2025). Increasingly, however, emphasis has shifted to preharvest, feed-based interventions aimed at reducing pathogen colonization in the GIT. These supplements include vaccines, phytochemicals, organic acids, bacteriophages, and a wide range of biotic compounds. Biotic feed additives include prebiotics, probiotics, synbiotics, and postbiotics, with prebiotics and postbiotics consisting of various carbohydrate sources or fermentate-derived metabolites and probiotics that contain live bacteria or yeast. Synbiotics combine both components to enhance microbial balance and pathogen exclusion (Swanson et al., 2025).

However, limiting *Salmonella* in live broiler or layer hen operations can be a challenge. Because *Salmonella* is widespread in the poultry environment, management and environmental factors that disrupt gut microbiota can create conditions conducive to *Salmonella* colonization (Park et al., 2008; Ricke, 2017; Gast et al., 2024). In addition, *Salmonella's* ability to adapt to environmental stresses, such as desiccation, extends the opportunity for contact and transmission (Park et al., 2008). These extended opportunities for contamination exist not only among birds but also within their shared spaces, such as dry feed and dust. Although disinfection and other hygiene practices reduce risk, environmental persistence indicates that contamination pathways remain. Among these resources, feed has been one of the most extensively investigated sources of *Salmonella* introduction into poultry flocks, making it a critical focus for control strategies.

7.3 Detection, monitoring and control of **Salmonella** *in animal feeds*

Animal feed can become contaminated in a variety of ways through fertilizers in the field, dust, coproducts, wild animals, and equipment (Tomičić et al., 2019). *Salmonella* has been isolated from cattle feed and silage (Nazareth et al., 2021). However, poultry feed has been the most extensively surveyed over the years for *Salmonella*, as summarized in several reviews (Williams, 1981a; Maciorowski et al., 2004; Wales and Davies, 2010; Jones, 2011). Certain feed ingredients, such as animal proteins, present a greater risk for *Salmonella* contamination (Williams, 1981a). Although thermal processing eliminates most of the indigenous *Salmonella,* the risk of recontamination remains. Similarly, pelleting is an effective decontaminant; however, exposure to environmental sources and condensation can serve as conduits for recontamination (Jones and Richardson, 2004; Jones, 2011; Tomičić et al., 2019). Therefore, inclusion of chemical antimicrobials, such as formic acid or formaldehyde, has been used to reduce the risk of recontamination (Jones, 2011; Ricke et al., 2019, 2020).

Limiting *Salmonella* in feeds is considered cost-effective, particularly for high-risk feeds (Wierup and Widell, 2014). Control measures can be categorized as preventing contamination, reducing multiplication, or complete elimination (Jones 2011; Wierup and Widell, 2014). The feed mill is considered the primary source of contamination, where different ingredients are brought in, mixed, and processed to make the final product. Opportunities for cross-contamination from ingredients, equipment, and other sources can significantly impact product contamination and transmission throughout the facility. Different factors can add to this complexity. Serovars can vary in feed mills by geographical location (Shariat et al., 2021). Seasons can impact *Salmonella* prevalence in feed mills, with higher prevalence occurring in the fall versus early spring/summer. The age of the feed mill can also be a factor (Magossi et al., 2019, 2020). Therefore, multi-hurdle approaches are warranted to achieve realistic reductions.

Identifying critical control points via HACCP has been used to address these complex concerns (Jones and Ricke, 1994; Wierup and Kristoffersen, 2014). Cultural selection and enrichment have been the mainstay of feed analyses and have remained essential for determining viable cells (Williams, 1981b; Maciorowski et al., 2004, 2005, 2006; Feye et al., 2021). However, certain feed amendments, such as organic acids and formaldehyde, can reduce *Salmonella* levels during enrichment, leading to an underestimation of the actual levels in the feed (Carrique-Mas et al., 2007). Animal feeds with low water activities may also be problematic for detection. For example, Li et al. (2020) showed that viable nonculturable *Salmonella* cells can persist in dry matrices, such as flour, and are difficult to culture. Additionally, feeds can be a challenge for selective enrichment because of the diverse microbial and fungal communities that can outcompete stringent selective conditions. Ha et al. (1995a, 1995b) used an antifungal agent to inhibit fungal growth on general media selecting for *Salmonella*. Obtaining representative samples that accurately reflect *Salmonella* levels and distribution in the presence of large bulk quantities of feed is also a challenge (Jones and Ricke, 1994).

Alternative nonculture methods have also been explored for detecting *Salmonella* in feed, including immunological and molecular-based techniques (Maciorowski et al., 2005, 2006). Polymerase chain reaction (PCR) assays have received the most attention in animal feeds (Maciorowski et al., 2005). Advancements in molecular technologies have increased the depth of information gained from these assays. For example, quantitation of expressed virulence genes helped to understand the phenotypic responses of heat-treated *Salmonella* in feed (Park et al., 2011). Simultaneous detection of multiple serovars using multiplex PCR has provided new and comprehensive insights into the extent and nature of feed contamination (Beaubrun et al., 2017; Borowiak et al., 2020). DNA microarrays and WGS increased the ability to track sources of *Salmonella* contamination in feed mill-connected sites, such as transportation and farms (Alvarez et al., 2003). Finally, clustered regularly interspaced short palindromic repeat (CRISPR) typing has been introduced as a molecular approach for *Salmonella* serotyping and subsequently applied to identify *Salmonella* serovars from feed mills (Shariat et al., 2021). Overall, the complexity of feed contamination underscores the need for integrated control measures supported by sensitive, modern detection methods.

8 Other microorganisms and the feed microbiome

While this review has focused on foodborne pathogens in feeds, a wide range of microorganisms inhabit feeds, some of which may not be pathogenic but can affect animal performance and health. Various microbial populations have been detected in corn-based poultry feed, including genera such as *Klebsiella*, *Kosakonia*, *Pantoea*, *Stenotrophomonas*, and *Enterococcus* (Olson et al., 2020; Olson et al., 2021). A comprehensive effort has been made to develop standardized procedures for conducting microbiome sequencing of poultry feeds. Olson et al. (2024b) compared five DNA extraction methods for 16S rRNA gene sequencing of a wide range of poultry feeds to determine which approaches provided the most consistent results for practical, routine analyses. They concluded that introducing mechanical disruption and adding lysozyme to commercial extraction kits improved the representation of the feed microbiota. In a follow-up study using these approaches, Olson et al. (2024c) compared the microbiomes of nine poultry feed types. They identified a 30% common core microbiome shared across all nine feeds, including microorganisms associated with the chicken gut. Feeds labeled as probiotics contained the most distinct microbial populations, indicating that probiotic supplementation apparently altered the baseline feed microbiome.

As more comparisons emerge, the influence of feed on shaping gut microbial composition may prove even more critical to animal health than the presence of specific pathogens alone. Since feed harbors a variety of microorganisms, it may potentially influence the composition and development of early microbiota in a young animal (Olson et al., 2022). In support of this, Olson et al. (2024c) observed several core microbial taxa in poultry feed that had been previously identified in poultry GIT. Understanding how this impacts the maturation of the microbial gut community and immune function can serve as guidance for new feed formulations and hygienic management strategies that have not been previously considered. Additionally, microbiome profiles can help evaluate the effectiveness of feed hygiene practices in reducing microbial contaminants. For example, comparing pretreatment and posttreatment communities can reveal how chemical or physical interventions shift microbial populations and how those shifts affect feed quality. Similarly, microbiome profiling offers a new tool to define indicator microorganisms that can approximate foodborne pathogen patterns in the respective feed or feed ingredient. Finally, comprehensive profiling enables more precise tracking of potential sources of cross-contamination and recontamination during feed mill processing.

9 Conclusion and future trends in research

Feed remains a critical vehicle for animals' overall exposure to foodborne pathogens. While much of the focus of feed microbial contamination has been on cereal grain-based diets, silages should not be overlooked, as fermentation conditions can be conducive to the growth of some foodborne pathogens, such as *L. monocytogenes*, and can cause diseases in susceptible ruminants. Spore formers, such as *Clostridia*, can persist during heat treatment and disinfection applications and germinate after feed processing. Pathogenic *E. coli* has been a concern from both a food safety perspective in ruminants and as a disease-causing organism in poultry. Foodborne *Salmonella* has received most of the research attention on feed contamination, in part because of its prominence as a foodborne diseases. This is especially true for poultry, which is among the primary sources of salmonellosis. While poultry can acquire *Salmonella* from a wide range of sources, feed has historically been a focus of concern. *Salmonella* has been identified in several feed ingredients, which can be introduced at the feed mill during mixing and processing. Feed mills are a significant contributor to feed contamination, not only during the initial stages of mixing raw feed but also through recontamination after post-thermal processes such as pelleting. Strategically identifying potential contamination sites in the feed mill and implementing HACCP plans has been proposed to limit the risk of recontamination.

While some progress has been made, foodborne pathogen contamination of animal feeds still has unresolved issues. Improved molecular detection and quantitation technologies offer an opportunity to identify pathogens and track individual serovars and strains during processing. However, given the erratic occurrence and low levels of *Salmonella* in large bulk feeds, optimizing representative sampling strategies remains a challenge. Additional insight is needed on the factors that make animals most susceptible to pathogen colonization of the GIT. This knowledge can guide cost-effective approaches to producing feeds that are less likely to be contaminated. Finally, overall feed-associated microbial populations should not be ignored. Feed microbiota may significantly influence gut microbial structure and, consequently, animal health. Studies based on 16S rRNA sequencing has revealed that many feedborne organisms are also potential colonizers of the GIT. This interaction requires further exploration, as it may reshape how feed microbial quality is evaluated and warrant a reappraisal of feed processing practices.

10 Where to look for further information

- Driehuis, F., et al. 2018. Silage review: animal and human health risks from silage. *Journal of Dairy Science* 101: 4093–4110. https://doi.org/10.3168/jds.2017–13836.
- Ferrari, R.G., et al. 2019. Worldwide epidemiology of *Salmonella* serovars in animal-based foods: meta-analysis. *Applied and Environmental Microbiology* 85: e00591-19. https://doi.org/10.1128/AEM00591-19.
- Jones, F.T. 2011. A review of practical *Salmonella* control measures in animal feed. *Journal of Applied Poultry Research 20:* 102–113. https://doi.org/10.3382/japr.2010–00217.
- Maciorowski, K.G., et al. 2007. Effects of poultry and livestock feed with bacteria and fungi. *Animal Feed Science and Technology* 133: 109–136. https://doi.org/10.1016/j.anifeedsci.2006.08.006.
- Maciorowski, K.G., et al. 2004. Incidence and control of food-borne *Salmonella* spp. in poultry feeds: a review. *World's Poultry Science Journal* 60: 446–457. https://doi.org/10.1079/WPS200428.
- Olson, E.G., et al. 2023. Chapter 27. Microbial pathogen contamination of animal feed. In: Anelich, L., et al. (Eds.), *Present Knowledge in Food Safety: A Risk-Based Approach Through the Food Chain*. International Life Sciences Institute, Academic Press, Elsevier. pp. 378–393. https://doi.org/10/1016/B978-0-0-12-819470–6.00023–8.
- Olson, E.G., et al. 2022. Application of microbial analyses to feeds and potential implications for poultry nutrition. *Poultry Science* 101: 101789. https://doi.org/10.1016/j.psj.2022.101789.
- Ricke, S.C. 2005. Chapter 7. Ensuring the safety of poultry feed. pp. 174–194 In: Mead, G.C. (Ed.), *Food Safety Control in the Poultry Industry*, Woodhead Publishing, Cambridge, UK. https://doi.org/10.1533/9781845690236.174.
- Sargeant, J.M., et al. 2021. *Salmonella* in animal feeds: a scoping review. *Frontiers in Veterinary Science* 8:727495. https://doi.org/10.3389/fvets.2021.727495.
- Wales, A.D., et al. 2010. Chemical treatment of animal feed and water for the control of *Salmonella*. *Foodborne Pathogens and Disease* seven: 3–15. https://doi.org/10.1089/fdp.2009.0373.

11 References

Abrego, A., et al. 2025. Summarizing the current knowledge and existing knowledge gaps for preharvest and postharvest *Salmonella* contamination in pork. *Journal of Food Protection* 88: 100609. https://doi.org/10.1016/j.jfp.2025.100609.

Adhikari, B., et al. 2009. The role of animal movement, including off-farm rearing of heifers, in the interherd transmission of multidrug resistant *Salmonella*. *Journal of Dairy Science* 92: 4229–4238. https://doi.org/3168/jds.2008–1494.

Aljuwayd, M., et al. 2025. Disinfectants and one health review: the role of reactive oxygen species in the bactericidal activity of chlorine against *Salmonella*. *One Health* 20: 100989. https://doi.org/10.1016/j.onehlt.2025.100989.

Alvarez, J.S., et al. 2003. Detection of a *Salmonella enterica* serovar California spreading in Spanish feed mills and genetic characterization with DNA microarrays. *Applied and Environmental Microbiology* 69: 7531–7534. https://doi.org/10.1128/AEM.69.12.7531–7534.2003.

Alves, M.B.R., et al. 2017. Feed can be a source of *Campylobacter jejuni* infection in broilers. *British Poultry Science* 58(1): 46–49. https://doi.org/10.1080/00071668.2016.1258691.

Amin, H., et al. 2022. Cow farmers' homes host more diverse airborne bacterial communities than pig farmers' homes and suburban homes. *Frontiers in Microbiology* 13: 883991. https://doi.org/10.3389/fmicb.2022.883991.

Anniballi, F., et al.. 2013. Management of animal botulism outbreaks: from clinical suspicion to practical countermeasures to prevent or minimize outbreaks. *Biosecurity and Bioterrorism: Biodefense Strategy, Practice, and Science* 11(Supplement 1): S191–S199. https://doi.org/10.1089/bsp.2012.0089.

Arcari, T., et al. 2020. Comparative review of the responses of *Listeria monocytogenes* and *Escherichia coli* to low pH stress. *Genes* 11: 1330. https://doi.org/10.3390/genes11111330.

Arnold, A.N., et al. 2023. Longitudinal evaluation of *Salmonella* in environmental and peripheral lymph nodes of fed cattle from weaning to finish in three distinct feeding locations. *Journal of Food Protection* 86: 1000062. https://doi.org/10.1016/j.jfp.2023.100062.

Arthur, T.M., et al. 2008. Source tracking of *Escherichia coli* O157:H7 and *Salmonella* contamination in the lairage environment at commercial U.S. beef processing plants and identification of an effective intervention. *Journal of Food Protection* 71: 1752–1760.

Awad, W.A., et al. 2018. Re-thinking the chicken: *Campylobacter jejuni* interaction: a review. *Avian Pathology* 47(4): 352–363. https://doi.org/10.1080/03079457.2018.1475724.

Awad, W.A., et al. 2016. Age-related differences in the luminal and mucosa-associated gut microbiome of broiler chickens and shifts associated with *Campylobacter jejuni* infection. *Frontiers in Cellular Infection Microbiology* 6: 154. https://doi.org/10.3389/fcimb.2016.00154.

Baker, C.A., et al. 2016a. Shiga toxin-producing *Escherichia coli* in food: incidence, ecology, and detection strategies in the United States. *Food Control* 59: 407–419. https://doi.org/10.1016/j.foodcont.2015.06.011.

Baker, C.A., et al. 2016b. Immuno-based detection of Shiga toxin-producing pathogenic *Escherichia coli:* a review on current approaches and potential strategies for optimization. *Critical Reviews in Microbiology* 42: 656–675. https://doi.org/10.3109/1040841X.2015.1009824.

Barham, A.R., et al. 2002. Effects of the transportation of beef cattle from the feedyard to the packing plant on prevalence levels of *Escherichia coli* O157 and *Salmonella* spp. *Journal of Food Protection* 65: 280–283.

Beales, N. 2004. Adaptation of microorganisms to cold temperatures, weak acid preservatives, low pH, and osmotic stress: a review. *Comprehensive Reviews in Food Science and Food Safety* 3: 1–20.

Beaubrun, J.J.-G., et al. 2017. Evaluation of a multiplex PCR method to serotype *Salmonella* in animal feeds pre-enrichment broth cultures. *MethodsX* 4: 335–345. https://doi.org/10.1016/j.mex.2017.09.003.

Bierne, H., et al. 2018. To be cytosolic or vacuolar: the double life of *Listeria monocytogenes*. *Frontiers in Cellular Infection Microbiology* 8: 136. https://doi.org/10.3389/fcimb.2018.00136.

Blank, G., et al. 1996. Microbiological decontamination of poultry feed: evaluation of steam conditioners. *Journal of the Science of Food and Agriculture* 72: 299–305.

Bodie, A.R., et al. 2023. Natural antimicrobials for *Listeria monocytogenes* in ready-to-eat meats: current challenges and future prospects. *Microorganisms* 11: 1301. https://doi.org/10.3390/microorganisms11051301.

Borowiak, M., et al. 2020. Development of a novel *mcr-6* to *mcr-9* multiplex PCR and assessment of *mcr-1* to *mcr-9* occurrence in colistin-resistant *Salmonella enterica* isolates from environment, feed, animals and food (2011–2018) in Germany. *Frontiers in Microbiology* 11: 80. https://doi.org/10.3389/fmicb.2020.00080.

Brar, P.K., et al. 2015. Survival of *Salmonella*, *Escherichia coli* O157:H7, and *Listeria monocytogenes* on raw peanut and pecan kernels stored at -24, 4, and 22°C. *Journal of Food Protection* 78: 323–332. https://doi.org/10.4315/0362-028X.JFP-14–327.

Broom, L.J. 2017. Necrotic enteritis: current knowledge and diet-related mitigation. *World's Poultry Science Journal* 73: 281–292. https://doi.org/10.1017/S0043933917000058.

Bucur, F.I., et al. 2018. Resistance of *Listeria monocytogenes* to stress conditions encountered in food and food processing environments. *Frontiers in Microbiology* 9: 2700. https://doi.org/10.3389/fmicb.2018.02700.

Callaway, T.R., et al. 2014. Ecological and dietary impactors of foodborne pathogens and methods to reduce fecal shedding in cattle. *Journal of Animal Science* 92: 1356–1365. https://doi.org/10.2527/jas.2013–7308.

Callaway, T.R., et al. 2006. Fecal prevalence of *Escherichia coli* O157, *Salmonella*, *Listeria*, and bacteriophage infecting *E. coli* O157:H7 in feedlot cattle in the Southern Plains Region of the United States. *Foodborne Pathogens and Disease* 3: 234–244.

Cao, X., et al. 2018. Isolation and characterization of *Listeria monocytogenes* from the black-headed gull feces in Kunming. *Journal of Infection and Public Health* 11: 59–63. https://doi.org/10.1016/j.jiph.2017.03.003.

Carrique-Mas, J.J., et al. 2007. Organic acid and formaldehyde treatment of animal feeds to control *Salmonella*: efficacy and masking during culture. *Journal of Applied Microbiology* 103: 88–96. https://doi.org/10.1111/j.1365–2672.2006.03233.x.

Chlebicz, A. and Śliżewska, K. 2018. Campylobacteriosis, Salmonellosis, Yersiniosis, and Listeriosis as zoonotic foodborne diseases: a review. *International Journal of Environmental Research in Public Health* 15: 863. https://doi.org/10.3390/ijerph15050863.

Costello, M.K., et al. 2025. The gastrointestinal tract microbiome of Holstein x Angus cross cattle is negatively impacted by the pre-harvest process. *Applied and Environmental Microbiology* 91(5): https://doi.org/10.1128/aem.02599–24 1.

Davis, M.L., et al. 2019. Establishment of *Listeria monocytogenes* in the gastrointestinal tract. *Microorganisms* 7: 75. https://doi.org/10.3390/microorganisms7030075.

Dittoe, D.K., et al. 2025. Chapter 11. Use of antimicrobials in poultry processing plants: the case of the United States. In: Ricke, S.C. (Ed.), *Achieving sustainable production of poultry meat Volume 1: Safety, quality and sustainability*, Burleigh Dodds Science Publishing, Cambridge, UK. pp. 274–300. http://dx.doi.org/10.19103/AS.2024.0146.13.

Doyle, M.E., et al. 2001. Heat resistance of *Listeria monocytogenes. Journal of Food Protection* 64: 410–429.

Driehuis, F., et al. 2018. Silage review: animal and human health risks from silage. *Journal of Dairy Science* 101: 4093–4110. https://doi.org/10.3168/jds.2017–13836.

Dunkley, K.D., et al. 2009. Foodborne *Salmonella* ecology in the avian gastrointestinal tract. *Anaerobe* 15: 26–35. https://doi.org/10.1016/j.anaerobe.2008.05.007.

Dziva, F. and Stevens, M.P. 2008. Colibacillosis in poultry: unravelling the molecular basis of virulence of avian pathogenic *Escherichia coli* in their natural hosts. *Avian Pathology* 37: 355–366. https://doi.org/10.1080/03079450802216652.

Farag, M.E.L.H., et al. 1990. Radiation sensitivity of *Listeria monocytogenes* in phosphate buffer, trypticase soy broth, and poultry feed. *Journal of Food Protection* 53: 648–651.

Ferrari, R.G., et al. 2019. Worldwide epidemiology of *Salmonella* serovars in animal-based foods: meta-analysis. *Applied and Environmental Microbiology* 85: e00591-19. https://doi.org/10.1128/AEM00591-19.

Feye, K.M., et al. Chapter 2. Isolation of *Salmonella* spp. from animal feed. In: Schatten, H. (Ed.), *Salmonella Methods and Protocols: Methods in Molecular Biology*, 3rd edition. Humana Press, Springer Science+Business Media LLC, pp. 7–16. https://doi.org/10.1007/978-1-0716-0791-6_2.

Foley, S.L., et al. 2013. *Salmonella* pathogenicity and host adaptation in chicken-associated serovars. *Microbiology and Molecular Biology Reviews* 77: 582–607. https://doi.org/10.1128/MMBR.00015–13.

Foley, S., et al. 2008. *Salmonella* challenges: prevalence in swine and poultry and potential pathogenicity of such isolates. *Journal of Animal Science* 86(suppl. 14): E149–E162. https://doi.org/10.2527/jas.2007–0464.

Gast, R.K., et al. 2024. *Salmonella* in eggs and egg-laying chickens: pathways to effective control. *Critical Reviews in Microbiology* 50: 39–63. https://doi.org/10.1080/1040841X.2022.2156772.

Ge, B., et al. 2025. *Applied and Environmental Microbiology* 91: e02230–24. https://doi.org/10.1128/aem.02230–24.

Ge, B., et al. 2013. Retrospective analysis of *Salmonella, Campylobacter*, *Escherichia coli*, and *Enterococcus* in animal feed ingredients. *Foodborne Pathogens and Disease* 10: 684–691. https://doi.org/10.1089/fpd.2012.1470.

Golden, C.E., et al. 2019a. Comparison between random forest and gradient boosting machine methods for predicting *Listeria* spp. prevalence in the environment of pastured poultry farms. *Food Research International* 122: 47–55. https://doi.org/10.1016/j.foodres.2019.03.062.

Golden, C.E., et al. 2019b. Using farm practice variables as predictors of *Listeria* spp. prevalence in pastured poultry farms. *Frontiers in Sustainable Food Systems* 3: 15. https://doi.org/10.3389/fsufs.2019.00015.

Gosling, R.J., et al. 2021. Control of *Salmonella* and pathogenic *E. coli* contamination of animal feed using alternatives to formaldehyde-based treatments. *Microorganisms* 9: 263. https://doi.org/10.3390/microorganisms9020263.

Ha, S.D., et al. 1995a. Cycloheximide as a media amendment for enumerating bacterial populations in animal feeds. *Journal of Rapid Methods and Automation in Microbiology* 4: 95–105.

Ha, S.D., et al. 1995b. Growth response of *Salmonella* spp. to cycloheximide amendment in media. *Journal of Rapid Methods and Automation in Microbiology* 4: 77–85.

Hancock, D.D., et al. 1998. Multiple sources of *Escherichia coli* O 157 in feedlots and dairy farms in the Northwestern USA. *Preventive Veterinary Medicine* 35: 11–19.

Hanning, I., et al. 2009. Salmonellosis outbreaks due to fresh produce: sources and potential intervention measures. *Foodborne Pathogens and Disease* 6: 635–648.

Hanning, I., et al. 2008. Pre-cut lettuce pre-packaged lettuce: a risk for listeriosis? *Foodborne Pathogens and Disease* 5: 731–746. https://doi.org/10.1089/fpd.2008.0142.

Hatheway, C.L. 1990. Toxigenic clostridia. *Clinical Microbiology Reviews* 3: 66–98.

Henzler, D.J. and Opitz, H.M. 1992. The role of mice in the epizootiology of Salmonella enteritidis infection on chicken layer farms. *Avian Diseases* 36: 625–631.

Hofreuter, D. 2014. Defining the metabolic requirements for the growth and colonization capacity of *Campylobacter jejuni*. *Frontiers in Cellular and Infection Microbiology* 4: 137. https://doi.org/10.3389/fcimb.2014.00137.

Holt, P.S. 2021. A revisiting of hen welfare and egg safety consequences of mandatory outdoor access for organic production. *Poultry Science* 100: 101436. https://doi.org/10.1016/j.psj.2021.101436.

Horn, N. and Bhunia, A.K. 2018. Food-associated stress primes foodborne pathogens for the gastrointestinal phase of infection. *Frontiers in Microbiology* 9: 1962. https://doi.org/10.3389/fmicb.2018.01962.

Horrocks, S.M., et al. 2009. Incidence and ecology of *Campylobacter* in animals. *Anaerobe* 15: 18–25. https://doi.org/10.1016/j.anaerobe.2008.09.001.

Huff, G.R., et al. 2009. Co-infection of market-age turkeys with *Escherichia coli* and *Listeria monocytogenes* in two stress models. *Avian Diseases* 53: 495–501. https://doi.org/10.1637/8675-030309-Reg.1.

Huff, G.R., et al. 2008. Pathogenicity of Listeria monocytogenes Scott A after oral and oculonasal challenges of day-old turkey poults. *Avian Diseases* 52: 444–450. https://doi.org/10.1637/8244-013008-Reg.1.

Indikova, I, et al. 2015. Survival with a helping hand: *Campylobacter* and microbiota. *Frontiers in Microbiology* 6: 1266. https://doi.org/10.3389/fmicb.2015.01266.

Jones, F.T. 2011. A review of practical *Salmonella* control measures in animal feed. *Journal of Applied Poultry Research* 20: 102–113. https://doi.org/10.3382/japr.2010–00217.

Jones, F.T. and Richardson, K.E. 2004. *Salmonella* in commercially manufactured feeds. *Poultry Science* 83: 384–391. https://doi.org/10.1093/ps/83.3.384.

Jones, F.T. and Ricke, S.C. 1994. Researchers propose tentative HACCP plan for feed manufacturers. *Feedstuffs* 66(18): 32, 36–38, 40–42.

Kalmokoff, M.L., et al. 2001. Adsorption, attachment and biofilm formation among isolates of *Listeria monocytogenes* using model conditions. *Journal of Applied Microbiology* 91: 725–734.

Kolenda R., et al. 2015. A systematic review and meta-analysis of the epidemiology of pathogenic *Escherichia coli* of calves and the role of calves as reservoirs for human pathogenic *E. coli*. *Frontiers in Cellular and Infection Microbiology* 5: 23. https://doi.org/10.3389/fcimb.2015.00023.

Lakicevic, B.Z., et al. 2022. Landscape of stress response and virulence genes among *Listeria monocytogenes* strains. *Frontiers in Microbiology* 12: 738470. https://doi.org/10.3389/fmicb.2021.738470.

Lee J.T., et al. 2023. Functional properties of amino acids: improve health status and sustainability. *Poultry Science* 102: 102288. https://doi.org/10.1016/j.psj.2022.102288.

Li, Y., et al. 2020. Reduction, prevention, and control of *Salmonella enterica* viable but non-culturable cells in flour food. *Frontiers in Microbiology* 11: 1859. https://doi.org/10.3389/fmicb.2020.01859.

Liegsalz, T.M., et al. 2025. Characterizing framework for broiler chicken's bedding materials for focusing on preventative health. *BMC Poultry Science and Management* 2: 6. https://doi.org/10.1186/s44364-025-00009-y.

Lindström, M., et al. 2010. *Clostridium botulinum* in cattle and dairy products. *Critical Reviews in Food Science and Nutrition* 50: 281–304. https://doi.org/10.1080/10408390802544405.

Locatelli, A., et al. 2017. The distribution of *Listeria* in pasture-raised broiler farm soils is potentially related to University of Vermont medium enrichment bias toward *Listeria innocua* over *Listeria monocytogenes*. *Frontiers in Veterinary Science* 4: 227. https://doi.org/10.3389/fvets.2017.00227.

Looper, M.L., et al. 2006. Prevalence of *Escherichia coli* O157:H7 and *Salmonella* in beef steers consuming different forage diets. *Letters in Applied Microbiology* 42: 583–588.

Lungu, B., et al. 2009. Growth, survival, proliferation and pathogenesis of *Listeria monocytogenes* under low oxygen or anaerobic conditions: a review. *Anaerobe* 15: 7–17. https://doi.org/10.1016/j.anaerobe.2008.08.001.

Maciorowski, K.G., et al. 2007. Effects of poultry and livestock feed with bacteria and fungi. *Animal Feed Science and Technology* 133: 109–136. https://doi.org/10.1016/j.anifeedsci.2006.08.006.

Maciorowski, K.G., et al. 2006. Cultural and immunological detection methods for *Salmonella* spp. in animal feeds: a review. *Veterinary Research Communications* 30: 127–137.

Maciorowski, K.G., et al. 2005. Polymerase chain reaction detection of foodborne *Salmonella* spp. in animal feeds. *Critical Reviews in Microbiology* 31: 45–53. https://doi.org/10.1080/10408410590912970.

Maciorowski, K.G., et al. 2004. Incidence and control of food-borne *Salmonella* spp. in poultry feeds: a review. *World's Poultry Science Journal* 60: 446–457. https://doi.org/10.1079/WPS200428.

Magossi, G., et al. 2020. Potential risk-factors affecting *Salmonella* sp. and *Escherichia coli* occurrence and distribution in midwestern United States swine feed mills. *Journal of Applied Microbiology* 129: 1744–1750. https://doi.org/10.1111/jam.14758.

Magossi, G., et al. 2019. Seasonal presence of *Salmonella* spp., *Salmonella* Typhimurium and its monophasic variant serotype 1 4,[5], 12:i:-, in selected United States feed mills. *Foodborne Pathogens and Disease* 16: 276–281. https://doi.org/10.1089/fpd.2018.2504.

Mason, T.G. and Richardson, G. 1981. A review: *Escherichia coli* and the human gut: some ecological considerations. *Journal of Applied Bacteriology* 1: 1–16.

Milillo, S.R., et al. 2012a. A review of the ecology, genomics and stress response of *Listeria innocua* and *Listeria monocytogenes*. *Critical Reviews in Food Science and Nutrition* 52: 712–725. https://doi.org/10.1080/10408398.2010.507909.

Milillo, S.R., et al. 2012b. *Listeria monocytogenes* and hemolytic *Listeria innocua* in poultry. *Poultry Science* 91: 2158–2163. https://doi.org/10.3382/ps.2012–02292.

Miller, M.F., et al. 2008. Environmental dust exposure as a factor contributing to an increase in *Escherichia coli* O157 and *Salmonella* populations on cattle hides in feedyards. *Journal of Food Protection* 71: 2078–2081.

Moore, R.J. 2016. Necrotic enteritis predisposing factors in broiler chickens. *Avian Pathology* 45: 275–281. https://doi.org/10.1080/03079457.2016.1150587.

Munns, K.D., et al. 2015. Perspectives on super-shedding of *Escherichia coli* O157:H7 by cattle. *Foodborne Pathogens and Disease* 12: 89–103. https://doi.org/10.1089/fpd.2014.1829.

Munoz, L.R., et al. 2021. Evaluation of commercially manufactured animal feeds to determine presence of *Salmonella*, *Escherichia*, and *Clostridium perfringens*. *Journal of Applied Poultry Research* 30: 100142. https://doi.org/org/10.1016/j.japr.2021.100142.

Nataro, J.P. and Kaper, J.B. 1998. Diarrheagenic *Escherichia coli*. *Clinical Microbiology Reviews* 11: 142–201.

Nazareth, J., et al. 2021. Food safety considerations in integrated organic crop: livestock systems: prevalence of *Salmonella* spp. and *E. coli* O157:H7 in organically raised cattle and organic feed. *Renewable Agriculture and Food Systems* 36: 8–16. https://doi.org/10.1017/S1742170519000292.

Nightingale, K.K., et al. 2004. Ecology and transmission of *Listeria monocytogenes* infecting ruminants and in the farm environment. *Applied and Environmental Microbiology* 70: 4458–4467. https://doi.org/10.1128/AEM.70.8.2004.

Notermans, S., et al. 1981. Persistence of *Clostridium botulinum* type B on a cattle farm after an outbreak of botulism. *Applied and Environmental Microbiology* 41: 179–183.

O'Bryan, C.A., et al. 2022. Public health impact of *Salmonella* spp. on raw poultry: current concepts and future prospects in the United States. *Food Control* 132: 108539. https://doi.org/10.1016/j.foodcont.2021.108539.

Okelo, P.O., et al. 2008. Improvements in reduction of feed contamination: an alternative monitor of bacterial killing during feed extrusion. *Journal of Applied Poultry Research* 17: 219–228.

Olafson, P.U., et al. 2016. Assessing transmission of *Salmonella* to bovine peripheral lymph nodes upon horn fly feeding. *Journal of Food Protection* 79: 1135–1142. https://doi.org/10.4315/0362-028X.JFP-15–414.

Olson, E.G., et al. 2025. Deaminase inhibitor impact on poultry cecal microbiota inoculated with *Campylobacter jejuni* in an *in vitro* model supplemented with casein hydrolysates. *Journal of Applied Microbiology* 136(3): lxaf046. https://doi.org/10.1093/jambio/lxaf046.

Olson, E.G., et al. 2024a. *Campylobacter jejuni* and casein hydrolysate addition: impact on poultry *in vitro* cecal microbiota and metabolome. *PLoS One* 19(5): e0303856. https://doi.org/10.1371/journal.pone.0303856.

Olson, E.G., et al. 2024b. Microbiome analyses of poultry feeds: part I. Comparison of five different DNA extraction methods. *Journal of Environmental Science and Health B* 59: 378–389. https://doi.org/10.1080/03601234.2024.2353002.

Olson, E.G., et al. 2024c. Microbiome analyses of poultry feeds: part II. Comparison of different poultry feeds. *Journal of Environmental Science and Health B* 59: 448–482. https://doi.org/10.1080/03601234.2024.2361596.

Olson, E.G., et al. 2020. Draft genome sequences of 11 bacterial strains isolated from commercial corn-based poultry feed. *Microbiology Resource Announcements* 9: e00170-20. https://doi.org/10.1128/ MRA.00170–20.

Olson, E.G., et al. 2021. Identification of bacterial isolates from commercial poultry feed via 16S rDNA. *Journal of Environmental Science and Health B* 56: 272–281. https://doi.org/10.1080/03601234.2020.1868236.

Olson, E.G., et al. 2022. Application of microbial analyses to feeds and potential implications for poultry nutrition. *Poultry Science* 101: 101789. https://doi.org/10.1016/j.psj.2022.101789.

Olson, E.G., et al. 2023. Chapter 27. Microbial pathogen contamination of animal feed. In: Anelich, L., et al. (Eds.), *Present Knowledge in Food Safety: a Risk-Based Approach Through the Food Chain*. International Life Sciences Institute, Academic Press, Elsevier. pp. 378–393. https://doi.org/10/1016/B978-0-0-12-819470–6.00023–8.

Painter, J.A., et al. 2013. Attribution of foodborne illnesses, hospitalizations, and deaths to food commodities by using outbreak data, United States, 1998–2008. *Emerging Infectious Diseases* 19: 407–415. https://doi.org/10.3201/eid1903.111866.

Pandey, P., et al. 2020. Improving biosecurity procedures to minimize the risk of spreading pathogenic infections agents after carcass recycling. *Frontiers in Microbiology* 11: 623. https://doi.org/10.3389/fmicb.2020.00623.

Park, S.H., et al. 2011. Detection of *Salmonella* spp. survival and virulence in poultry feed by targeting the *hilA* gene. *Journal of Applied Microbiology* 111: 426–432. https://doi.org/10.1111/j.1365–2672.2011.05054.x.

Park, S.Y., et al. 2008. Environmental dissemination of foodborne *Salmonella* in preharvest poultry production: reservoirs, critical factors and research strategies. *Critical Reviews in Environmental Science Technology* 38: 73–111. https://doi.org/10.1080/10643380701598227.

Perez, H.G., et al. 2025. Chapter 3. *Escherichia coli* in poultry meat. In: Ricke,S.C. (Ed.), *Achieving sustainable production of poultry meat Volume 1: Safety, quality and sustainability*, Burleigh Dodds Science Publishing, Cambridge, UK. pp. 53–77. http://dx.doi.org/10.19103/AS.2024.0146.03.

Pillai, S.D. and S.C. Ricke. 2002. Aerosols from municipal and animal wastes: Background and contemporary issues, *Can. J. Microbiol.* 48: 681-696. doi: 10.1139/W02-070.

Rasetti-Escargueil, C., et al. 2020. Public health risk associated with botulism as foodborne zoonoses. *Toxins* 12(1): 17. https://doi.org/10.3390/toxins12010017.

Ricke, S.C. 2005. Chapter 7: ensuring the safety of poultry feed. In: Mead, G.C. (Ed.), Food Safety Control in the Poultry Industry, Woodhead Publishing, Cambridge, UK. pp. 174–194. https://doi.org/10.1533/9781845690236.174.

Ricke, S.C. 2017. Insights and challenges of *Salmonella* infections in laying hens. *Current Opinions in Food Science* 18: 43–49. https://doi.org/10.1016/j.cofs.2017.10.012.

Ricke, S.C. 2018. Chap. 8. Feed Hygiene. In: Dewulf, J. and Van Immerseel, F. (Eds.), *Biosecurity in Animal Production and Veterinary Medicine: From Principles to Practice*. ACCO (Academische Coöperative Vennootschap cvba), Leuven, Belgium, pp. 177–209.

Ricke, S.C., et al. 2018. Molecular-based identification and detection of *Salmonella* in food production systems: current perspectives. *Journal of Applied Microbiology* 125: 313–327. https://doi.org/10.1111/jam.13888.

Ricke, S.C., et al. 2019. Formaldehydes in feeds and interaction with the poultry gastrointestinal tract microbial community. *Frontiers in Veterinary Science* 6: 188. https://doi.org/10.3389/fvets.2019.00188.

Ricke, S.C., et al. 2020. Formic acid as an antimicrobial for poultry production: a review. *Frontiers in Veterinary Science* 7: 563. https://doi.org/10.3389/fvets.2020.00563.

Ricke, S.C., et al. 2023. *Listeria* occurrence in conventional and alternative egg production systems. *Microorganisms* 11: 2164. https://doi.org/10.3390/microorganisms11092164.

Roberts, B.N., et al. 2020. *Listeria monocytogenes* response to anaerobic environments. *Pathogens* 9: 210. https://doi.org/10.3390/pathogens9030210.

Rocha, P.R.D.A., et al. 2013. Ruminant rhombencephalitis-associated *Listeria monocytogenes* strains constitute a genetically homogeneous group related to human outbreak strains. *Applied and Environmental Microbiology* 79: 3059–3066. https://doi.org/10.1128/AEM.00219-13.

Rothrock Jr., M.J., et al. 2017. *Listeria* occurrence in poultry flocks: detection and potential implications. *Frontiers in Veterinary Science* 4: 125. https://doi.org/10.3389/fvets.2017.00125.

Rothrock Jr., M.J., et al. 2019. *Listeria* occurrence and potential control strategies in alternative and conventional poultry processing and retail. *Frontiers in Sustainable Food Systems* 3: 33. https://doi.org/10.3389/fsufs.2019.00033.

Rukambile, E., et al. 2019. Infection, colonization and shedding of *Campylobacter* and *Salmonella* in animals and their contribution to human disease: a review. *Zoonoses and Public Health* 66: 562–578. https://doi.org/10.1111/zph.12611.

Ryser, E.T., et al. 1997. Effects of pH on distribution of *Listeria* ribotypes in corn, hay, and grass silage. *Applied and Environmental Microbiology* 63: 3695–3697.

Sahin, O., et al. 2015. *Campylobacter* in poultry: ecology and potential interventions. *Avian Diseases* 59: 185–200. https://doi.org/10.1637/11072-032315-Review.

Santos, R.A.N.D., et al. 2022. *Clostridium perfringens* associated with dairy farm systems show diverse genotypes. *International Journal of Food Microbiology* 382: 109933. https://doi.org/10.1016/j.ijfoodmicro.2022.109933.

Sargeant, J.M., et al. 2021. *Salmonella* in animal feeds: a scoping review. *Frontiers in Veterinary Science* 8: 727495. https://doi.org/10.3389/fvets.2021.727495.

Scallan, E., et al. 2011. Foodborne illness acquired in the United States: major pathogens. *Emerging Infectious Diseases* 17: 7–15. https://doi.org/10.3201/ eid1701.p11101.

Shariat, N.W., et al. 2021. Incidence of *Salmonella* serovars isolated from commercial animal feed mills in the United States and serovar identification using CRISPR analysis. *Journal of Applied Microbiology* 130: 2141–2146. https://doi.org/10.1111/jam.14933.

Shi, X. and Zhu, X. 2009. Biofilm formation and food safety in food industries. *Trends in Food Science & Technology* 20: 407–413.

Sibanda, N., et al. 2018. A review of the effect of management practices on *Campylobacter* prevalence in poultry farms. *Frontiers in Microbiology* 9: 2002. https://doi.org/10.3389/fmicb.2018.02002.

Silva, J., et al. 2011. *Campylobacter* spp. as a foodborne pathogen: a review. *Frontiers in Microbiology* 2: 100. https://doi.org/10.3389/fmicb.2011.00200.

Skovgaard, N. and Morgen, C.-A. 1988. Detection of *Listeria* spp. in faeces from animals, in feeds, and in raw foods of animal origin. *International Journal of Food Microbiology* 6: 229–242.

Stevens, M.P., et al. 2009. Molecular insights into farm animal and zoonotic *Salmonella* infections. *Philosophical Transactions of the Royal Society B: Biological Sciences* 364: 2709–2723. https://doi.org/10.1098/rstb.2009.0094.

Swanson, K.S., et al. 2025. Use of biotics in animals: impact on nutrition, health, and food production. *Journal of Animal Science* 103: ska061. https://doi.org/10.1093/jas/skaf061.

Tajkarimi, M., et al. 2008. Ammonia disinfection of animal feeds: laboratory study. *International Journal of Food Microbiology* 122: 23–28. https://doi.org/10.1016/j.ijfoodmicro.2007.11.040.

Tessari, E.N.C., et al. 2014. Analysis of the presence of *Clostridium perfringens* in feed and raw material used in poultry production. *Food and Nutrition Sciences* 5: 614–617. https://doi.org/10.4236/fns.2014.57072.

Tomičić, Z., et al. 2019. *Salmonella* in the feed industry: problems and potential solutions. *Journal of Agronomy, Technology and Engineering Management* 2(1): 130–137.

Van Immerseel, F., et al. 2009. Rethinking our understanding of the pathogenesis of necrotic enteritis in chickens. *Trends in Microbiology* 17: 32–36. https://doi.org/10.1016/j.tim.2008.09.005.

Udhayavel, S., et al. 2017. Occurrence of *Clostridium perfringens* contamination in poultry feed ingredients: isolation, identification and its antibiotic sensitivity pattern. *Animal Nutrition* 3: 309–312. https://doi.org/10.1016/j.animu.2017.05.006.

Van Immerseel, F., et al. 2004. *Clostridium perfringens* in poultry: an emerging threat for animal and public health. *Avian Pathology* 33(6): 537–549. https://doi.org/10.1080/03079450400013162.

Villagrán-de la Mora, Z., et al. 2020. *Clostridium perfringens* as foodborne pathogen in broiler production: pathophysiology and potential strategies for controlling necrotic enteritis. *Animals* 10: 1718. https://doi.org/10.3390/ani10091718.

Vongkamjan, K., et al. 2012. Silage collected from dairy farms harbours an abundance of listeriaphages with considerable host range and genome size diversity. *Applied and Environmental Microbiology* 78: 8666–8675. https://doi.org/10.1128/AEM.01859–12.

Wales, A.D., et al. 2010. Chemical treatment of animal feed and water for the control of *Salmonella*. *Foodborne Pathogens and Disease* 7: 3–15. https://doi.org/10.1089/fdp.2009.0373.

Welch, C.B., et al. 2021.The impact of pre-slaughter fasting on the ruminal microbial population of commercial Angus steers. *Microorganisms* 9: 2625. https://doi.org/10.3390/microorganisms9122625.

Whyte, P., et al. 2003. A survey of the prevalence of *Salmonella* and other enteric pathogens in a commercial poultry feed mill. *Journal of Food Safety* 23: 13–24.

Wierup, M. and Kristoffersen, T. 2014. Prevention of *Salmonella* contamination of finished soybean meal used for animal feed by a Norwegian production plant despite frequent *Salmonella* contamination of raw soy beans, 1994–2012. *Acta Veterinaria Scandinavica* 56: 41. http://www.actavetscand.com/content/56/1/41.

Wierup, M. and Widell, S. 2014. Estimation of costs for control of *Salmonella* in high-risk feed materials and compound feed. *Infection, Ecology & Epidemiology* 4: 23496. https://doi.org/10.3402/iee.v4.23496.

Wigley, P. 2024. *Salmonella* and the chicken: reflections on salmonellosis and its control in the United Kingdom. *BMC Poultry Science and Management* 1: 1. https://doi.org/10.1186/s44364-024-00001-y.

Williams, J.E. 1981a. Salmonellas in poultry feeds: a worldwide review. Part I. Introduction. *World's Poultry Science Journal* 37: 6–19.

Williams, J.E. 1981b. Salmonellas in poultry feeds: a worldwide review. Part II. Methods in isolation and identification. *World's Poultry Science Journal* 37: 19–25.

Williams, J.E. 1981c. Salmonellas in poultry feeds: a worldwide review. Part III. Methods in control and elimination. *World's Poultry Science Journal* 37: 97–105.

Wilson, J., et al. 2005. Manifestations of *Clostridium perfringens* and related bacterial enteritides in broiler chickens. *World's Poultry Science Journal* 61: 435–449. https://doi.org/10.1079/WPS200566.

Xylouri, E., et al. 1997. Rapid identification of *Clostridium perfringens* in animal feedstuffs. *Anaerobe* 3: 191–193.

Chapter 2

Controlling bacterial pathogen contamination of animal feed products

Elena G. Olson and Steven C. Ricke, University of Wisconsin-Madison, USA

1 Introduction

Despite substantial advances in food safety, foodborne illness continues to pose a persistent public health threat in the United States and worldwide (Jones and Yackley, 2018; Franklin et al., 2020; Sarno et al., 2021; Grace, 2023). The primary zoonotic foodborne pathogens of public health interest include nontyphoidal *Salmonella*, pathogenic *Escherichia coli*, *Campylobacter*, and *Listeria monocytogenes*, among others (Baker et al., 2016a; Omer et al., 2018). Agricultural commodities, such as fresh produce and food products, represent potential reservoirs of foodborne pathogens and contribute to varying degrees of risk of foodborne illness (Hanning et al., 2008, 2009; Carstens et al., 2019; Machado-Moreira et al., 2019; Zwietering et al., 2021). Food animals can serve as asymptomatic carriers, facilitating transmission of pathogens to susceptible individuals (Chlebicz and Śliżewska, 2018; Omer et al., 2018; Conrad et al., 2017). Animal-derived food products such as milk, meat, and eggs, as well as further processed retail foods and ingredients, can become contaminated at various stages of production and handling (Wu et al., 2017; Chlebicz and Śliżewska, 2018; Omer et al., 2018; Gast et al., 2024). Regardless of the source, foodborne infections result in millions of illnesses annually and impose an economic burden of billions of U.S. dollars (Chlebicz and Śliżewska, 2018).

http://dx.doi.org/10.19103/AS.2026.0172.09

Efforts to control foodborne pathogens in animal products have traditionally focused on antimicrobial use at meat processing facilities (O'Bryan et al., 2022; Dittoe et al., 2025). Compounds such as peracetic acid, organic acids, and other antimicrobial agents have historically been used and continue to be used in processing operations to reduce microbial contamination (Dittoe et al., 2025). However, there has been a growing emphasis on pre-harvest intervention strategies before animals enter the processing plant. These pre-harvest interventions have included numerous biotic feed additives, such as prebiotics, probiotics, postbiotics, and synbiotics (Swanson et al., 2025). The concept behind biotic feed additives is to either introduce beneficial bacteria capable of colonizing or transiently influencing the GIT gastrointestinal tract (GIT) or provide dietary supplements that select for indigenous GIT beneficial bacteria. Other approaches have included bacteriophage, botanicals, and vaccines (Olson et al., 2022).

Effective implementation of live-animal intervention strategies relies on identifying the primary sources of foodborne pathogens and determining the critical points in an animal's life cycle at which susceptibility to colonization is highest. For instance, young animals, such as newly hatched chicks, can be particularly vulnerable to *Salmonella* colonization because their immature GIT microbiota has not yet developed a protective community structure that can serve as a barrier to pathogen establishment (Stavric, 1987). Conditions that induce GIT dysbiosis, such as feed withdrawal imposed before harvest or during molt induction in laying hens, can disrupt the local microbiota and create an environment conducive to pathogen proliferation (Corrier et al., 1999; Ricke, 2003a, 2017; Gast et al., 2024; Costello et al., 2025). Consequently, identifying potential pathogen sources throughout all animal production stages is essential. However, this task is challenging, as pathogens can arise from various environmental reservoirs, including aerosols, water, insects, wild birds, rodents, and other biological and mechanical vectors.

One pathogen source in animal production that has received increased attention over the years is animal feed. Beyond its role in supporting animal nutrition and performance, feed can also be an important vehicle for pathogen exposure (Maciorowski et al., 2004; Obe et al., 2023; Olson et al., 2022, 2023). Contamination can occur at various stages, before ingredients enter the feed mill and during milling. Primary feed components, such as corn or other cereal grains, may carry pathogens from the field, while additional ingredients and protein supplements, including meat and bone meal, can further contribute to the overall microbial contamination of feed (Allred et al., 1967; Williams et al., 1969; Ha et al., 1998a). Consequently, contaminated feed can directly transmit bacterial pathogens to animals that consume it (Maciorowski et al., 2004; Jones, 2011; Obe et al., 2023; Olson et al., 2023). Therefore, minimizing or preventing pathogen contamination of feed is a critical control point in livestock production systems. This chapter explores current and emerging strategies for mitigating bacterial contamination in animal feed, emphasizing practical, regulatory, and technological considerations relevant to feed safety and animal health.

2 Microbial ecology and contamination pathways in feed production

2.1 Feed ingredients and feed mills as interconnected sources of contamination

Over the years, considerable effort has been devoted to identifying and characterizing the microorganisms associated with animal feeds (Ricke, 2005; Maciorowski et al., 2007; Olson et al., 2023). Much of this work has relied on culture-based methods, employing either nonselective or selective media, depending on the study objectives. Research has primarily focused on organisms of concern for animal health and food safety. Among these, mycotoxin-producing fungi are well-documented contaminants of feed and have been the subject of numerous studies aimed at identifying toxigenic members, determining the timing of mycotoxin production, and elucidating the environmental conditions that promote toxin synthesis (Maciorowski et al., 2007; Oguz, 2011; Chavez et al., 2020). This is a crucial consideration for feed treatment applications aimed at protecting animal health by preventing mycotoxin formation and/or eliminating the fungi that produce them. From a food safety perspective, *Salmonella* has been the primary target organism in feed microbiology research. For nearly half a century, numerous studies have sought to identify contamination sources, characterize individual serovars, and develop effective control measures (Williams, 1981a, 1981b, 1981c; Maciorowski et al., 2004). More recently, the emergence of advanced molecular and sequence-based techniques has provided greater insight into the complex nature of microbial communities that inhabit both feed and animal GIT ecosystems, revealing intricate ecological interactions that influence pathogen persistence and transmission (Ricke et al., 2017; Weinroth et al., 2022; Ge et al., 2025; Lyte et al., 2025). Recent 16S rRNA gene sequencing studies have demonstrated that diverse microbial populations, including commensals and pathogens, are present across various feed types, with composition varying according to formulation and supplementation (Olson et al., 2024).

Microbial contamination can occur at multiple time points throughout the feed production chain. Feed mills can serve as reservoirs for microbial persistence and transmission (Jones and Ricke, 1994). The complexity of this issue is underscored by the detection of distinct *Salmonella* serovars in different feed mills across the United States (Shariat et al., 2021). During feed processing, cross-contamination may occur through equipment surfaces, inadequate sanitation practices, or lapses in personnel hygiene. The risk of contamination often begins with raw ingredients, particularly those sourced from animal byproducts, improperly stored grains, or bulk commodities exposed to environmental sources of pathogens. Such contaminated materials can introduce microorganisms, including pathogens, into the mill environment, where they may persist and spread (Jones and Ricke, 1994; Crump et al., 2002; Ge et al., 2013). Each step in milling – grinding, mixing, and pelleting – offers opportunities for cross-contamination, especially when equipment is inadequately cleaned or improperly maintained (Jones and Ricke, 1994). Feed mills also face challenges from environmental vectors, such as rodents, birds, and insects, while moisture and elevated temperatures can further promote microbial proliferation. Storage and transport conditions, such as contaminated containers or exposure to humid environments, add to overall contamination risk (Jones, 2011). Certain ingredients, especially animal-derived protein sources, are frequently associated with higher levels of pathogens such as *Salmonella* (Crump et al., 2002). Studies have highlighted that even commercial feeds intended for early poultry stages can harbor potential pathogens, including *Campylobacter*, *Helicobacter*, and *Listeria*, as part of their core microbiomes (Olson et al., 2024). This knowledge is particularly critical when managing feed for young animals, as early exposure may increase the risk of colonization in immunologically immature hosts with underdeveloped gastrointestinal microbiota (Olson et al., 2022).

2.2 Post-processing microbial dynamics and antimicrobial resistance in stored feeds

Once feed leaves the feed mill, the risk of microbial contamination does not diminish. Environmental factors, such as dust, insects, rodents, and birds, are common vectors for introducing microorganisms and pathogens into feed (Maciorowski et al., 2004; Park et al., 2011). Moreover, airborne transmission may facilitate the spread of pathogens over geographic distances, complicating efforts to trace contamination sources and increasing the potential for exposure across diverse populations (Pillai and Ricke, 2002). Feed additives, including probiotics or organic acid blends, while beneficial in modulating gut microbiota, may also influence the microbial ecology of feed in unpredictable ways. For example, probiotic-amended feeds have been reported to exhibit reduced microbial richness and increased dominance by Alphaproteobacteria – a class that includes taxa associated with early poultry-gut colonization (Olson et al., 2024).

Storage and transport introduce additional challenges to feed safety, as factors such as elevated or fluctuating moisture levels, temperature variations, and the use of previously contaminated storage bins or transport containers can contribute to microbial proliferation. Microbiome-based analyses have revealed that microbial community composition shifts significantly with storage duration, storage conditions, and feed matrix type, highlighting the influence of physical and compositional factors on microbial persistence and potential pathogen survival. Importantly, emerging metagenomic analyses have revealed that in addition to pathogens, animal feeds may also carry antimicrobial resistance (AMR) genes. A recent shotgun metagenomic study examining cattle, poultry, and dog feeds identified up to ten AMR gene families within feed matrices (Ge et al., 2025). These findings underscore that, beyond traditional pathogens, animal feed may act as a reservoir for AMR determinants, contributing to the broader *One Health* challenge of AMR dissemination. However, the extent to which feed composition and exposure to antimicrobial treatments influence the development or persistence of AMR within indigenous microbial populations remains to be fully elucidated (Ricke et al., 2012; Olson et al., 2023).

Efforts to minimize microbial contamination in animal feeds have been of interest for several decades. Given the complexity of potential sources of contamination throughout the feed production chain, intervention methods must be effective from the initial stages of feed processing at the feed mill and maintain efficacy through storage, transport, and delivery to the farm. Consequently, control measures should address the visible contamination risks and underlying microbial community dynamics, including the influence of feed formulation and the presence of genomic markers associated with resistance or virulence. Even after interventions are applied, the potential for recontamination persists across all production stages. Moreover, depending on community composition and ecological interactions, the diverse microbial communities in different feed matrices may respond variably to treatment approaches. Detecting potential pathogens associated with specific diet transitions, such as *Campylobacter* in grower diets, further underscores the need for targeted microbiological surveillance across the feed production continuum to optimize intervention strategies for particular feed types and pathogens. The following sections will examine current and emerging interventions to reduce microbial contamination in animal feeds.

3 Physical strategies for feed decontamination

3.1 Rendering: thermal recycling of animal byproducts

Physical treatment of animal feed generally involves exposure to nonchemical external processes such as heat or various forms of radiation. In the animal feed industry, thermal processing is widely used across multiple feed types and ingredients to enhance nutritional quality and microbial safety. Thermal rendering is one of the earliest and most established forms of recycling food and animal byproducts (Meeker and Meisinger, 2015). This process converts animal tissues and other meat-processing components not intended for human consumption into usable materials through large-scale cooking systems that remove moisture and separate proteins from fats (Morawicki, 2010; Ricke et al., 2025). The resulting dried products have diverse commercial applications. Rendered fats and proteins are used in both feed and non-feed applications, depending on product specifications and market requirements (Meeker and Meisinger, 2015; Ricke et al., 2025). Although the high temperatures and extended processing times employed during rendering are generally effective in reducing bacterial and viral contaminants, *Salmonella* and other microorganisms have occasionally been detected in rendered byproducts, potentially due to cross-contamination within rendering facilities or recontamination occurring after processing (Bender et al., 1997; Kinley et al., 2010; Gong and Jiang, 2017).

3.2 Pelleting: conventional thermal processing for microbial reduction

Pelleting is a long-established feed processing technique in which small feed particles are combined into larger, more uniform pellets by applying moisture, heat, and pressure (Thomas and van der Poel, 1996; Rowe et al., 1999). Pelleting generally occurs in three main stages: 1) conditioning, in which steam is introduced to the mash form of the feed, 2) pelleting, where the conditioned feed is forced through a metal die, and 3) cooling, during which the formed pellets are air-cooled to stabilize their structure (Jones, 2011; Huss et al., 2017a). The overarching objective is to produce pellets that maintain structural integrity throughout production, transportation, and handling (Thomas et al., 1997). The physical and nutritional properties of the resulting pellets – such as size, density, and durability – offer several advantages. These advantages include improved feed efficiency, more consistent digestion rates, and reduced time spent by animals sorting feed (Rowe et al., 1999; Boroojenia et al., 2016). However, grain type, chemical composition, and harvest conditions can significantly influence pellet formation, physical quality, and operational parameters such as energy consumption during production (Fancher et al., 1996; Thomas et al., 1998).

The temperatures used in pelleting are generally sufficient to reduce *Salmonella* and other bacteria (Maciorowski et al., 2004; Jones, 2011; Huss et al., 2017a). However, the extent of microbial reduction varies considerably among feed types and processing conditions. For example, exposure to 71°C has been shown to reduce *Salmonella* counts, while higher temperatures in the 80–85°C range are typically required for reduction to below detectable levels under specified processing conditions. In contrast, much lower temperatures, approximately 54°C, can effectively inactivate porcine epidemic diarrhea virus

(Stott et al., 1975; Veldman et al., 1995; Jones and Richardson, 2004; Huss et al., 2017a). Absolute temperature alone, however, is not a reliable predictor of microbial mitigation efficacy, as inactivation depends on a combination of factors, including exposure time, moisture, and pressure (Himanthonkham et al., 1996; Jones and Richardson, 2004; Boroojenia et al., 2016). Moreover, heat resistance varies among microorganisms, with some, particularly spore formers, exhibiting substantially greater thermal tolerance (Cox et al., 1986; Maciorowski et al., 2004). Even among *Salmonella* strains, thermal tolerance varies, leading to differences in survival across serovars (Liu et al., 1969; Williams, 1981c; Andino et al., 2014; Jarvis et al., 2016). If heat temperature is sublethal, *Salmonella* may develop adaptive responses that increase thermotolerance and, in some cases, enhance virulence gene expression (Droffner and Yamamoto, 1991; Churi et al., 2010; Park et al., 2011; Jarvis et al., 2016; Dawoud et al., 2017; Ricke, 2018). Hygienic zoning and controlled airflow in feed mills are typically used to minimize recontamination after processing (Jones and Ricke, 1994). However, similar to rendering, recontamination after pelleting remains a concern (Jones and Richardson, 2004; Jones, 2011; Muñoz et al., 2021).

3.3 Extrusion: high-temperature, short-term processing for pathogen inactivation

Extrusion is a feed processing technique similar to pelleting and involves exposure to higher temperatures and moisture levels than those typically used in conventional pelleting (Williams, 1981c; Fancher et al., 1996). Extrusion aims to disrupt grain structure, enhance starch gelatinization, and improve pellet durability (Fancher et al., 1996; Rowe et al., 1999). In recent years, extrusion-based technologies have received considerable attention for animal feed production, food manufacturing, and polymer fabrication for packaging materials (Prabha et al., 2021; Altiparmak et al., 2022). Extrusion combines mechanical and thermal forces, wherein feed ingredients are subjected to intense heat, moisture, and pressure as they are forced through a die with specially shaped openings (Prabha et al., 2021). The process generally occurs within a barrel, where the material experiences increasing resistance, generating frictional heat and pressure as it advances toward the die (Rowe et al., 1999). Extrusion conditions can vary depending on process configuration.

Although extrusion uses higher temperatures than pelleting, exposure is relatively brief–typically 15–30 s – and is therefore classified as a high-temperature, short-time (HTST) treatment (Fancher et al., 1996; Rowe et al., 1999; Boroojenia et al., 2016). The combined effects of heat, pressure, and shear during extrusion enhance microbial inactivation compared with conventional pelleting (Fancher et al., 1996; Boroojenia et al., 2016). When properly optimized, extrusion can eliminate *Salmonella*, offering a higher level of feed decontamination than pelleting alone (Williams, 1981c; Boroojenia et al., 2016). Pelleting typically reduces *Salmonella* populations by approximately 50% or greater (Veldman et al., 1995; Jones and Richardson, 2004; Jones, 2011; Boroojenia et al., 2016), whereas extrusion has been shown to achieve reductions to below detectable levels (Okelo et al., 2006; Okelo et al., 2008; Boroojenia et al., 2016). The enhanced microbial reduction makes extrusion particularly valuable for feeds formulated for young or immunocompromised animals, which are more susceptible to infection. However, as with other thermal treatments, extrusion-processed feeds remain vulnerable to recontamination during cooling, storage, and handling. Therefore, stringent post-processing hygiene and feed mill management practices, similar to those used for pelleted feeds, are critical to maintaining microbial safety (Jones and Richardson, 2004; Jones, 2011; Muñoz et al., 2021).

3.4 Nonthermal treatments: radiation and innovative physical technologies

Radiation propagates through space or matter in electromagnetic or particulate form (Mshelia et al., 2022). Using ionizing radiation to preserve food products has been explored for over a century, primarily to eliminate contaminating microorganisms by disrupting their cells (Maciorowski et al., 2004; O'Bryan et al., 2008; Indiarto et al., 2023). With the availability of manufactured radioactive isotopes and the development of electron linear accelerators, practical applications of food irradiation became increasingly feasible (O'Bryan et al., 2008). These technologies differ in how ionizing energy is generated and in their penetration characteristics: gamma rays and X-rays are electromagnetic radiation, whereas e-beams consist of accelerated electrons (Indiarto et al., 2023). Gamma irradiation, typically generated by cesium-137 or cobalt-60 is the most commonly used isotope, rather than leaving the modifier ambiguous, is employed for cold pasteurization of foods (O'Bryan et al., 2008; Mshelia et al., 2022). Electron beam (e-beam) irradiation, by contrast, does not require radioisotopes. Instead, e-beam relies on an electron accelerator that can be switched on or off as needed, though e-beams exhibit lower penetration than gamma or X-ray radiation (O'Bryan et al., 2008; Mshelia et al., 2022). Similarly, X-rays are produced by high-energy electron streams rather than radioactive decay, offering an additional non-isotopic option for irradiation processing (O'Bryan et al., 2008; Mshelia et al., 2022).

These treatments can inactivate microorganisms and disrupt spoilage-associated proteins and enzymes (Indiarto et al., 2023). Irradiation dosage is expressed as the absorbed dose, or Gray (Gy), where 1 Gy equals 1 joule of energy absorbed per kilogram of target material. Commercial irradiation applications are classified according to dose: radurization (≤2.5 kGy) extends shelf life by reducing microbial loads; radicidation (2.5–5 kGy) targets non-spore-forming pathogenic bacteria and substantially reduces their populations; and radappertization (>10 kGy) destroys spore formers, producing a commercially sterile product (Mshelia et al., 2022). Historically, irradiation has been investigated as a potential method for improving feed hygiene, offering an advantage over thermal treatments by preserving the nutritional quality of proteins (Williams, 1981c; Leeson and Marcotte, 1993; Maciorowski et al., 2004). High radiation doses were originally considered desirable for sterilizing feeds for germ-free animal studies (Ley et al., 1969; Leeson and Marcotte, 1993). To control *Salmonella* contamination, irradiation doses of 10–15 kGy have been reported to reduce populations below detectable levels, whereas 15–35 kGy are typically required to produce *Salmonella*-free feed (Leeson and Marcotte, 1993; Maciorowski et al., 2004). Applying irradiation post-pelleting has been suggested to achieve microbial reduction with lower dosages (Mossel et al., 1967; Leeson and Marcotte, 1993). However, with advancements in molecular detection sensitivity for *Salmonella*, recommended dosage ranges may need to be reassessed. Although irradiation dosages for *Salmonella* control are relatively well established, lower doses can produce variable effects on total microbial loads and the composition of indigenous microbial populations in feeds (Maxcy, 1983; Szczawińska et al., 1991; Ha et al., 1997a). This variability is particularly relevant when feeds contain live probiotic cultures that may be sensitive to ionizing radiation, potentially diminishing their efficacy. Additionally, consumer acceptance remains a key challenge for food irradiation and continues to generate public debate (Maciorowski et al., 1999; O'Bryan et al., 2008; Indiarto et al., 2023). Whether such concerns extend to the irradiation of animal feeds remains unclear.

Other forms of radiation, such as infrared (IR), have also been explored for agricultural and food applications. IR radiation spans wavelengths from approximately 0.78 μm to 1000 μm, positioned between visible light and microwaves on the electromagnetic spectrum (Krishnamurthy et al., 2008; Indiarto et al., 2023). These wavelengths are categorized into near-IR (0.75–1.4 μm), short-wavelength IR (1.4–3 μm), mid-wavelength IR (3–8 μm), long-wavelength IR (8–15 μm), and far-IR (15–1000 μm), each with distinct physical properties and uses (Anumudu et al., 2024). In poultry production, IR technologies have been applied for animal welfare monitoring, precision beak treatment, supplemental heating, and feed quality assessment (Hayat et al., 2024).

Near-infrared spectroscopy (NIRS) has long been established as a robust analytical tool for evaluating the nutritive composition of forage and grain-based feeds, including protein, fiber, starch, and moisture content (Givens et al., 1997; Givens and Deaville, 1999; Hossain et al., 2024). NIRS is an analytical or monitoring technology, not an antimicrobial treatment. Beyond analytical applications, IR radiation has demonstrated antimicrobial properties in foods, inhibiting vegetative bacterial cells, spore formers, yeasts, and molds through energy absorption and molecular disruption (Anumudu et al., 2024). For instance, Oduola et al. (2020) reported that IR exposure inactivated bacteria and molds on rough rice, though bacterial populations appeared more resistant than molds. However, its application to animal feeds remains limited. Future research integrating microbiome-based 16S rRNA gene sequencing could provide valuable insights into how IR treatments influence the broader microbial ecology of feed ingredients and composite feed matrices. Further development of scalable, energy-efficient IR-based processing systems will be essential to translating this technology into practical, commercially viable feed safety interventions.

4 Chemical interventions for feed hygiene management

4.1 Organic acids: traditional yet variable antimicrobials

Organic acids are weak, mostly carboxylic acids with variable carbon chain lengths (Cherrington et al., 1991; Ricke, 2003b; Wales et al., 2010; Dittoe et al., 2018). They are commonly classified as short-chain or volatile fatty acids (SCFA; <6 C), medium-chain fatty acids (MCFA; C7–C10), and long-chain fatty acids (LCFA; >C10) (Dittoe et al., 2018). These definitions vary, with MCFAs commonly including C6 or broader ranges. Their antimicrobial properties depend on acid type and structure, and several mechanisms contribute to their bacteriostatic or bactericidal activity. These properties include uncoupling of electron transport, disruption of pH homeostasis, energy dissipation, and alteration of electrochemical gradients across cell membranes (Russell, 1992; Ricke, 2003b). The production of fermentation acids, such as acetate, lactate, propionate, and butyrate, by the GIT microbiota is well established (Dittoe et al., 2022; Chatman et al., 2024; Viquez-Umana et al., 2025). These metabolites can inhibit enteric pathogens, providing a natural barrier against colonization (Ricke, 2003b; Dittoe et al., 2018, 2022). Strategies to enhance endogenous acid production through supplementation with biotics have been identified as key mechanisms by which feed additives limit pathogen establishment in the GIT (Dittoe et al., 2018; Callaway and Ricke, 2023; Swanson et al., 2025).

Beyond their endogenous roles, organic acids have long been used directly in food and animal production systems due to their antimicrobial efficacy. In meat processing, lactic and acetic acids are widely used to reduce bacterial contamination on carcasses, extend shelf life, and improve product safety (Cherrington et al., 1991; Dittoe et al., 2025). In feed production, organic acids serve diverse functions depending on formulation and species. For example, propionate, acetate, sorbate, and benzoate are used to inhibit yeasts and molds in silage and improve aerobic stability (Muck et al., 2018; Ricke et al., 2020), while formate acts as a silage acidifier that suppresses *Clostridium* and spoilage organisms, preserving protein quality (Borreani et al., 2018; Ricke et al., 2020). Organic acidifiers have also been incorporated into anaerobically fermented high-moisture duck feeds to enhance feed quality and reduce the burden of GIT pathogens (Ibrahim et al., 2020).

Organic acids have a long history as feed additives in cereal-grain-based poultry diets (Williams, 1981c; Maciorowski et al., 2004; Dittoe et al., 2018). Organic acids are valued for controlling bacterial contamination during feed storage and post-processing. They also have a favorable safety profile because many occur naturally in the GIT, can be metabolized by the animal, and generally do not leave harmful residues in meat products (Wales et al., 2010). Most applications of organic acids have targeted *Salmonella* by reducing feed contamination and inhibiting intestinal colonization (Maciorowski et al., 2004; Wales et al., 2010). However, *in vivo* efficacy can be variable. Hume et al. (1993) reported that feeding propionate-treated diets to chicks did not significantly alter crop or cecal pH or reduce *S.* Typhimurium colonization, as most propionate was metabolized before reaching the lower GIT. More recent studies corroborate these findings. Oakley et al. (2014) observed minimal effects of propionate, formate, or MCFA supplementation on cecal microbiome composition or *Salmonella* prevalence in inoculated chicks.

The efficacy of organic acids in feed matrices is also variable. To achieve antimicrobial activity, relatively high concentrations are often required, which may cause palatability issues and equipment corrosion (Wales et al., 2010). Buffered formulations, such as neutralized propionate, tend to produce only modest reductions in *S.* Typhimurium (Ha et al., 1998b). Even commercial acid-based products show inconsistent performance depending on feed composition, moisture, temperature, and contamination level (Ricke et al., 2020; Gosling et al., 2021). Methodology can also confound results. For instance, the carryover of acid residues into unbuffered culture media may suppress *Salmonella* recovery, leading to underestimation of viable counts (Carrique-Mas et al., 2007). Buffering effects can influence community-level detection. Wages et al. (2022) demonstrated that microbial diversity profiles of poultry carcass rinsates varied depending on whether rinse solutions were buffered, suggesting that similar artifacts could occur in feed microbiome analyses. Although organic acid–treated feeds were not included in that study, Olson et al. (2024) observed significant microbiome shifts in probiotic-supplemented feeds, implying that comparable effects could be detected in acid-treated feed.

Several approaches have been investigated to improve efficacy. Higher dosages in high-risk feed ingredients have been proposed, though palatability and equipment concerns remain (Wales et al., 2010; Ricke et al., 2020).

Microencapsulation has been explored to deliver organic acids to the lower GIT, thereby enhancing targeted inhibition of pathogens (van Immerseel et al., 2006; Broom, 2015; Ricke et al., 2020). Combining organic acids with other antimicrobial agents, including MCFAs, SCFAs, inorganic acids, aldehydes, or botanicals, has been shown to enhance efficacy through complementary mechanisms (Cochrane et al., 2016; Ricke et al., 2019; Dee et al., 2021). For instance, modified derivatives such as butyrate glycerides may offer an improved delivery method to lower GIT compartments (Broom, 2015; Bedford and Gong, 2018). Physical–chemical combinations have also proven effective. For example, Milillo et al. (2011) demonstrated synergistic inhibition of *S.* Typhimurium when mild heat treatment was combined with SCFA exposure. Such multiple-hurdle strategies can exploit mechanistic diversity among compounds to achieve synergistic antimicrobial effects. Expanding these multifactorial approaches represents a promising direction for developing robust, feed-safe antimicrobial systems. However, validation in live-animal models remains essential.

4.2 Aldehydes and formaldehyde: potent but regulated feed disinfectants

Aldehydes, particularly formaldehyde, have been widely used for several decades as chemical treatments to control microbial contamination in feeds and animal production systems. These compounds have been extensively reviewed by Ricke et al. (2019) and are briefly summarized here. Aldehydes occur naturally in foods and have numerous industrial applications, including pharmaceuticals and chemical manufacturing (Ricke et al., 2019). Their antimicrobial activity is mainly attributed to their function as fixatives, which chemically modify proteins, a property that has been exploited for disinfection in clinical and agricultural settings (McLeod, 2018). Formaldehyde's ability to cross-link proteins stabilizes intact bacterial cells for vaccine development and standardization of detection assays (Baker et al., 2016b; Ricke and Schaefer, 1990; Ricke et al., 2019). This cross-linking activity and interactions involving other cellular macromolecules, such as nucleic acids, highlight its lethality toward microorganisms (Ricke et al., 2019).

Over the years, a variety of aldehydes, including glutaraldehyde and formaldehyde, have been applied as antimicrobial agents in livestock environments such as animal housing and transport equipment (Böhm, 1998; Reuter, 1998; Battersby et al., 2017; Jiang et al., 2018; Ricke et al., 2019). In the poultry industry, one of the earliest and most common uses of formaldehyde was as an eggshell fumigant in hatcheries (Williams, 1970; Sacco et al., 1989). Hatching eggs are susceptible to microbial contamination, including *Salmonella*, which can penetrate the eggshell and infect chicks post-hatch (Gast et al., 2024). Because formaldehyde is toxic to *Salmonella*, it has been considered an effective control measure for reducing hatchery contamination. However, concerns about occupational exposure and potential safety risks have prompted the exploration of alternative egg disinfection strategies (Scott and Swetnam, 1993a, 1993b; Scott et al., 1993).

Formaldehyde also has a long history as a feed disinfectant. It was initially applied as a fumigant to control molds in high-moisture corn and later demonstrated to be effective for *Salmonella* reduction in feeds (Williams, 1981c). Subsequent work showed that liquid formaldehyde formulations incorporated directly into feed could effectively reduce *Salmonella* levels in artificially contaminated feed (Williams, 1981c; Moustafa et al., 2002). Among chemical feed treatments, formaldehyde remains one of the most effective disinfectants (Carrique-Mas et al., 2007; Cochrane et al., 2016; Gosling et al., 2021). Nevertheless, factors such as moisture content and the timing of application to feed still need to be considered to maximize efficacy (Carrique-Mas et al., 2007; Jones, 2011; Ricke et al., 2019). For example, Jones (2011) suggested post-pelleting application of formaldehyde as a strategy to minimize *Salmonella* recontamination as a management option.

4.3 Emerging alternatives to formaldehyde treatments

Despite its proven effectiveness, formaldehyde's use in feed treatment has come under increased regulatory scrutiny, particularly in Europe, due to concerns about worker safety and potential residues. These regulatory and safety concerns prompted a gradual shift within the feed industry toward alternative antimicrobial agents and integrated control strategies (Gosling et al., 2021). Numerous compounds have been investigated as potential alternatives for feed decontamination. As discussed previously, organic and inorganic acids have received the most extensive attention. However, several other compounds have also been evaluated, including ethylene oxide, methyl bromide, trisodium phosphate, terpenes, botanicals, alcohols, chlorate, and surfactants (Maciorowski et al., 2004; Wales et al., 2010) some listed agents represent historical or highly regulated treatments and may not be approved for contemporary feed applications. Some agents exert antimicrobial activity directly within the feed matrix, while others act primarily as delivery vehicles, releasing active compounds that exhibit antimicrobial effects once ingested and present in the gut (Wales et al., 2010). As with organic acids, the efficacy of these compounds is influenced by factors such as concentration, timing of application, and feed composition. Additionally, the microbial community structure of feed contaminants may influence responsiveness. For example, Ha et al. (1997a, 1997b) reported that ethyl alcohol treatments reduced artificially inoculated *S.* Typhimurium and indigenous fungal populations but had limited effects on aerobic bacteria.

4.4 Phytogenic compounds and essential oils: plant-derived feed antimicrobials

Plant-derived phytogenic compounds, long used in the food industry for flavoring and preservation, have recently gained interest in animal agriculture for their potential to improve performance and feed hygiene (Applegate et al., 2010). Given the diversity of plant sources, phytochemical composition varies widely, encompassing a broad range of secondary metabolites with antimicrobial, antioxidant, or bioactive properties (Applegate et al., 2010; Hardin et al., 2010). Among these, essential oils represent a notable subcategory of naturally occurring antimicrobials composed of hydrocarbons, oxygenated compounds, and minor nonvolatile residues (Applegate et al., 2010; Hardin et al., 2010).

Essential oils and plant extracts such as thymol and eugenol have demonstrated antimicrobial activity against various foodborne pathogens, though Gram-negative bacteria generally exhibit greater resistance than Gram-positive species (Burt, 2004; Calo et al., 2015; O'Bryan et al., 2015). Chemically, essential oils can be divided into two major groups: terpene compounds and aroma compounds (Calo et al., 2015). However, their specific composition can vary with plant species, developmental stage, the anatomical origin of the extract (e.g. leaf, seed, or fruit), and the extraction method used (Calo et al., 2015).

Essential oils' antimicrobial mechanisms are multifaceted. Generally, their activity is attributed to their hydrophobicity, which allows them to penetrate bacterial membranes and disrupt intracellular processes. These include leakage of cytoplasmic contents, alteration of membrane potential, inhibition of aerobic respiration, and interference with quorum sensing (O'Bryan et al., 2015). Given their structural and chemical diversity, combinations of essential oil components can simultaneously target multiple cellular functions, leading to additive or synergistic antimicrobial effects (Bassolé and Juliani, 2012).

Although essential oils are increasingly accepted as alternative growth-promoting feed additives for poultry and swine, inconsistent efficacy and incomplete understanding of their mechanisms have limited widespread adoption (Biswas et al., 2024). Nevertheless, essential oil blends, while generally less potent than MCFA or formaldehyde, have demonstrated efficacy in reducing post-processing contamination with *S.* Typhimurium in rendered feed ingredients. However, these results vary with feed type (Cochrane et al., 2016). Moreover, commercial feed additives that combine essential oils with other antimicrobials such as organic acids and formaldehyde have been developed and shown to improve performance and health outcomes in swine under viral challenge models (Wales et al., 2010; Dee et al., 2021).

4.5 Integrating biosecurity, Hazard Analysis and Critical Control Point (HACCP) and risk-based management in feed safety

Biological hazards, including viruses, parasites, and pathogenic bacteria, within food production systems pose significant risks to animal health and public safety (Kumar et al., 2024). Consequently, identifying critical points along the farm-to-fork continuum where contamination can occur is essential for minimizing risk throughout the food supply chain. Equally important is controlling infectious agents that may not directly threaten humans but cause substantial economic losses in animal agriculture, such as viral pathogens affecting poultry and livestock (Huss et al., 2017b).

Managing these risks is challenging due to the inherent complexity of food production systems. Environmental factors, transmission routes, and vector ecology vary widely across operations. At the same time, pathogens differ in their ability to persist under adverse conditions and develop resistance to environmental stressors and antimicrobial interventions. Therefore, minimizing opportunities for contact between biological hazards and feed or food products is fundamental to effective risk management. Biosecurity provides a systematic framework for reducing the introduction and spread of biological hazards within animal production environments (Huss et al., 2017b). Farm-level biosecurity programs, emphasizing hygiene improvements, personnel education, and the strategic use of disinfectants, have been implemented for decades to mitigate the spread of of pathogens, including antibiotic-resistant organisms (Zwietering et al., 2021; Jimenez et al., 2023). A review by Jimenez et al. (2023) concluded that integrated approaches combining management, education, and sanitation protocols contribute substantially to overall farm hygiene and disease control.

In the context of feed safety, biosecurity extends beyond the farm itself. The feed mill represents a critical point in the production chain where microbial contamination can be introduced before feed reaches the animal. Multiple studies have identified feed mills as reservoirs for *Salmonella* and other pathogens, with risk influenced by factors such as mill age, ingredient sourcing diversity, and environmental hygiene (Huss et al., 2017b; Mogassi et al., 2020; Muñoz et al., 2021; Shariat et al., 2021). Although contamination may originate from a single entry point, such as dust or raw material intake, pathogens have been isolated from a wide range of mill environments, indicating extensive potential for dissemination (Jones and Richardson, 2004; Davies and Wales, 2010; Huss et al., 2017b). Identifying high-risk contamination sites within feed manufacturing facilities is essential to ensure that feed leaving the mill is not a source of infection at the farm level.

Table 1 Summary of feed pathogen control methods

Category	Representative methods	Primary mechanism	Advantages	Limitations
Physical	Rendering, pelleting, extrusion, irradiation	Heat, shear, radiation-induced damage	Effective, scalable	Recontamination post-processing
Chemical	Organic acids, aldehydes, phytochemicals	pH reduction, protein crosslinking, membrane disruption	Flexible, low residue	Variable efficacy, corrosion, safety
Biological	Probiotics, atoxigenic fungi, bacteriophages	Competitive exclusion, enzyme degradation	Sustainable, One Health alignment	Strain specificity, stability
Procedural	Biosecurity, HACCP	Risk mitigation, hygiene management	Systems-based, regulatory alignment	Implementation cost

The Hazard Analysis and Critical Control Point (HACCP) system provides a structured framework for identifying, monitoring, and managing biological risks in feed production (Jones and Ricke, 1994). According to Huss et al. (2017b), feed mill hazard analysis typically includes: identification of ingredients and processing steps; hazard identification; and hazard evaluation based on severity and probability. Once potential hazards are characterized, targeted mitigation measures can be implemented to reduce or eliminate the risk. HACCP programs can be explicitly adapted for feed manufacturing to manage critical control points, including ingredient reception, thermal processing, and post-production storage. However, as Jones and Ricke (1994) noted, assessing contamination levels of pathogens such as *Salmonella* is inherently difficult due to the vast volumes of feed produced and the statistical limitations of sampling strategies. One practical solution is to use indicator organisms. These are microorganisms that mimic the behavior of target pathogens but are more prevalent and abundant. Tracking indicator bacteria may enable efficient monitoring and trend detection (Jones and Richardson, 2004; Ricke, 2018). Overall, preventive and proactive strategies form the cornerstone of microbial control in feed production. A systems-based approach integrating hygiene, traceability, and continuous risk assessment can significantly reduce contamination events and improve the overall safety and sustainability of the food supply chain (Table 1).

5 Conclusion and future trends in research

Ongoing research is needed to improve feed pathogen control strategies in food production systems. A *One Health* perspective that integrates animal, human, and environmental health can guide risk assessment and intervention design, with the potential for mathematical modeling (Jimenez et al., 2023; Kumar et al., 2024). Although much emphasis has traditionally focused on interventions at the farm and processing plant levels, the repeated recovery of pathogens from animal feed and feed ingredients underscores the need to incorporate feed processing into comprehensive, longitudinal food safety strategies.

Microbial contamination of animal feed is a complex issue requiring multifaceted solutions. Physical, chemical, and procedural controls must be integrated into a cohesive management plan. Innovation, regulatory compliance, and proactive training are key to enhancing feed safety and protecting public health. Research continues to focus on developing novel methods with applications for feed analysis and interventions. For example, new molecular techniques to detect viable but nonculturable (VBNC) organisms – such as combining propidium monoazide (PMA) treatment with PCR – have been successfully applied to dry food matrices, such as flour, and may hold promise for pathogen detection in dry feeds (Li et al., 2020). In parallel, studies on prebiotic feed additives are expanding beyond conventional formulations toward innovative feed-processing–driven modifications (Callaway and Ricke, 2023; Swanson et al., 2025). Urango et al. (2024) demonstrated a nonenzymatic Maillard reaction (MR) between proteins and polysaccharides by adding prebiotic fibers to food matrices. The resulting products were subsequently shown to exert prebiotics effects by modulating the intestinal microbiota and promoting gut health. Whether similar MR products generated during feed processing can influence feed-associated microbiomes remains an open area for exploration. Advances in microbial ecology have facilitated the development of non-toxic fungal strains capable of colonizing and competitively excluding aflatoxin-producing fungal species, as well as bacterial species that degrade aflatoxins enzymatically (Atehnkeng et al., 2022; Nguyen et al., 2024). Adopting these technologies and discovering novel approaches for the feed industry will require cross-sector collaboration among academia, industry, and government agencies to address ongoing and emerging challenges.

6 Where to look for further information

Maciorowski, K.G., et al. 2007. Effects of poultry and livestock feed with bacteria and fungi. *Animal Feed Science and Technology* 133: 109–136. https://doi.org/10.1016/j.anifeedsci.2006.08.006.

Maciorowski, K.G., et al. 2004. Incidence and control of food-borne *Salmonella* spp. in poultry feeds: a review. *World's Poultry Science Journal* 60: 446–457. https://doi.org/10.1079/WPS200428.

Olson, E.G., et al. 2023. Chapter 27. Microbial pathogen contamination of animal feed. In: L. Anelich, et al. (Eds.), *Present Knowledge in Food Safety: A Risk-Based Approach Through the Food Chain*. International Life Sciences Institute, Academic Press, Elsevier. pp. 378–393. https://doi.org/10/1016/B978-0-0-12-819470-6.00023-8.

Olson, E.G., et al. 2022. Application of microbial analyses to feeds and potential implications for poultry nutrition. *Poultry Science* 101:101789. https://doi.org/10.1016/j.psj.2022.101789.

Ricke, S. C. 2005. Chapter 7. Ensuring the safety of poultry feed. In: G.C. Mead (Ed.), *Food Safety Control in the Poultry Industry*. Woodhead Publishing, Cambridge, UK. pp. 174–194. https://doi.org/10.1533/9781845690236.174.

Ricke, S.C., et al. Richardson. 2020. Formic acid as an antimicrobial for poultry production: a review. *Frontiers in Veterinary Science* 7: 563. https://doi.org/10.3389/fvets.2020.00563.

Ricke, S.C., et al. 2019. Formaldehydes in feeds and interaction with the poultry gastrointestinal tract microbial community. *Frontiers in Veterinary Science* 6: 188. https://doi.org/10.3389/fvets.2019.00188.

Wales, A.D., et al. 2010. Chemical treatment of animal feed and water for the control of *Salmonella*. *Foodborne Pathogens and Disease* 7: 3–15. https://doi.org/10.1089/fdp.2009.0373.

7 References

Allred, J.N., et al. 1967. A survey to determine the *Salmonella* contamination rate in livestock and poultry feeds. *Journal of the American Veterinary Medical Association* 151: 1857–1860.

Altiparmak, S.C., et al. (2022). Extrusion-based additive manufacturing technologies: State of the art and future perspectives. *Journal of Manufacturing Processes*. 83, 606–636. https://doi.org/10.1016/mappro2022.09.032.

Andino, A., et al. 2014. Survival of *Salmonella enterica* in poultry feed is strain dependent. *Poultry Science* 93: 441–447. https://doi.org/10.3382/ps.2013-03401.

Anumudu, C.K., et al. 2024. Advances in the Application of Infrared in Food Processing for Improved Food Quality and Microbial Inactivation. *Foods* 13: 4001. https://doi.org/10.3390/foods13244001.

Applegate, T.J., et al. Probiotics and phytogenics for poultry: myth or reality? *Journal of Applied Poultry Research* 19: 194–210. https://doi.org/10.3382/japr.2010-00168.

Atehnkeng, J., et al. 2022. Impact of frequency of application on the long-term efficacy of the biocontrol product Aflasafe in reducing aflatoxin contamination in maize. *Frontiers in Microbiology* 13: 1049013. https://doi.org/10.3389/fmicb.2022.1049013.

Baker, C.A., et al. 2016a. Shiga toxin-producing *Escherichia coli* in food: incidence, ecology, and detection strategies in the United States. *Food Control* 59: 407–419. https://doi.org/10.1016/j.foodcont.2015.06.011.

Baker, C.A., et al. 2016b. Formalin-fixed cells as an internal standard approach for the detection and quantitative assessment of shiga toxin-producing *Escherichia coli* (STEC). *Food Control* 63: 76–82. https://doi.org/10.1016/j.foodcont.2015.11.014.

Bassolé, I.H.N. and Juliani, H.R. 2012. Essential oils in combination and their antimicrobial properties. *Molecules* 17(4): 3989–4006. https://doi.org/10.3390/molecules17043989.

Battersby, T., et al. 2017. Evaluating and improving terminal hygiene practices on broiler farms to prevent *Campylobacter* cross-contamination between flocks. *Food Microbiology* 64: 18. https://doi.org/10.1016/j.fm.2016.11.018.

Bedford, A. and Gong, J. 2018. Implications of butyrate and its derivatives for gut health and animal production. *Animal Nutrition* 4: 151–159. https://doi.org/10.1016/j.aninu.2017.08.010.

Bender, J.B., et al. 1997. Animal by-products contaminated with *Salmonella* in the diets of lactating dairy cows. *Journal of Dairy Science*. 80: 3064–3067.

Biswas, S., et al. 2024. Assessing the potential of phytogenic feed additives: a comprehensive review on their effectiveness as a potent dietary enhancement for nonruminant in swine and poultry. *Journal of Animal Physiology and Animal Nutrition* 108: 711–723. https://doi.org/10.1111/jpn.13922.

Böhm, R. 1998. Disinfection and hygiene in the veterinary and disinfection of animal houses and transport vehicles. *International Biodeterioration and Biodegradation* 41: 217–224. https://doi.org/10.1016/S0964-8305(98)00030-4.

Boroojenia, F.G., et al. 2016. The effects of hydrothermal processing on feed hygiene, nutrient availability, intestinal microbiota and morphology in poultry: a review. *Animal Feed Science and Technology* 220: 187–215. https://doi.org/10.1016/j.anifeedsci.2016.07.010.

Borreani, G., et al. 2018. Silage review: factors affecting dry matter and quality losses in silages. *Journal of Dairy Science* 101: 3952–3979. https://doi.org/10.3168/jds.2017-13837.

Broom, L.J. 2015. Organic acids for improving intestinal health of poultry. *World's Poultry Science Journal* 71: 630–642. https://doi.org/10.1017/S0043933915002391.

Burt, S. 2004. Essential oils: their antibacterial properties and potential applications in foods: a review. *International Journal of Food Microbiology* 94(3): 223–253.

Callaway, T.R. and Ricke, S.C. (Eds.) 2023. *Direct Fed Microbials/Prebiotics for Animals: Science and Mechanisms of Action*, 2nd edition. Springer Science, New York, NY, 348 pp.

Calo, J.R., et al. 2015. Essential oils as antimicrobials in food systems: a review. *Food Control* 54: 111–119. https://doi.org/10.1016/j.foodcont.2014.12.040.

Carrique-Mas, J.J., et al. 2007. Organic acid and formaldehyde treatment of animal feeds to control *Salmonella*: efficacy and masking during culture. *Journal of Applied Microbiology* 103: 88–96.

Carstens, C.K., et al. 2019. Multistate outbreaks of foodborne illness in the United States associated with fresh produce from 2010 to 2017. *Frontiers in Microbiology* 10: 2667. https://doi.org/10.3389/fmicb.2019.02667.

Chatman, C.C., et al. 2024. Meat animal biologics discovery opportunities from the gut microbiome: Application of metabolomics. *Meat and Muscle Biology* 8(1): 18261, 1–11. https://doi.org/10.22175/mmb.18261.

Chavez, R.A., et al. 2020. A review of the methodology of analyzing aflatoxin and fumonisin in single corn kernels and the potential impacts of these methods on food security. *Foods* 9: 297. https://doi.org/10. 3390/foods9030297.

Cherrington, C.A., et al. 1991. Organic acids: chemistry, antibacterial activity and practical applications. *Advances in Microbial Physiology* 32: 87–108.

Chlebicz, A. and Śliżewska, K. 2018. Campylobacteriosis, salmonellosis, yersiniosis, and listeriosis as zoonotic foodborne diseases: A review. *International Journal of Environmental Research Public Health* 15: 863. https://doi.org/10.3390/ijerph15050863.

Churi, A., et al. 2010. Increased temperature influences *hilA* gene fusion expression in a *Salmonella* Typhimurium poultry isolate. *Food Biotechnology* 24: 51–61.

Cochrane, R., et al. 2016. Evaluating chemical mitigation of *Salmonella* Typhimurium ATCC 14028 in animal feed ingredients. *Journal of Food Protection* 79: 672–676. https://doi.org/10.4315/0362-028X.JFP-15-320.

Conrad, C.C., et al. 2017. Farm fairs and petting zoos: a review of animal contact as a source of zoonotic enteric disease. *Foodborne Pathogens and Disease* 14: 59–73. https://doi.org/10.1089/fpd.2016.2185.

Corrier, D.E., et al. 1999. Presence of *Salmonella* in the crop and ceca of broiler chickens before and after preslaughter feed withdrawal. *Poultry Science* 78: 45–49.

Costello, M.K., et al. 2025. The gastrointestinal tract microbiome of Holstein x Angus cross cattle is negatively impacted by the pre-harvest process. *Applied Environmental Microbiology* 91 (5): e02599-24. https://doi.org/10.1128/aem.02599-241.

Cox, N.A., et al. 1986. Effect of steam conditioning nd pelleting process on the microbiology and quality of commercial-type poultry feeds. *Poultry Science* 65: 704–709.

Crump, J.A., et al. 2002. Bacterial contamination of animal feed and its relationship to human foodborne illness. *Clinical Infectious Diseases* 355: 859–865.

Davies, R.H. and Wales, A.D. 2010. Investigations into *Salmonella* contamination in poultry feedmills in the United Kingdom. *Journal of Applied Microbiology* 109: 1430–1440. https://doi.org/10.1111/j.1365-2672.2010.04767.x

Dawoud, T.M., et al. 2017. *Salmonella* thermal resistance - Molecular responses. *Frontiers in Veterinary Science* 4: 93. https://doi.org/10.33889/fvets.2017.00093.

Dee, S.A., et al. 2021. An evaluation of additives for mitigating the risk of virus contaminated feed using an ice-block challenge model. *Transboundary Emerging Diseases* 68: 833–845. https://doi.org/10.1111/tbed.13749

Dittoe, D.K, et al. 2018. Organic acids and potential for modifying the avian gastrointestinal tract and reducing pathogens and disease. *Frontiers in Veterinary Science* 5: 216. https://doi.org/10.3389/fvets.2018.00216.

Dittoe, D.K., et al. 2022. Impact of the gastrointestinal microbiome and fermentation metabolites on broiler performance. *Poultry Science* 101: 101786. https://doi.org/10.1016/j.psj.2022.101786.

Dittoe, D.K., et al. 2025. Chapter 11. Use of antimicrobials in poultry processing plants: the case of the United States. In: S.C. Ricke, (Ed.), *Achieving sustainable production of poultry meat Volume 1: Safety, quality and sustainability*, Burleigh Dodds Science Publishing, Cambridge, UK. pp. 274–300. https://doi.org/10.19103/AS.2024.0146.13.

Droffner, M.L. and Yamamoto, N. 1991. Prolonged environmental stress via a two step process selects mutants of *Escherichia*, *Salmonella* and *Pseudomonas* that grow at 54 C. *Archives of Microbiology* 156(4): 307e311.

Fancher, B.I., et al. 1996. Feed processing using the annular gap expander and its impact on poultry performance. *Journal of Applied Poultry Research*. 5: 386–394.

Franklin, N., et al. 2020. Describing the epidemiology of foodborne outbreaks in New South Wales from 2000 to 2017. *Foodborne Pathogens and Disease* 17: 701–711. https://doi.org/10.1089/fpd.2020.2806.

Gast, R.K., et al. 2024. *Salmonella* in eggs and egg-laying chickens: pathways to effective control. *Critical Reviews in Microbiology* 50: 39–63. https://doi.org/10.1080/1040841X.2022.2156772.

Ge, B., et al. 2025. Exploring animal food microbiomes and resistomes via 16S rRNA gene amplicon sequencing and shotgun metagenomics. *Applied and Environmental Microbiology* 91(2): e02230–24.

Ge, B., et al. 2013. Retrospective analysis of *Salmonella*, *Campylobacter*, *Escherichia coli*, and *Enterococcus* in animal feed ingredients. *Foodborne Pathogens and Disease* 10: 684–691. https://doi.org/10.1089/fpd.2012.1470.

Givens, D.I. and Deaville, E.R. 1999. The current and future role of near infrared reflectance spectroscopy in animal nutrition: a review. *Australian Journal of Agricultural Research* 50: 1131–1145.

Givens, D.I, et al. 1997. The principles, practices and some future applications of near infrared spectroscopy for predicting the nutritive value of foods for animals and humans. *Nutrition Research Reviews* 10: 83–114.

Gong, C. and Jiang, X. 2017. Characterizing *Salmonella* contamination in two rendering processing plants. *Journal of Food Protection* 80: 265–270. https://doi.org/10.4315/0362-028XJFP-16-210.

Gosling, R.J., et al. 2021. Control of *Salmonella* and Pathogenic *E. coli* contamination of animal feed using alternatives to formaldehyde-based treatments. *Microorganisms* 9: 263. https://doi.org/10.3390/microorganisms902 0263

Grace, D. 2023. Burden of foodborne disease in low-income and middle-income countries and opportunities for scaling food safety interventions. *Food Security* 15: 1475–1488. https://doi.org/10.1007/s12571-023-01391-3.

Ha, S.D., et al. 1997a. Survival of an unirradiated *Salmonella typhimurium* marker strain inoculated in poultry feeds after irradiation. *Journal of Rapid Methods and Automation in Microbiology* 5: 47–59.

Ha, S.D., et al. 1997b. Ethyl alcohol reduction of *Salmonella typhimurium* in poultry feed. *Journal of Rapid Methods and Automation in Microbiology* 5: 75–85.

Ha, S.D., et al. 1998a. Indigenous feed microflora and *Salmonella typhimurium* marker strain survival in poultry feed with varying levels of protein. *Animal Feed Science and Technology* 76: 23–33.

Ha, S.D., et al. 1998b. Survivability of indigenous feed microflora and a *Salmonella typhimurium* marker strain in poultry feed treated with buffered propionic acid. *Animal Feed Science and Technology* 75: 145–155.

Hanning, I., et al. 2008. Pre-cut lettuce pre-packaged lettuce: a risk for listeriosis? *Foodborne Pathogens and Disease* 5: 731–746.

Hanning, I., et al. 2009. Salmonellosis outbreaks due to fresh produce: sources and potential intervention measures. *Foodborne Pathogens and Disease* 6: 635–648.

Hardin, A., et al. 2010. Essential oils and antioxidants derived from citrus by-products in food protection and medicine: an introduction and review of recent literature. *Journal of Agricultural & Food Information* 11: 99–122. https://doi.org/10.1080/10496501003680680.

Hayat, K., et al. 2024. Beyond the spectrum: unleashing the potential of infrared radiation in poultry industry advancements. *Animals* 14: 1431. https://doi.org/10.3390/ani14101431

Himanthonkham, S., et al. 1996. Heat destruction of *Salmonella* in poultry feeds. *Avian Diseases* 40: 72–77.

Hossain, M.E., et al. 2024. Near-infrared spectroscopy for analysing livestock diet quality: a systematic review. *Heliyon* 10: e40016. https://doi.org/10.1016/j.heliyon.2024.e40016.

Hume, M.E., et al. 1993. Metabolism of [14C] propionic acid in broiler chicks. *Poultry Science* 72: 786–793.

Huss, A., et al. 2017a. Chapter 5. Physical and chemical methods for the reduction of biological hazards in animal feeds. In: S.C. Ricke, et al. (Eds.), *Food and Feed Safety Systems and Analysis*. Elsevier Inc., San Diego, CA. pp. 83–95. https://doi.org/10.1016/B978-0-12-811835-1.00019-1.00005-1.

Huss, A., et al. 2017b. Chapter 4. Animal feed mill biosecurity: prevention of biological hazards. In: S.C. Ricke, et al. (Eds.), *Food and Feed Safety Systems and Analysis*. Elsevier Inc., San Diego, CA. pp. 63–81. https://doi.org/10.1016/B978-0-12-811835-1.00019-1.00004-x.

Ibrahim, D., et al. 2020. Impact of feeding anaerobically fermented feed supplemented with acidifiers on its quality and growth performance, intestinal villi and enteric pathogens of mulard ducks. *Livestock Science*. 242: 104299. https://doi.org/10.1016/j.livsci.2020.104299

Indiarto, R., et al. 2023. Meat irradiation: a comprehensive review of its impact on food quality and safety. *Foods* 12: 1845. https://doi.org/10.3390/foods12091845.

Jarvis, N.A., et al. 2016. An overview of *Salmonella* thermal destruction during food processing and preparation. *Food Control* 68: 280–290. https://doi.org/10.1016/B978-0-12-81135-1.00005-1.

Jiang, L., et al. 2018. Effect of different disinfectants on bacterial aerosol diversity in poultry houses. *Frontiers in Microbiology* 9: 2113. https://doi.org/10.3389/fmicb.2018.02113.

Jimenez, C.E.P., et al. 2023. Biosecurity and water, sanitation, and hygiene (WASH) interventions in animal agricultural settings for reducing infection burden, antibiotic use, and antibiotic resistance: a one health systematic review. *Lancet Planet Health* 7: e418–434.

Jones, F.T. 2011. A review of practical *Salmonella* control measures in animal feed. *Journal of Applied Poultry Research* 20: 102–113. https://doi.org/10.3382/japr.2010-00217.

Jones, F.T. and Richardson, K.E. 2004. *Salmonella* in commercially manufactured feeds. *Poultry Science* 83: 384–391. https://doi.org/10.1093/ps/83.3.384.

Jones, F.T. and Ricke, S.C. 1994. Researchers propose tentative HACCP plan for feed manufacturing. *Feedstuffs* 66(18): 32, 36–38, 40–42.

Jones, T.F. and Yackley, J. 2018. Foodborne disease outbreaks in the United States: a historical overview. *Foodborne Pathogens and Disease* 15: 11–15. https://doi.org/10.1089/fpd.2017.2388.

Kinley, B., et al. 2010. Analysis of *Salmonella* and enterococci isolated from rendered animal products. *Canadian Journal of Microbiology* 56: 65–73. https://doi.org/10.1139/W09-108.

Krishnamurthy, K., et al. 2008. Infrared heating in food processing: an overview. *Comprehensive Reviews in Food Science and Food Safety* 7: 2–13.

Kumar, V., et al. 2024. Advancing microbial food safety and hazard analysis through predictive modeling. *The Microbe* 2: 100049. https://doi.org/10.1016.j.microb.2024.100049.

Leeson, S. and Marcotte, M. 1993. Irradiation of poultry feed I. Microbial status and bird response. *World's Poultry Science Journal* 49: 19–33.

Ley, F.J., et al. 1969. Sterilisation of laboratory animal diets using gamma radiation. *Laboratory Animals* 3: 221–254.

Li, Y., et al. 2020. Reduction, prevention, and control of *Salmonella enterica* viable but non-culturable cells in flour food. *Frontiers in Microbiology* 11: 1859. https://doi.org/10.3389/fmicb.2020.01859.

Liu, T.S., et al. 1969. Thermal resistance of Salmonella senftenberg 775W in dry animal feeds. *Avian Diseases* 13: 611–631.

Lyte, J.M., et al. 2025. Do we need a standardized 16S rRNA gene amplicon sequencing analysis protocol for poultry microbiota research? *Poultry Science* 104: 105242. https://doi.org/10.1016/j.psj.2025.105242.

Machado-Moreira, B., et al. 2019. Microbial contamination of fresh produce: what, where, and how? *Comprehensive Reviews in Food Science and Food Safety* 18: 1727–1759. https://doi.org/10.1111/1541-4337.12487.

Maciorowski, K.G., et al. 1999. Consumer food safety awareness and acceptance of irradiated raw poultry in three Texas cities. *Dairy, Food and Environmental Sanitation* 19: 18–28.

Maciorowski, K.G., et al. 2004. Incidence and control of food-borne *Salmonella* spp. in poultry feeds: a review. *World's Poultry Science Journal* 60: 446–457. https://doi.org/10.1079/WPS200428.

Maciorowski, K.G., et al. 2007. Effects of poultry and livestock feed with bacteria and fungi. *Animal Feed Science and Technology* 133: 109–136. https://doi.org/10.1016/j.anifeedsci.2006.08.006.

Maxcy, R.B. 1983. Significance of residual organisms in foods after substerilizing doses of gamma radiation: a review. *Journal of Food Safety* 5: 203–211.

McLeod V. 2018. Sterilizing safely: closer look at glutaldehyde based chemical agents. *Lab Manager* 13: 30–33.

Meeker, D.L. and Meisinger, J.L. 2015. Rendered ingredients significantly influence sustainability, quality, and safety of pet food. *Journal of Animal Science* 93: 835–847. https://doi.org/10.2527/jas2014-8524.

Milillo, S.R., et al. 2011. Immediate reduction of *Salmonella enterica* serotype Typhimurium following exposure to multiple-hurdle treatments with heated, acidified organic acid salt solutions. *Applied and Environmental Microbiology* 77: 3765–3772. https://doi.org/10.1128/AEM.02839-10.

Mogassi, G., et al. 2020. Potential risk-factors affecting *Salmonella* sp. and *Escherichia coli* occurrence and distribution in midwestern United States swine feed mills. *Journal of Applied Microbiology* 129: 1744–1750. https://doi.org/10.1111/jam.14758.

Morawicki, R. 2010. Chapter 19. Coproducts and by-products from poultry processing. In C.M. Owens, et al. (Eds.), *Poultry Meat Processing*. 2nd edition. CRC Press, Boca Raton, FL, pp. 381–396.

Mossel, D.A.A., et al. 1967. Comparative study on decontamination of mixed feeds by radicidation and by pelletisation. *Journal of Science Food and Agriculture*. 18: 362–367.

Moustafa, G.Z., et al. 2002. Hygienic control of *Salmonella* in artificially contaminated feed. *Veterinary Medical Journal* (Giza) 50: 239–246.

Mshelia, R.D., et al. 2022. Food irradiation: an effective but under-utilized technique for food preservations. *Journal of Food Science and Technology* 60(10): 2517–2525. https://doi.org/10.1007/s13197-022-05564-4.

Muck, R.E., et al. 2018. Silage review: recent advances and future uses of silage additives. *Journal of Dairy Science*. 101: 3980–4000. https://doi.org/10.3168/jds.2017-13839

Muñoz, L.R., et al. 2021. Evaluation of commercially manufactured animal feeds to determine presence if *Salmonella*, *Escherichia coli*, and *Clostridium perfringens*. *Journal of Applied Poultry Research* 30: 100142. https://doi.org/10.1016/j.japr.2021.100142.

Nguyen, T., et al. 2024. Mycotoxin biodegradation by *Bacillus* bacteria: a review. *Toxins* 16: 478. https://doi.org/10.3390/toxins16110478.

Oakley, B.B., et al. 2014. Successional changes in the chicken cecal microbiome during 42 days of growth are independent of organic acid feed additives. *BMC Veterinary Research* 10: 282. http://www.biomedcentral.com/1746-6148/10/282.

Obe, T., et al. 2023. Controlling *Salmonella*: strategies for feed, the farm, and the processing plant. *Poultry Science* 102: 103086. https://doi.org/10.1016/j.psj.2023.103086.

O'Bryan, C.A., et al. 2008. Impact of irradiation on the safety and quality of poultry and meat products: A review. *Critical Reviews in Food Science and Nutrition* 48: 442–447. https://doi.org/10.1080/10408390701425698.

O'Bryan, C.A., et al. 2015. Potential of plant essential oils and their components in animal agriculture – *in vitro* studies on antibacterial mode of action. *Frontiers in Veterinary Science* 2: 35. https://doi.org/10.3389/fvets.2015.00035.

O'Bryan, C.A., et al. 2022. Public health impact of *Salmonella* spp. on raw poultry: Current concepts and future prospects in the United States. *Food Control* 132: 108539. https://doi.org/10.1016/j.foodcont.2021.108539.

Oduola, A.A., et al. 2020. Impacts of broadband and selected infrared wavelength treatments on inactivation of microbes on rough rice. *Journal of Food Safety* 40: e12764. https://doi.org/10.1111/jfs.12764.

Oguz, H. 2011. A review from experimental trials on detoxification of aflatoxin in poultry feed. *Eurasian Journal of Veterinary Science* 27: 1–12.

Okelo, P.O., et al. 2008. Improvements in reduction of feed contamination: an alternative monitor of bacterial killing during feed extrusion. *Journal of Applied Poultry Research* 17: 219–228. https://doi.org/10.3382/japr.2007-00060.

Okelo, P.O., et al. 2006. Optimization of extrusion conditions for elimination of mesophilic bacteria during thermal processing of animal feed mash. *Animal Feed Science and Technology* 129: 116–137. https://doi.org/10.1016/j.anifeedsci.2005.12.011.

Olson, E.G., et al. 2022. Application of microbial analyses to feeds and potential implications for poultry nutrition. *Poultry Science* 101: 101789. https://doi.org/10.1016/j.psj.2022.101789.

Olson, E.G., et al. 2023. Chapter 27. Microbial pathogen contamination of animal feed. In: L. Anelich, et al. (Eds.), *Present Knowledge in Food Safety: A Risk-Based Approach Through the Food Chain*. International Life Sciences Institute, Academic Press, Elsevier. pp. 378–393. https://doi.org/10/1016/B978-0-0-12-819470-6.00023-8.

Olson, E.G., et al. 2024. Microbiome analyses of poultry feeds: part II. Comparison of different poultry feeds. *Journal of Environmental Science and Health, Part B* 59(8): 448–482. https://doi.org/10.1080/03601234.2024.2361596.

Omer, M.K., et al. 2018. A systematic review of bacterial foodborne outbreaks related to red meat and meat products. *Foodborne Pathogens and Disease* 15: 598–611. https://doi.org/10.1089/fpd.2017.2393.

Park, S.H., et al. 2011. Detection of *Salmonella* spp. survival and virulence in poultry feed by targeting the *hilA* gene. *Journal of Applied Microbiology* 111: 426–432.

Pillai, S.D. and Ricke, S.C. 2002. Aerosols from municipal and animal wastes: Background and contemporary issues, *Canadian Journal of Microbiology* 48: 681–696.

Prabha, K., et al. 2021. Recent development, challenges, and prospects of extrusion technology. *Future Foods* 3: 100019. https://doi.org/10.1016/j.fufo.2021.100019.

Reuter, G. 1998. Disinfection and hygiene in the field of food of animal origin. *International Biodeterioration and Biodegradation* 41: 209–215. https://doi.org/10.1016/S0964-8305(98)00029-8.

Ricke, S.C. 2003a. The gastrointestinal tract ecology of *Salmonella* Enteritidis colonization in molting hens. *Poultry Science*. 82: 1003–1007.

Ricke, S.C. 2003b. Perspectives on the use of organic acids and short chain fatty acids as antimicrobials. *Poultry Science*. 82: 632–639.

Ricke, S.C. 2005. Chapter 7. Ensuring the safety of poultry feed. In: G.C. Mead (Ed.), Food Safety Control in the Poultry Industry, Woodhead Publishing, Cambridge, UK. pp. 174–194. https://doi.org/10.1533/9781845690236.174.

Ricke, S.C. 2017. Insights and challenges of *Salmonella* infections in laying hens. *Current Opinions in Food Science* 18: 43–49. https://doi.org/10.1016/j.cofs.2017.10.012.

Ricke, S.C. 2018. Chap. 8. Feed hygiene. In: J. Dewulf and F. Van Immerseel (Eds.), *Biosecurity in Animal Production and Veterinary Medicine – From Principles to Practice*. ACCO (Academische Coöperative Vennootschap cvba), Leuven, Belgium. pp. 177–209.

Ricke, S.C., et al. 2012. Chapter 16 Antimicrobials in feed: Benefits and limitations. In: J. Fink-Gremmels (Ed.), *Feed and Fodder Contamination: Effects on Livestock and Food Safety*. Woodhead Publishing Ltd., Cambridge, UK. pp. 411–431.

Ricke, S.C., et al. 2017. Chap. 19. Unravelling food production microbiomes: concepts and future directions. In: S.C. Ricke, et al. (Eds.), *Food and Feed Safety Systems and Analysis*. Elsevier Inc., San Diego, CA. pp. 347–374. https://doi.org/10.1016/B978-0-12-811835-1.00019-1.

Ricke, S.C., et al. 2019. Formaldehydes in feeds and interaction with the poultry gastrointestinal tract microbial community. *Frontiers in Veterinary Science* 6: 188. https://doi.org/10.3389/fvets.2019.00188.

Ricke, S.C., et al. 2020. Formic acid as an antimicrobial for poultry production: a review. *Frontiers in Veterinary Science* 7: 563. https://doi.org/10.3389/fvets.2020.00563.

Ricke, S.C., et al. 2025. Chapter 16. Valorising byproducts from broiler processing: Biologics and sustainability. In: S.C. Ricke (Ed.), *Achieving sustainable production of poultry meat Volume 1: Safety, quality and sustainability*, Burleigh Dodds Science Publishing, Cambridge, UK. pp. 403–428. https://doi.org/10.19103/AS.2024.0146.20.

Ricke, S.C. and Schaefer, D.M. 1990. Characterization of egg yolk antibodies for detection and quantitation of rumen selenomonads using an enzyme - linked immunosorbent assay. *Applied and Environmental Microbiology* 56: 2795–2800.

Rowe, J.B., et al. 1999. Processing cereal grains for animal feeding. *Australian Journal of Agricultural Research*. 50: 721–736.

Russell, J.B. 1992. Another explanation for the toxicity of fermentation acids at low pH: anion accumulation versus uncoupling. *Journal of Applied Bacteriology* 73: 363–370.

Sacco, R.E., et al. 1989. Effect of hatching egg sanitizers on embryonic survival and hatchability of turkey eggs from different lines and on egg shell bacterial populations. *Poultry Science* 68: 1179–1184. https://doi.org/10.3382/ps.0681179.

Sarno, E., et al. 2021. A review of significant European foodborne outbreaks in the decade. *Journal of Food Protection* 84: 2059–2070. https://doi.org/10.4315/JFP-21-096.

Scott, T.A. and Swetnam, C. 1993a. Screening sanitizing agents and methods for hatching eggs I. Environmental and user friendliness. *Journal of Applied Poultry Research* 2: 1–6. https://doi.org/10.1093/japr/2.1.1.

Scott, T.A. and Swetnam, C. 1993b. Screening sanitizing agents and methods for hatching eggs II. Effectiveness against microorganisms on the egg shell. *Journal of Applied Poultry Research* 2:7–11. https://doi.org/10.1093/japr/2.1.7.

Scott, T.A., et al. 1993. Screening sanitizing agents and methods for hatching eggs III. Effect of concentration and exposure time embryo viability. *Journal of Applied Poultry Research* 2: 12–18. https://doi.org/10.1093/japr/2.1.12.

Shariat, N.W., et al. 2021. Incidence of *Salmonella* serovars isolated from commercial animal feed mills in the United States and serovar identification using CRISPR analysis. *Journal of Applied Microbiology* 130: 2141–2146. https://doi.org/10.1111/jam.14933.

Stavric, S. 1987. Microbial colonization control of chicken intestine using defined cultures. *Food Technology* 41: 93–98.

Stott, J.A., et al. 1975. Incidence of salmonellae in animal feed and the effect of pelleting on content of enterobacteriaeae. *Journal of Applied Bacteriology* 39: 41–46.

Swanson, K.S., et al. 2025. Use of biotics in animals: impact on nutrition, health, and food production. *Journal of Animal Science* 103: ska061. https://doi.org/10.1093/jas/skaf061.

Szczawińska, M.E., et al. 1991. Fate of unirradiated *Salmonella* in irradiated mechanically deboned chicken meat. *International Journal of Food Microbiology* 14: 313–324.

Thomas, M. and van der Poel, A.F.B. 1996. Physical quality of pelleted animal feed 1. Criteria for pellet quality. *Animal Feed Science Technology* 61: 89–112.

Thomas, M., et al. 1997. Physical quality of pelleted animal feed. 2. Contribution of processes and its conditions. *Animal Feed Science Technology* 64: 173–192.

Thomas, M., et al. 1998. Physical quality of pelleted animal feed. 3. Contribution of feedstuff components. *Animal Feed Science Technology* 70: 59–78.

Urango, A.C.M., et al. 2024. Maillard conjugates produced from proteins and prebiotic dietary fibers: technological properties, health benefits and challenges. *Trends in Food Science & Technology* 147: 104438. https://doi.org/10.1016/j.tifs.2024.104438.

van Immerseel, F., et al. 2006. Strategies to control *Salmonella* in the broiler production chain. *World's Poultry Science Journal* 62: 367–382. https://doi.org/10.1017/S0043933906001088.

Veldman, A., et al. 1995. A survey of the incidence of Salmonella species and Enterobacteriaceae in poultry feeds and feed components. *Veterinary Record* 136: 169–172.

Viquez-Umana, F.L., et al. 2025. Meat animal biologics discovery from the rumen microbiome: application of genomics. *Meat and Muscle Biology* 9(1): 1–17. https://doi.org/10.22175/mmb.19820.

Wages, J.A., et al. 2022. Consequences of implementing neutralizing buffered peptone water in commercial poultry processing on the microbiota of whole bird carcass rinses and the subsequent microbiological analyses. *Frontiers in Microbiology* 13: 813461. https://doi.org/10.3389/fmicb.2022.813461.

Wales, A.D., et al. 2010. Chemical treatment of animal feed and water for the control of *Salmonella*. *Foodborne Pathogens and Disease* 7: 3– 15. https://doi.org/10.1089/fdp.2009.0373.

Weinroth, M.D., et al. 2022. Considerations and best practices in animal science 16S rRNA gene sequencing microbiome studies. *Journal of Animal Science* 100: 1–18. https://doi.org/10.1093/jas/skab346.

Williams, J.E. 1981a. Salmonellas in poultry feeds: worldwide review. Part I. Introduction. *World's Poultry Science Journal* 37: 6–19.

Williams, J.E. 1981b. Salmonellas in poultry feeds: worldwide review. Part II. Methods in isolation and identification. *World's Poultry Science Journal* 37: 19–25.

Williams, J.E. 1981c. Salmonellas in poultry feeds: a worldwide review. Part III. Methods in control and elimination. *World's Poultry Science Journal* 37: 97–105.

Williams, J.E. 1970. Effect of high-level formaldehyde fumigation on bacterial populations on the surface of chicken hatching eggs. *Avian Diseases* 14: 386–392. https://doi.org/10.2307/1588482

Williams Jr., L.P., et al. 1969. A ten-month study of *Salmonella* contamination in animal protein meals. *Journal of the American Veterinary Medical Association* 155: 167–174.

Wu, S., et al. 2017. Food safety hazards associated with ready-to-eat cookie dough and its ingredients. *Food Control* 73: 986–993. https://doi.org/10.1016/j.foodcont.2016.10.010.

Zwietering, M.H., et al. 2021. All food processes have a residual risk, some are small, some very small and some are extremely small: zero risk does not exist. *Current Opinions in Food Science* 39: 83–92. https://doi.org/10.1016/j.cofs.2020.12.017.

Chapter 3

Viral pathogen contamination of animal feed products and their effects[1]

Scott Dee, Pipestone Veterinary Services, USA

1 Introduction

Effective biosecurity protocols are essential toward protecting the health status of swine farms. In the US, tremendous resources, with costs per pig weaned ranging from US$1.50 to US$3.00 US, have been invested to reduce the risk of viral pathogens, such as porcine reproductive and respiratory virus entry into susceptible populations. Protocols including shower in-shower out, transport sanitation, quarantine and testing of incoming genetics, and the filtration of incoming air are commonplace throughout the US swine industry, particularly at the level of the sow farm (Silva et al., 2018). In contrast, prior to the entry of

1 This chapter is based on and updates the following article published under an Open Access CC BY License: Dee, S.A., et al. (2020) 'The risk of viral transmission in feed: What do we know, what do we do?', *Transbound Emerg Dis*, 67(6):2365–2371.

http://dx.doi.org/10.19103/AS.2026.0172.10

porcine epidemic diarrhea virus (PEDV) in the US swine population during May 2013 (Chen et al., 2014), the role of feed as a vehicle for pathogen transport and transmission had not been considered, even though feed is delivered to swine farms daily in the absence of any biosecurity protocols. Since the identification of this novel risk factor, scientists around the world have conducted numerous studies to understand its relevance. The chapter will review these studies in detail.

This chapter will review the literature on the risk of feed and feed ingredients for the transport and transmission of viral pathogens, provide information on means currently in place to mitigate this risk, and summarize how this new risk factor has changed behavior throughout the global agriculture. Finally, as much of the published literature on feed risk has centered on viral diseases of pigs, the chapter will primarily focus on swine but will also reference recent work in poultry, regarding the risk of feed for the transmission of avian influenza virus (AIV). The goal of the chapter is to provide a comprehensive summary of the literature on the risk of feed for viral transmission to animals and to educate readers on current actions taken by the global swine industry to mitigate it.

2 Background: tracking transmission of porcine epidemic diarrhea virus

To truly appreciate the risk of feed for viral transport and transmission to livestock, it is imperative to start the discussion with the story of PEDV, an alphacoronavirus that prior to 2013 had never been detected in the US, but was known to be widespread in Europe and Asia. In early 2013, breeding herds throughout the midwestern US, experienced epidemics of high morbidity and mortality in sows and lactating piglets characterized by vomiting and diarrhea, with up to 100% mortality in suckling piglets less than 7days of age (Lee, 2015). Upon its entry, PEDV spread rapidly throughout the country at an unprecedented rate (Niederwerder and Hesse, 2018). Following phylogentic analysis, it was determined that it most likely originated from China (Huang et al., 2013). During the initial outbreak, the American Association of Swine Veterinarians, the National Pork Producers Council and the United States Department of Agriculture Center for Epidemiology and Animal Health conducted an epidemiological investigation involving porcine epidemic diarrhea (PED) affected case and control herds, and of the 100 variables surveyed, seven were significantly associated with acquiring PEDV during the process of feeding animals (http://www.aasv.org).

In 2014, the risk was confirmed when it was proven that ingestion of contaminated complete feed could serve as a vehicle for PEDV transmission to naïve pigs (Dee et al., 2014a). This study involved the detection of the virus in feed dust samples from the interior walls of feed bins that had provided feed for the index cases of PED in sows across a subset of farms, followed by a demonstration of virus viability using a bioassay model involving pigs consuming this material via natural feeding behavior. Within 3–4 days post ingestion, clinical signs of PED (vomiting and diarrhea) and the detection of PEDV RNA in rectal swabs along with lesions in the gastrointestinal tract suggestive of PED were documented. Shortly thereafter, the minimum infectious dose of PEDV in feed was determined to be 5.6×10^1 Tissue Culture Infectious $Dose_{50}$/g using the 10-day-old piglet bioassay (Schumacher et al., 2016).

Another important finding was the potential for widespread PEDV contamination of surfaces in an animal food manufacturing facility (Schumacher et al., 2017). In this study, a virulent US PEDV isolate was used to inoculate 50 kg of swine feed, which was then mixed, conveyed, and discharged into bags using pilot-scale feed manufacturing equipment. Subsequent collection of environmental swabs demonstrated widespread distribution of virus via feed dust, with the presence of PEDV RNA in 100% of dust samples collected from animal food-contact surfaces and in 89% of dust samples from non-animal food-contact surfaces.

Once contamination of the feed mill environment was demonstrated, the question of whether viral survival would differ across the various feed ingredients found in a milling environment was investigated (Dee et al., 2015, Trudeau et al., 2017a). Interestingly, viable PEDV was detected out to 180 days post-inoculation (DPI) in conventional (high-protein/low-fat) soybean meal, as well as out to 30 DPI in Distillers Dried Grains with Solubles, meat and bone meal, red blood cells, lysine HCL, D/L methionine, choice white grease, choline chloride, and out to 7 DPI in limestone and 14 DPI in threonine. In contrast, viable PEDV was not present in several other ingredients, including corn, various animal protein sources, and vitamin/trace mineral mixes.

These data, along with observations from the field, posed the question of whether certain ingredients could serve as vehicles for the movement of PEDV between countries. This issue was raised in January 2014, when PEDV was detected for the first time in Ontario, Canada (Pasma et al., 2016). Following extensive epidemiologic investigation, the source of the virus appeared to be contaminated samples of spray-dried plasma protein originating from the US (Perri et al., 2018, 2019). However, while infectious PEDV was demonstrated in samples of case-specific plasma by swine bioassay, transmission of the virus to pigs following consumption of feed containing PEDV-positive plasma was not successful (Pascik et al., 2014).

Building on the potential for transboundary movement of PEDV, a Trans-Pacific model, simulating the movement of cargo from Beijing, China, to the Anquin terminal in Shanghai, China, to port of San Francisco, CA, US, and then to Des Moines, IA, US, was developed (Dee et al., 2016). The model utilized transport times, environmental conditions, and feed ingredients (www.hs

.usitc.gov) representative of transport of cargo from China to the US. Under the conditions of this study, PEDV survived the trans-oceanic simulation in soy-based products, including lysine, choline, and vitamin D. Surprisingly, the virus did not survive in the absence of a feed matrix, suggesting that survival is dependent upon the presence of the ingredient, not the container (tote) per se.

3 Tracking the transmission of other viruses

Based on these collective data involving PEDV, the transboundary experiment was repeated across 11 other viruses, including Foot and Mouth Disease Virus (FMDV), Classical Swine Fever Virus (CSFV), African Swine Fever Virus (ASFV), Influenza A Virus of Swine (IAV-S), Pseudorabies virus (PRV), Nipah Virus (NiV), Porcine Reproductive and Respiratory Syndrome Virus (PRRSV), Swine Vesicular Disease Virus (SVDV), Vesicular Stomatitis Virus (VSV), Porcine Circovirus Type 2 (PCV2), and Vesicular Exanthema of Swine Virus (VESV) (Dee et al., 2018). In certain cases, surrogate viruses were employed, including Senecavirus A (SVA) for FMDV, Bovine Viral Diarrhea Virus (BVDV) for CSFV, Bovine Herpesvirus Type 1 (BHV-1) for PRV, Canine Distemper Virus (CDV) for Nipah Virus, Porcine Sapelovirus (PSV) for SVDV, and Feline Calicivirus (FCV) for VESV.

Since ASFV had not been reported in China at the time of the study and was actively circulating in Eastern Europe, a Trans-Atlantic model was developed involving representative feed ingredients, transport times and environmental conditions simulating shipment from Warsaw, Poland to the port of LeHarve, France, to the port of New York City, NY, US, and then to Des Moines, IA, US (Dee et al., 2018). Under the conditions of this study, SVA (representing FMDV), FCV (representing VESV), BHV-1 (representing PRV), PRRSV, PSV (representing SVDV), ASFV, and PCV2 maintained infectivity, while BVDV (representing CSFV), VSV, CDV (representing NiV), and IAV-S did not. As before, the majority of viruses survived in conventional soybean meal, lysine hydrochloride, choline chloride, vitamin D, and pork sausage casings. These collective results led to the development of the concept of 'high-risk combination' (the right virus in the right ingredient) and contributed to the growing body of evidence that contaminated feed ingredients may represent a risk for the transport of pathogens at domestic and global levels.

These findings became significant following the announcement of the initial cases of ASFV in China's pig population during August 2018 for several reasons:

1. Fifty percent of the world's pigs were in China at the time.
2. The Chinese national herd was historically naïve to ASFV.
3. ASFV had survived in 9 of 12 of the ingredients utilized in the study, including three varieties of soy-based products, choline chloride, three

types of pet food, pork sausage casings and complete feed (Dee et al., 2018).

4. The US imported approximately 2 M metric tons of agricultural products, including 55 000 metric tons of soy-based products from China, along with 45 000 metric tons and 3 000 metric tons of soy-based products from the Ukraine and Russia, respectively, in 2018 (Patterson et al., 2021).

4 Understanding the role of feed in virus transmission

This crisis accelerated the research efforts to better understand the risk of feed, specifically as it pertained to ASFV. This resulted in the seminal work of Niederwerder and others, who documented transmission of ASFV to naïve pigs following natural consumption of contaminated feed and water (Niederwerder et al., 2019). This study determined the minimum infectious dose of ASFV in liquid (10^1 $TCID_{50}$) and in feed (10^4 $TCID_{50}$) following a one-time exposure. However, further analysis indicated that the more frequent the exposure (3×, 10×, 30×) to virus in feed or water, the higher the probability of infection, even in the presence of lower doses (10^2 $TCID_{50}$).

Another significant finding was the calculation of ASFV half-life in feed ingredients. Original estimates, based on limited (n = 2) data points derived from the trans-Atlantic model indicated that half-life ranged from 4–5 days across all nine of the virus-positive ingredients (Dee et al., 2018). Niederwerder and Stoian used the trans-Atlantic model to conduct a more comprehensive half-life evaluation, incorporating data from all four sampling points in the model, resulting in half-live values that ranged from 9.6 days to 14.2 days (inclusive of 95% confidence intervals) across all 9 supportive ingredients (Stoian et al., 2019), suggesting that ASFV survival could occur far beyond the 30-day transport period used in the model.

Furthermore, Stenfeldt and others reported that FMDV may remain infectious in pig feed ingredients for durations compatible with transoceanic transport (Stenfeldt et al., 2022). Although the observed minimum infectious dose was relatively high, it was concluded that variations in feeding conditions and biophysical characteristics of different virus strains may alter the probability of infection. Finally, the transoceanic model was also applied to SVA, PRV, and CSFV, once again demonstrating survival of these pathogens in feed and feed ingredients (Stoian et al., 2020, Caserta et al., 2022). In conclusion, these lab-based studies provided the initial evidence that specific viral agents in combination with the proper ingredient can survive long-distance transport under simulated transboundary conditions. As this was a new information to the veterinary profession and animal industries, following publication of these data, attempts to quantify this risk were undertaken.

Using a hazard analysis, Jones determined that there was a measurable risk for feed-based transmission of porcine delta coronavirus, SVA, FMDV, CSFV, Chinese PRV, PEDV, and ASFV through the US feed supply chain (Jones et al., 2020). In another assessment, Shurson and others concluded that no system nor standardized analytical procedures at commercial laboratories existed to facilitate and accurately measure potential viable virus contamination in feed ingredients and complete feeds globally, and as a result, there was high uncertainty of the extent of swine virus contamination in global feed supply chains (Shurson et al., 2024).

However, in 2022, SVA was detected for the first time in the national swine herd in the country of Chile, following the importation and feeding of soybean meal from endemically infected countries including the USA, China, and Brazil (Dee et al., 2022). This was a significant case, as SVA RNA was detected in the imported meal and pigs consuming diets formulated with this soybean meal became infected with SVA, and affected animals demonstrated vesicular lesions on the snout and feet. As this was the initial report of a vesicular disease in its national swine herd, immediate efforts were put in place by the Chilean governmental to eliminate the virus. While it was fortunate that the virus in question was SVA, and not ASFV or FMDV, this case proved that while the risk of viral transmission via feed may be low, it could occur.

Based on this case, a quantitative risk assessment model to estimate the probability that one or more corn or soybean meal ocean vessels (25 000 tons) contaminated with ASFV would be imported into the United States annually was developed (Schambow et al., 2022). The model's probability estimate was conditionally based on five likelihoods: the probability of initial ASFV contamination, ASFV inactivation during processing, transport recontamination, and ASFV inactivation while awaiting customs clearance at United States entry. Results indicated that days of transport, virus survival during transport, and number of vessels shipped were the parameters most influential for increased likelihood of a vessel with ASFV-contaminated soybean meal or corn entering the United States.

To further bridge the gap from the laboratory to the field, Dee and others tested whether three significant viruses of swine could survive in feed ingredients during long-distance commercial transport across the continental US (Dee et al., 2023). One-metric ton totes of soybean meal (organic and conventional) and complete feed were spiked with PRRSV, PEDV, and SVA and transported for 23 days across the US in a commercial semi-trailer truck, crossing 29 states, and 10 183 km. In addition, the same feed material was also spiked with *Emiliania huxleyi* virus (EhV), a surrogate virus for ASFV recovered from marine algae (*Emiliania huxleyi*) [30]. At the end of the journey, viable PRRSV, PEDV, SVA, and EhV were detected in both soy products and viable PEDV, SVA, and EhV in complete feed. These results provided additional evidence that viral pathogens of pigs and a novel surrogate for ASFV can survive in representative volumes of feed and feed ingredients during long-distance commercial transport across the continental United States (Palowski et al., 2022).

In further support of the risk of feed and/or feed transport, the introduction of PEDV into a 10 000-sow farrow-to-wean farm located in northwest Mexico was published (Garrido-Mantilla et al., 2022). Based on data collected during the outbreak investigation, it was concluded that the most probable route for PEDV incursion into this breeding herd was contaminated feed or a contaminated feed transport vehicle.

Finally, besides the potential of the feed transport vehicle to move viral contaminated feed, ASFV DNA persistence in feed manufacturing surfaces was assessed. This novel study involved the use of a model feed mill housed in a BSL-3 laboratory facility purposefully contaminated with the virus in feed (Houston et al., 2022). Results indicated that ASFV DNA was detectable on feed manufacturing surfaces for at least 180 days with no overt evidence of reduction, highlighting the importance of bio-exclusion of ASFV within feed manufacturing facilities and the need for thorough/effective decontamination and other mitigation processes in affected areas.

In closing, these studies suggest that while the risk of viral transport and transmission via feed and feed ingredients may be of low-to-moderate risk, it is both a domestic issue and a transboundary threat. Therefore, due to its significant impact on animal health and welfare, as well as farm economics, it is important to understand how to manage and prevent this risk.

5 Assessing feed risks for swine and other livestock species

So far the risk of viral transmission via feed has focused on plant-based/non-porcine feed ingredients. However, as the original route of entry of PEDV from the US to Canada was proposed to be through importation of porcine plasma, it is important to review the literature on this topic. In January 2014, approximately 9 months following the initial detection of PEDV in the US, the first case of PED was confirmed in a swine herd in southwestern Ontario, Canada, and a follow-up epidemiological investigation pointed to feed as a common risk factor. Several lots of feed and spray-dried porcine plasma (SDPP) were tested for the presence of PEDV RNA by PCR and found to be positive, supporting the notion that contaminated feed may have been responsible for the introduction of PEDV into Canada. These findings led to conduct a bioassay experiment in which three PEDV-positive SDPP samples (from a single lot) and two PEDV-positive feed samples supplemented with this SDPP were used to orally inoculate 3-week-old piglets. Although the feed-inoculated piglets did not show any significant excretion of PEDV, the SDPP-inoculated piglets shed PEDV at a relatively high level for ≥9 days, indicating the possibility for this ingredient to harbor infectious virus and to transmit it to pigs following ingestion (Pascik, 2014).

Regarding the risk of ASFV infection through porcine plasma, Fischer and others demonstrated that SDPP contaminated with ASFV and stored at 4°C remains infectious for at least 5 weeks. In contrast, spiked SDPP stored at room temperature displayed a distinct ASFV titer reduction after 1 week and complete inactivation after 2 weeks (Fischer et al., 2021). In conclusion, they concluded that the residual risk of ASFV transmission through externally contaminated SDPP is low if SDPP is stored at room temperature for a period of at least 2 weeks before feeding.

As viral risk of animal feed has primarily been a focus of research in the swine industry, little information currently exists regarding other species, such as poultry or cattle. A recent paper by Azeem et al. described the potential role of feedstuffs for the area spreading of avian influenza virus (AIV). Thirty feed samples collected across commercial layer facilities that experienced highly pathogenic avian influenza (H5N2) in 2014–2015 and four feed mills were tested by PCR and virus isolation. Twenty-one of the 30 samples were PCR positive, but all were negative by virus isolation. Secondly, feed materials obtained from an AIV-negative farm were spiked with various concentrations of a low pathogenic AIV H5N2. In the spiked feeds, the AIV titer was 1–3 logs lower than a cell culture media control, even when tested immediately after spiking, suggesting that feed might have a negative impact on the virus or PCR detection. Collectively, the study observations indicated that AIV survivability in feed is relatively low, thus rendering it a low risk.

6 Current strategies to mitigate feed risks

Since the discovery of PEDV in the US and the role that feed appears to play in the epidemiology of the disease, there has been extensive effort put forth to evaluate the efficacy of multiple protocols and products to reduce risk. Current attempts to mitigate this risk have centered on one of four approaches: mechanical reduction (flushing and sequencing of feed batches), heat treatment, chemical mitigation, and storage time of feed prior to feeding (Stewart et al., 2020). Regarding ASFV, the notifiable disease is considered a major threat to the pig industry, animal health, and food security worldwide. According to the European Food Safety Authority, ASF virus (ASFV) survival and transmission in feed and feed materials is a major research gap that represents a massive threat to animal health and the pig industry worldwide (Fischer, 2021). As ASFV is efficiently transmitted via blood and meat from infected animals and can be highly stable in the environment, there is a concern about the potential role of contaminated raw materials used for feed in the spread of ASFV, especially crops and derived products originating from areas with ASF in wild boar and thus with high environmental ASFV contamination. However, little is known about the stability of ASFV on contaminated crops and possible inactivation methods.

To measure the efficacy of these methods, approaches to sampling and testing feed for the presence of viral nucleic acid and viable virus have been developed (Jones et al., 2019, Shurson et al., 2024). The findings of Jones and others indicated that the sampling of bulk ingredients stored in totes or containers for PEDV should include a composite of ten individual samples, collected in a designated pattern (Jones et al., 2019), followed by testing by PCR to detect nucleic acid and if possible, swine bioassay to determine viral viability. During this evaluation, it was reported that the sensitivity of detection was dependent upon dose, and a consistent reduction in viral load during the extraction methods involving feed samples can be expected. In addition, the use of surrogates such as the previously described algal-based EhV could be successfully recovered from spiked samples, viral load quantified, and tested for viability (Shurson et al., 2024). This novel approach is a safe surrogate for ASFV and could also provide opportunities to further investigate sampling of feed and feed ingredients. While more work is needed, these are good first steps to not only conduct surveillance and measure the risk of potentially contaminated feed but validate the efficacy of mitigation protocols as well, which will now be discussed.

7 Mechanical reduction: flushing and sequencing

Several experiments have been conducted to assess the efficacy of decontaminating PEDV-positive feed and feed manufacturing facilities through the physical process of mixing, using repeated sequencing of clean feed following known contaminated batches or using chemically treated rice hulls. Regarding sequencing, results demonstrated that sequenced batches of feed had reduced quantities of PEDV RNA, although sequenced feed without detectable PEDV RNA was still infectious (Schumacher et al., 2018). Therefore, this protocol can reduce but not eliminate the risk of producing infectious PEDV carryover from the first batch of feed.

Another study tested whether flushing feed with rice hulls, treated with formaldehyde or medium chain fatty acid blends, reduced the quantity of detectable RNA present after mixing a batch of PEDV-positive feed (Gebhardt et al., 2018). Elijah and others demonstrated that sequencing batches of feed was shown to decrease the concentration of ASFV contamination but did not eliminate it (Elijah et al., 2022). This study also supported previously described sampling methods for PEDV-contaminated feed, showing that bulk ingredients can be accurately evaluated for ASFV contamination by collecting ten subsamples using the sampling method described in the paper.

8 Heat treatment

Several studies have demonstrated a positive effect of temperature on the survival of PEDV in feed. Early work on the effect of heat-treatment by Trudeau et al. indicated that heating swine feed at temperatures over 130°C effectively reduced PEDV survival (Trudeau et al., 2016, 2017b). Finally, regarding pelleting, conditioning, and pelleting temperatures above 54.4°C were effective in reducing the quantity and infectivity of PEDV in swine feed. In contrast, viable virus was present following exposure to lower (37.8°C and 46.1°C) conditioning temperatures (Cochrane et al., 2017).

Regarding ASFV, Songkasupa et al. determined the thermal inactivation of ASFV in non-animal and animal origin feed ingredients. The rate of thermal inactivation was represented by decimal reduction time (D_T) or the time required to reduce ASFV per 1 log at temperature T. The mean D_{60}, D_{70}, D_{80}, and D_{85} of meat and bone meal (MBM), soybean meal (SBM), and maize grain (MZ) are in the ranges 5.11–6.78, 2.19–3.01, 0.99–2.02, and 0.16–0.99 min, respectively, which were determined not to be statistically different from each other (Songkasupa et al., 2022). In addition, work by Azeem et al. demonstrated that heating (37°C) had a negative effect on the survivability of AIV in feed.

Regarding porcine plasma, a spray-drying process used for commercial preparation has been tested for its ability to neutralize PEDV, involving an inlet air temperature of 166°C, an aspiration rate of 0.6 m^3/min, an outlet temperature of 80 °C, and an 820 mL/h sample flow rate under 0.1 MPa of pressure. The resulting spray-dried plasma powder was found to be non-infectious when fed to pigs (Gerber et al., 2014).

9 Chemical mitigation

Extensive studies have been conducted to evaluate the effect of chemical mitigation on virus-contaminated feed, with a focus on PEDV and ASFV. The initial work revolved around a formaldehyde-based product (Sal CURB® Kemin Industries, Des Moines, IA, USA), an FDA-approved liquid antimicrobial used to control Salmonella contamination in poultry and swine diets. In groups of pigs fed Sal CURB®-treated feed spiked with PEDV versus non-treated feed, clinical signs of PEDV infection (vomiting and diarrhea) and viral shedding in feces were observed 2–3 days post-consumption of non-treated feed. In contrast, no evidence of infection was observed in pigs fed formaldehyde-treated feed (Dee et al., 2014b). Expanding the application of chemical mitigation, Niederwerder and others evaluated the efficacy of both medium-chain fatty acid (monoglycerides) and formaldehyde-based feed additives in inactivating ASFV (Niederwerder et al., 2021). Feed additives were tested in cell culture and in feed ingredients under a transoceanic shipment model, and both chemical additives reduced ASFV infectivity in a dose-dependent manner.

Building on the role of monoglycerides, Phillips et al. conducted a dose-titration with all monoglyceride-treated groups of pigs (1.5, 2.5, and 3.5 kg/t) exhibiting 100% protection from PEDV (Phillips et al., 2022). In another study, feed samples were spiked with PEDV and mixed with either organic acid mixtures, sugar, or salt was incubated at room temperature for up to 21 days. All additives tested were effective in reducing the survival of PEDV as compared to controls (Trudeau et al., 2016). To expand the portfolio, Jackman et al. evaluated the antiviral activities of selected Medium Chain Fatty Acids (MCFAs), namely caprylic, capric, and lauric acids, and a related monoglyceride, glycerol monolaurate (GML), to inhibit ASFV in liquid and feed conditions were investigated (Jackman et al., 2020). The findings from this study indicate that MCFA and GML inhibit ASFV in liquid conditions and that GML is also able to reduce ASFV infectivity in feed.

Dee evaluated the ability of 15 different feed additives to mitigate the risk of virus-contaminated feed using an 'iceblock' challenge model, containing equal concentrations of PRRSV, SVA, and PEDV (Dee and others, 2021). In 14 of the additives tested, pigs on supplemented diets had significantly greater average daily gain (ADG), significantly lower clinical signs and infection levels, and numerically lower mortality rates compared to non-supplemented controls. In conclusion, most of the additives evaluated mitigated the effects of PRRSV 174, PEDV, and SVA in contaminated feed, resulting in improved health and performance.

Similarly, regarding AIV, Toro et al., 2016, demonstrated that a blend of formaldehyde, propionic acid, terpenes, and surfactant or a blend of approved phytochemicals and carboxylic acids effectively inactivated AIV in chicken feed.

10 Extended storage

The final control strategy to discuss is extended storage. With the generation of new knowledge on viral half-life in feed, it now became possible to store feed in temperature-controlled facilities for pre-determined periods of time to facilitate viral decay and reduce viability. Niederwerder evaluated the ability of ASFV survive in three feed matrices (complete feed, soybean meal, and ground corncobs) exposed to three environmental storage temperatures (40°F, 68°F, 95°F) for up to 365 days (Niederwerder et al., 2022). Results indicated ASFV DNA was highly stable and detectable by qPCR in almost all feed matrices through the conclusion of each study. Infectious ASFV was most stable in soybean meal, maintaining infectivity for at least 112 days at 40°F, at least 21 days at 68°F, and at least 7 days at 95°F. As a follow-up, Houston demonstrated that the quarantining of feed may help reduce but does not eliminate the presence of ASFV DNA in feed over time when products are held at room temperature (Houston et al., 2023). In addition, ASFV DNA was detectable on feed manufacturing surfaces for at least 180 days with no overt evidence of reduction.

From the European perspective, Fischer et al., 2020, tested the effect of drying and heat treatment on the inactivation of ASFV on six different types of field crops: wheat, barley, rye, triticale, corn, and peas, which were artificially contaminated with infectious blood. Samples were analyzed for the presence of viral DNA by PCR and infectious virus after 2-h drying at room temperature or after drying and 1 h exposure to moderate heat at a specific temperature between 40°C and 75°C. ASFV genome was detected in all samples by PCR, including samples that had been dried for 2 h and incubated for 1 h at 75°C. On the other hand, no infectious virus could be detected after 2 h of drying using virus isolation. Based on these data, investigators concluded that the risk of ASFV transmission via contaminated crops is most likely low following incubation for at least 2 h at room temperature. Nonetheless, to minimize the risk of transmission as much as possible crops from ASF-affected zones should not be used for pig feed (Fischer et al., 2020).

Looking at the relationship of storage time and temperature, Dee and others inoculated 1-metric-ton totes of conventional soybean meal with PRRSV and SVA, stored totes for 30 days at 23.9°C, 15.5°C, or 10°C, and subsequently fed them to pigs. Results indicated that inactivation of both viruses occurred in soy stored at 23.9°C. In contrast, SVA infectivity was observed in soy stored at both 15.5°C and 10°C, while PRRSV 144 L1C variant infectivity was only observed in soy stored at 10°C. These results suggest that a storage period of 30 days and a temperature of 23.9°C may assist in the reduction of the risk of virus-contaminated plant-based feed ingredients, such as soybean meal (Dee et al., 2022).

11 Managing imported feed

In summary, there is a growing body of scientific evidence suggesting that contaminated feed and feed ingredients are risk factors for the spread of viral diseases at the domestic and the transboundary levels. This information has stimulated collaborative efforts across North America between livestock and grain commodity groups, governmental agencies, and the veterinary profession to manage this risk. In the US, the application of a 'Responsible Imports' approach has been adopted across the swine industry (Patterson et al., 2019). Responsible Imports, a science-based protocol to safely introduce essential feed ingredients from high-risk countries, is based on the following principles:

- **Necessity**: is importation of the ingredient an absolute necessity?
- **Alternatives:** can the ingredient be obtained from a country free from foreign animal diseases?

- **Virus:** which virus is causing the concern?
- **Viral half-life:** is there published information on the half-life of the virus in the designated ingredient?
- **Transport time:** what is the projected time for delivery of the ingredient from the source to its destination?
- **Viral load:** are there safe products that can be added to the ingredient to reduce viral load during transport?
- **Storage period:** is there published information on storage time and temperature that will eliminate residual virus from the ingredient prior to use?

Therefore, as production companies across the US develop storage facilities for incoming products, a new way of thinking is taking shape, one that is based on 'feed quarantine' that brings together information across several fronts, including feed science, microbiology, and oceanic transport logistics, to understand how to minimize risk. This approach is intriguing as it is non-regulatory in nature and does not negatively impact trade.

Along these lines, the Canadian Food Inspection Agency has implemented a national policy-based program using designated secondary control zones to manage the introduction of high-risk feed ingredients, such as grains, oilseeds, and meals from 44 ASFV-positive countries into Canada, using extended storage in designated facilities (Calvin et al., 2022).

While a limited number of Canadian seaports allows for efficient allocation of resources for this effort, this is not the case in the US. However, new analytic tools have been developed to track where high-risk imports, i.e., soy-based products from China, are entering the US. For example, a recent analysis by Patterson and others indicated that the vast majority (>80%) of soy-based imports entered the US at 2 of 13 seaports: San Francisco/Oakland, CA (60.4%) and Seattle, WA (20.5%), and this knowledge clearly could enhance the allocation of resources to these specific high-risk sites (Patterson et al., 2021).

Furthermore, following passage of resolutions from the National Pork Producers Council requesting collaborative efforts across North America to reduce the risk of foreign animal disease entry via the risk of feed ingredients by rapidly adapting the use of feed additives as chemical mitigants and the Responsible Imports. In addition, a US feed safety task force involving representatives from governmental agencies, as well as industry stakeholder groups has been formed to develop a national plan to manage this risk. This group will provide guidance to another new initiative based on the National Poultry Improvement Plan, known as the US Swine Health Improvement Plan during the development of a national standard for feed biosecurity for the prevention of ASFV and CSFV incursion into the US swine herd.

In addition, the US Food and Drug Association has modified the language of GFI #245, entitled 'Hazard Analysis and Risk-Based Preventative Controls for Feed for Animals,' inserting the statement that 'Biological hazards in animal feed could also include viral pathogens.' Finally, based on the risk of viral incursion through feed imports, a questionnaire-based program is in place in Australia for FMDV and Denmark has recently enacted the requirement of heat-treating soy-based products imported from ASFV-positive countries (Dahl, 2024).

12 Conclusion and future trends

However, despite the progress that has been made, significant research gaps still exist regarding the risk of feed. For example, many of the published papers are based on experimental inoculation and models and more field-based information, via research and case studies is required. Furthermore, despite early efforts, a universally validated method to test feed is not available and routine testing of feed and feed ingredients is not permitted in certain countries such as the US. In addition, more information on risk management and how to logistically apply large-scale mitigation strategies is needed.

In closing, while challenging, the science supports the need to manage the risk of feed across the local farm, the large-scale production system, or the country. Based on this new information, it is hoped that all the efforts reviewed in this chapter will continue to stimulate communication and collaboration between the feed and livestock industries, resulting in the emerging concept of "global feed biosecurity". Ideally, current and future information regarding the risk of pathogen spread in feed will enhance the accuracy of risk assessments, drive the continual development of efficacious feed-based mitigation strategies, and ultimately change the philosophy regarding the global trade of feed ingredients from one that is based on price to one where country of origin health status is a major consideration.

13 Where to look for further information

To close the chapter, here are some resources on the topic of feed risk. Please check the representative websites of the organizations listed for contact information and meeting dates.

13.1 Seminal papers

Dee, S., et al. (2014a), 'An evaluation of contaminated complete feed as a vehicle for porcine epidemic diarrhea virus infection of naïve pigs following consumption via natural feeding behavior: proof of concept', *BMC Vet Res*, 10:176.

Niederwerder, M.C., et al. (2019), 'Infectious dose of African swine fever virus when consumed naturally in liquid or feed', *Emerg Inf Dis*, 25(5):891–897. https://doi.org/10.3201/eid2505.181495.

Niederwerder, M.C., et al. (2022), 'Stability of African swine fever virus in feed during environmental storage', *Transbound Emerg Dis*, 69(6):3216–3224. https://doi.org/10.1111/tbed.14666.

Dee, S., et al. (2016), 'Modeling the transboundary risk of feed ingredients contaminated with porcine epidemic diarrhea virus', *BMC Vet Research*, 12:51. https://doi.org/10.1186/s12917-016-0674-z.

Dee, S.A., et al. (2017), 'Survival of viral pathogens in animal feed ingredients under transboundary shipping models, *PLoS One*, 13(3):e0194509. https://doi.org/10.1371/journal.pone.0194509.

Dee, S.A., et al. (2022), 'Detection of Senecavirus A in pigs from a historically negative national swine herd and associated with feed imports from endemically infected countries', *Transbound Emerg Dis*, 1–3, https://doi.org/10.1111/tbed.14684.

Feed Risk Special Issue: Transboundary and Emerging Diseases, Volume 69, number 1, Jan 2022.

13.2 Organizations

- Midwest Animal Scientists.
- US Animal Health Association.
- American Association of Swine Veterinarians.
- American Veterinary Medical Association.

14 References

American Association of Swine Practitioners, National Pork Producers Council and USDA Center of Epidemiology and Animal Health. PEDV investigation (2013), http://www.aasv.org. (accessed 22 July 2015).

Azeem, S., et al. (2022), 'Evaluation of feedstuffs as a potential carrier of avian influenza virus between feed mills and poultry farms', *Pathogens*, 11(7):755. https://doi.org/10.3390/pathogens11070755.

Calvin, S., et al. (2022), 'African swine fever risk and plant-based feed ingredients: Canada's approach to risk management of imported feed products', *Transbound Emerg Dis*, 69(1):176–181. https://doi.org/10.1111/tbed.14281.

Caserta, L., et al. (2022), 'Stability of Senecavirus A in animal feed ingredients and infection following consumption of contaminated feed', *Transbound Emerg Dis*, 69(1):88–96. https://doi.org/10.1111/tbed.14310.

Chen, Q., et al. (2014), 'Isolation and characterization of porcine epidemic diarrhea viruses associated with the 2013 disease outbreak among swine in the United States', *J Clin Microbiol*, 52:234–243.

Cochrane, R.A., et al. (2017), 'Effect of pelleting on survival of porcine epidemic diarrhea virus-contaminated feed', *J Anim Sci*, 95(3):1170–1178. https://doi.org/10.2527/jas.2016.0961.

Dahl, J. (2024), 'Risk assessments and risk mitigation to prevent the introduction of African Swine fever into the Danish Pig Population', *Animals*, 14(17):2491. https://doi.org/10.3390/ani14172491.

Dee, N., et al. (2022), 'Evaluating the effect of temperature on viral survival in plant-based feed during storage', *Transbound Emerg Dis*, 69:e2105–e2110. https://doi.org/10.1111/tbed.14546.

Dee, S.A., et al. (2014a), 'An evaluation of contaminated complete feed as a vehicle for porcine epidemic diarrhea virus infection of naïve pigs following consumption via natural feeding behavior: proof of concept', *BMC Vet Res*, 10:176.

Dee, S.A., et al. (2014b), 'An evaluation of a liquid antimicrobial (Sal CURB®) for reducing the risk of porcine epidemic diarrhea virus infection of naïve pigs during consumption of contaminated feed', *BMC Vet Res*, 10:220. https://doi.org/10.1186/s12917-014-0220-9.

Dee, S.A., et al. (2015), 'An evaluation of porcine epidemic diarrhea virus survival in individual feed ingredients in the presence or absence of a liquid antimicrobial', *Porcine Health Manag*, 1:9. https://doi.org/10.1186/s40813-015-0003-0.

Dee, S.A., et al. (2016), 'Modeling the transboundary risk of feed ingredients contaminated with porcine epidemic diarrhea virus', *BMC Vet Research*, 12:51. https://doi.org/10.1186/s12917-016-0674-z.

Dee, S.A., et al. (2018), 'Survival of viral pathogens in animal feed ingredients under transboundary shipping models, *PLoS One*, 13(3):e0194509. https://doi.org/10.1371/journal.pone.0194509.

Dee, S.A., et al. (2021), 'An evaluation of additives for mitigating the risk of virus-contaminated feed using an ice-block challenge model', *Transbound Emerg Dis*, 68(2):833–845. https://doi.org/10.1111/tbed.13749.

Dee, S.A., et al. (2022), 'Detection of Senecavirus A in pigs from a historically negative national swine herd and associated with feed imports from endemically infected countries', *Transbound Emerg Dis,* 1–3. https://doi.org/10.1111/tbed.14684.

Dee, S.A., et al. (2023), 'Evidence of viral survival in representative volumes of feed and feed ingredients during long-distance commercial transport across the continental United States', *Transbound Emerg Dis*, 2023:3455128. https://doi.org/10.1155/2023/3455128.

Elijah, C.G., et al. (2022), 'Effect of mixing and feed batch sequencing on the prevalence and distribution of African swine fever virus in swine feed', *Transbound Emerg Dis*, 69(1):115–120. https://doi.org/10.1111/tbed.14177.

Fischer, M., et al. (2020), 'Stability of African swine fever virus on heat-treated field crops', *Transbound Emerg Dis*, 67(6):2318–2323. https://doi.org/10.1111/tbed.13650.

Fischer, F., et al. (2021), 'Stability of African swine fever virus on spiked spray-dried porcine plasma', *Transbound Emerg Dis*, 68(5):2806–2811. https://doi.org/10.1111/tbed.14192.

Garrido-Mantilla, J., et al. (2022), 'Feed or feed transport as a potential route for a porcine epidemic diarrhoea outbreak in a 10,000-sow breeding herd in Mexico', *Transbound Emerg Dis*, 69(1):66–71. https://doi.org/10.1111/tbed.14354.

Gebhardt, J.T., et al. (2018), 'Evaluation of the effects of flushing feed manufacturing equipment with chemically treated rice hulls on porcine epidemic diarrhea virus cross-contamination during feed manufacturing', *J Anim Sci*, 29;96(10):4149–4158. https://doi.org/10.1093/jas/sky295.

Gerber, P.F., et al. (2014), 'The spray-drying process is sufficient to inactivate infectious porcine epidemic diarrhea virus in plasma', *Vet Microbiol*, 174(1):86–92. https://doi.org/10.1016/j.vetmic.2014.09.008.

Houston, G.E., et al. (2022), 'Detection of African swine fever virus in feed and feed mill environment following extended storage', *Transbound Emerg Dis*, 69(1):149–156. https://doi.org/10.1111/tbed.14057.

Houston, G.E., et al. (2023), 'Detection of African swine fever virus in feed and feed mill environment following extended storage', *Transbound Emerg Dis*, 2023:3455128. https://doi.org/10.1155/2023/3455128.

Huang, Y.W., et al. (2013), 'Origin, evolution, and genotyping of emergent porcine epidemic diarrhea virus strains in the United States', American Association of Swine Practitioners, National Pork Producers Council and USDA Center for Epidemiology and Animal Health. PEDV Investigation. http://www.aasv.org. (accessed 22 July 2015).

Jackman, J.A., et al. (2020), 'Inhibition of African swine fever virus in liquid and feed by medium-chain fatty acids and glycerol monolaurate', *Anim Sci Biotechnol*, 8;11(1):114. https://doi.org/10.1186/s40104-020-00517-3.

Jones, C.K., et al. (2019), 'Validation of sampling methods in bulk feed ingredients for detection of swine viruses', *Trans Emerg Dis*, 67:1–5. https://doi.org/10.1111/tbed.13326.

Jones, C.K., et al. (2020), 'Reviewing the risk of feed as a vehicle for swine pathogen transmission', *Vet Med Sci*, 6(3):527–534. https://doi.org/10.1002/vms3.227.

Lee, C. (2015), 'Porcine epidemic diarrhea virus: an emerging and re-emerging epizootic swine virus', *Virol J*, 12:193. https://doi.org/10.1186/s12985-015-0421-2.

Niederwerder, M.C. and Hesse, R.A. (2018), 'Swine enteric coronavirus disease: a review of 4 years with porcine epidemic diarrhea virus and porcine deltacoronavirus in the United States and Canada', *Transbound Emerg Dis*, 65:660–675. https://doi.org/10.1111/tbed.12823.

Niederwerder, M.C., et al. (2019), 'Infectious dose of African swine fever virus when consumed naturally in liquid or feed', *Emerg Inf Dis*, 25(5):891–897. https://doi.org/10.3201/eid2505.181495.

Niederwerder, M.C., et al. (2021), 'Mitigating the risk of African swine fever virus in feed with anti-viral chemical additives', *Transbound Emerg Dis*, 68(2):477–486. https://doi.org/10.1111/tbed.13699.

Niederwerder, M.C., et al. (2022), 'Stability of African swine fever virus in feed during environmental storage', *Transbound Emerg Dis*, 69(6):3216–3224. https://doi.org/10.1111/tbed.14666.

Palowski, A., et al. (2022), 'Survival of a surrogate African swine fever virus-like algal virus in feed matrices using a 23-day commercial United States truck transport model', *Front Microbiol*, 13:1059118. https://doi.org/10.3389/fmicb.2022.1059118.

Pascik, J., et al. (2014), 'Investigation into the role of potentially contaminated feed as a source of the first-detected outbreaks of porcine epidemic diarrhea in Canada', *Transbound Emerg Dis*, 61:397–410.

Pasma, T., et al. (2016), 'Outbreak investigation of porcine epidemic diarrhea in swine in Ontario' *Can Vet J*, 57(1):84–89.

Patterson, G., et al. (2019), 'Risks to animal health associated with imported feed ingredients', *J Am Vet Med Assn*, 254(7):790–791. https://doi.org/10.2460/javma.254.7.790.

Patterson, G., et al. (2021), 'Quantification of soy-based feed ingredient entry from ASFV-positive countries to the United States by ocean freight shipping and associated seaports', *Transbound Emerg Dis.* 2021, Jul; 68(4): 2603-2609. https://doi.org/10.111/tbed.13881.

Perri, A.M., et al. (2018), 'An epidemiological investigation of the early phase of the porcine epidemic diarrhea (PED) outbreak in Canadian swine herds in 2014: a case-control study', *Prev Vet Med*, 150:101–109. https://doi.org/10.1016/j.prevetmed.2017.12.009.

Perri, A.M., et al. (2019), 'Network analyses using case-control data to describe and characterize the initial 2014 incursion of porcine epidemic diarrhea (PED) in Canadian swine herds', *Prev Vet Med*, 162:18–28. https://doi.org/10.1016/j.prevetmed.2018.11.001.

Phillips, F.C., et al. (2022), 'Monoglyceride reduces viability of porcine epidemic diarrhoea virus in feed and prevents disease transmission to post-weaned piglets', *Transbound Emerg Dis*, 69(1):121–127. https://doi.org/10.1111/tbed.14353.

Schambow, R.A., et al. (2022), 'Rethinking the uncertainty of African swine fever virus contamination in feed ingredients and risk of introduction into the United States', *Transbound Emerg Dis*, 69(1):157–175. https://doi.org/10.1111/tbed.14358.

Schumacher, L.L., et al. (2016), 'Evaluation of the minimum infectious dose of porcine epidemic diarrhea virus in virus-inoculated feed', *Am J Vet Res*, 77(10):1108–1113. https://doi.org/10.2460/ajvr.77.10.1108.

Schumacher, L.L., et al. (2017), 'Characterizing the rapid spread of porcine epidemic diarrhea virus (PEDV) through an animal food manufacturing facility', *PLoS One*, 12(11):e0187309. https://doi.org/10.1371/journal.pone.0187309.

Schumacher, L.L., et al. (2018), 'Feed batch sequencing to decrease the risk of porcine epidemic diarrhea virus (PEDV) cross-contamination during feed manufacturing', *J Anim Sci*, 21;96(11):4562–4570. https://doi.org/10.1093/jas/sky320.

Shurson, G.C., et al. (2024), 'Stability of a surrogate African swine fever-like algal virus in corn- and soybean-based feed ingredients during extended storage and *in vitro* digestion processes', *Front Vet Sci*, 11:1498977. https://doi.org/10.3389/fvets.2024.1498977.

Silva, G.S., et al. (2018), 'Development and validation of a scoring system to assess the relative vulnerability of swine breeding herds to the introduction of PRRS virus', *Prev Vet Med*, 160:116–122. https://doi.org/10.1016/j.prevetmed.2018.10.004.

Songkasupa, T., et al. (2022), 'Thermal inactivation of African swine fever virus in feed ingredients', *Sci Rep*, 12(1):15998. https://doi.org/10.1038/s41598-022-20290-9.

Stewart, S.C., et al. (2020), 'A review of strategies to impact swine feed biosecurity', *Anim Health Res Rev*, 21(1):61–68. https://doi.org/10.1017/S146625231900015X.

Stenfeldt, C., et al. (2022), 'The risk and mitigation of foot and mouth disease virus infection of pigs through consumption of contaminated feed', *Transbound Emerg Dis*, 69:72–87. https://doi.org/10.1111/tbed.14230.

Stoian, A.M.M., et al. (2019), 'Half-life of African swine fever virus in shipped feed', *Emerg Infect Dis*, 25(12):2261–2263. https://doi.org/10.3201/eid2512.191002.

Stoian, A.M.M., et al. (2020), 'Stability of classical swine fever virus and pseudorabies virus in animal feed ingredients exposed to transpacific shipping conditions', *Transbound Emerg Dis* 2020 Jul;67(4): 1623-1632. https://doi.org/10.1111/tbed.13498.

Toro, H., et al. (2016), 'Inactivation of Avian Influenza virus in nonpelleted chicken feed', *Avian Dis*, 60(4):846–849. https://doi.org/10.1637/11465-070616.

Trudeau, M.P., et al. (2016), 'Comparison of thermal and non-thermal processing of swine feed and the use of selected feed additives on inactivation of porcine epidemic diarrhea virus (PEDV)', *PLoS One*, 11(6):e0158128. https://doi.org/10.1371/journal.pone.0158128.

Trudeau, M.P., et al. (2017a), 'Environmental persistence of porcine coronaviruses in feed and feed ingredients', *PLoS One*, 12(5):e0178094. https://doi.org/10.1371/journal.pone.0178094.

Trudeau, M.P., et al. (2017b), 'Survival of porcine epidemic diarrhea virus (PEDV) in thermally treated feed ingredients and on surfaces' *Porcine Health Manag*, 3:17. https://doi.org/10.1186/s40813-017-0064-3.

Part 2

Other feed contaminants risks

Chapter 4

Assessing risks from mycotoxins in animal feed and animal-derived products

Stoycho D. Stoev, Trakia University, Bulgaria

1 Introduction

Mycotoxins are fungal metabolites that frequently contaminate feeds or food commodities, posing a serious threat to global animal and human health. Many ailments and diseases in animals and humans can be caused by such food/feedstuffs contaminated by mycotoxins. Cereals are invaded by fungi in both the field and following harvest, and such invasion and subsequent mycotoxin production are often unavoidable due to various environmental factors, such as excessive rain at harvest time, increased humidity, and inadequate storage conditions. Of the more than 400 types of natural fungal metabolites, only 10–14 are primarily responsible for foodborne ailments, such as ochratoxin A (OTA) (especially toxic for pigs, turkeys, ducklings, chicks, dogs, and rabbits); aflatoxins (AFs), among which aflatoxin B1 (AFB1) (especially toxic for young animals and poultry, calves, pigs, dogs, and fish) and aflatoxin M1 (AFM1) (especially toxic for calves) are of particular importance; fumonisins (FUMs), among which fumonisin B1 (FB1) is the most toxic (especially for horses and pigs); ergot alkaloids (especially toxic for horses but can also be a problem

http://dx.doi.org/10.19103/AS.2026.0172.16

for cows and sows); zearalenone (ZEA) (especially toxic for pigs and dogs); patulin (PAT) (especially toxic for rodents and cattle but also for humans); deoxynivalenol (DON) (especially toxic for pigs); diacetoxyscirpenol (DAS) (especially toxic for pigs and poultry); and T-2 and HT-2 (especially toxic for pigs and poultry). The food chain, the method of detoxification, the dosage, the half-life, and bioavailability of each mycotoxin should be taken into account for each animal species. For example, low doses of mycotoxins (especially DON and FB1) typically provoke an inflammatory response, whilst higher doses cause immunosuppression. Mycotoxins often contaminate feedstuffs or animal/poultry products, including eggs, meat, or milk in concentration, which can compromise human health and animal wellbeing (Vlachou et al., 2022; Marc, 2022; SCOOP, 2002).

If exposed to mycotoxins via feedstuffs, animals typically display changes in behavior including nervous signs or refusal to eat such feedstuffs as well as poor feed conversion, all of which are seen in the acute stages of mycotoxin exposure. Some foodborne diseases or an increased number of secondary bacterial infections can be seen in the acute stage of such mycotoxicoses in animals or poultry (due to immunosuppression, which is the first pronounced toxic effect of mycotoxin exposure), whilst decreased reproductive capacities and reduced body weight gain are often seen in chronic stages of the same mycotoxicoses (Stoev, 2015; Gashaw, 2015). The known toxic effects of mycotoxins include nephrotoxic (mainly OTA and slightly FB1), hepatotoxic (mainly AFB1), neurotoxic (mainly FB1), immunosuppressive (mainly OTA, AFB1, DON, and T-2 toxin), carcinogenic (mainly AFB1, OTA, and FB1), estrogenic (ZEA), and genotoxic (mainly AFB1, OTA, T-2) effects (Stoev, 2023, 2024a, 2024b) (Table 1).

Multiple mycotoxin contamination involving several mycotoxins often occurs in feedstuffs or food commodities. The mycotoxin content in these cases is often below the maximum permitted levels and within European requirements, but such multiple mycotoxin contamination at low levels can be harmful to animals and humans, bearing in mind the additive or synergistic effects between some key mycotoxins, which are often responsible for the appearance of various foodborne ailments. Such joint mycotoxin interactions should be carefully investigated via in vitro or in vivo studies, and appropriate conclusions should be made with regard to necessary hygiene control and required risk assessment. By doing so, the potential harm to animals or humans can be analyzed in depth and adequate preventive measures proposed for each particular case.

According to the Food and Drug Administration (FDA), the economic costs of crop losses in the US due to mycotoxin contamination and subsequent destruction of feeds or food commodities amount to nearly $930 million annually (CAST Report, 2003). The Food and Agriculture Organization (FAO)

Table 1 Target toxicity, species, age, and sex sensitivity to different mycotoxins (Akinmoladun et al., 2025)

Mycotoxins	Organs damaged	Target toxicity	Order of species sensitivity	Age and sex sensitivity
AFB1	Liver, kidneys	Hepatotoxic, nephrotoxic, carcinogenic, teratogenic, immunotoxic	Poultry > rabbits > pigs > ruminants	Young and female animals with increased sensitivity
OTA	Kidneys, liver	Nephrotoxic, hepatotoxic, carcinogenic, teratogenic, immunotoxic	Pigs > poultry > rabbits > ruminants	Male animals with increased sensitivity
DON	Liver, kidney, lymphocytes	Gastrointestinal toxicity, immunotoxic	Pigs > poultry > ruminants	Male and older animals with increased sensitivity
ZEA	Reproductive system (ovaries, testicles, uterus, vulva, vagina, mammary gland), liver, kidneys	Estrogenic	Pigs > dairy cattle > poultry > ruminants	Young prepubertal females with increased sensitivity
FB1 + FB2	Brain, liver, kidneys, small intestine	Neurotoxic, hepatotoxic, nephrotoxic, carcinogenic, teratogenic	Horses > pigs > poultry > ruminants > fish	Females with increased sensitivity
T-2 + HT-2	Liver, kidneys, lymphocytes	Hepatotoxic, gastrointestinal toxicity, teratogenic, immunotoxic	Poultry > pigs > ruminants	Young animals with increased sensitivity

reports that nearly 25% of the world's crops are contaminated by mycotoxins each year, which contributes to nearly 1 billion tons of annual losses of feedstuffs and foods (Smith et al., 2016). There are different kinds of losses attributed to mycotoxins, including loss or decreased livestock production, illness or death of animals and humans, increased necessity for medical care and veterinary services, increased costs for regulatory and preventive measures or mycotoxin detoxification, and economic losses caused by food/feed scrapping (Marc, 2022; Stoev, 2024a).

The European Union (EU) has put in place maximum permissible limits and guidance values for the most dangerous mycotoxins in human foods and animal feedstuffs (EC Directive 2002/32/EC, EC Directive 2003/100/EC, EC Recommendations 2013/165/EU and 2006/576/EC) (Table 2), but these

Table 2 Maximum permitted levels or guidance values of target mycotoxins in various feeds designed for animals, according to EU legislation and regulation (EC Directive 2002/32/EC, EC Directive 2003/100/EC, EC Regulations No 2023/915 and No 2024/1808, EC Recommendations 2013/165/EU and 2006/576/EC), United States Department of Agriculture (USDA-GB 2761-2017), and various other countries (Akinmoladun et al., 2025; Abdallah et al., 2015)

Mycotoxins	Cereals or cereal products and animal products	Maximum permitted levels or guidance values µg/kg (country)
AFB1	• Complete feedstuffs for pigs and poultry (except young animals)	• 20 (EU – EC Directive 2002/32/EC)
	• Complete feedstuffs for cattle, sheep, and goats with exception of	• 20 (EU – EC Directive 2002/32/EC)
	✓ Complete feedstuffs for dairy animals	• 5 (EU – EC Directive 2002/32/EC)
	✓ Complete feedstuffs for calves and lambs	• 10 (EU – EC Directive 2002/32/EC)
	• Complementary feedstuffs for cattle, sheep, and goats (except dairy animals, calves, and lambs)	• 20 (EU – EC Directive 2002/32/EC)
	• Feed for dairy cattle, breeding animals	• 20 (USA – FDA Compliance Policy Guide Sec. 683.100, 2022)
	• Feed for finishing cattle, swine, and poultry	• 100 (USA – FDA Compliance Policy Guide Sec. 683.100, 2022)
	• Animal feed	• 10 (Egypt – Abdallah et al., 2015)
	• Animal feed	• 50 (Turkey – Abdallah et al., 2015)
	• Animal feed	• 20 (Canada, Egypt, Iran – Abdallah et al., 2015)
Total AFs		
	• Animal feed	• 50 (Brazil – Abdallah et al., 2015)
	• Cereal feedstuffs	• 200 (Mexico – Abdallah et al., 2015)
	• Feedstuff ingredients	• 20 (Japan, Korea, USA – Abdallah et al., 2015)
	• Raw milk, heat-treated milk, and milk-based products	• 0.05 (EU – EC Regulation No 2023/915)
AFM1	• Milk and dairy products	• 0.5 (USA – USDA-GB 2761-2017)
	• Milk and dairy products	• 0.5 (Egypt – Abdallah et al., 2015)
	• Milk and dairy products	• 0.05 (Turkey – Abdallah et al., 2015)

OTA	• Cereals and cereal products	• 250 (EU – EC Recommendations 2006/576/EC)
	• Complementary and complete feedstuffs with exception of:	• 5000 (EU – EC Recommendations 2006/576/EC)
	• Complementary and complete feedstuffs for pigs	• 50 (EU – EC Recommendations 2006/576/EC)
	• Complementary and complete feedstuffs for poultry	• 100 (EU – EC Recommendations 2006/576/EC)
	• General feed (guideline)	• 250 (Canada – CFIA Guidance for OTA in Livestock Feed – 2009)
DON	• Cereals and cereal products without maize by-products	• 8000 (EU – EC Recommendations 2006/576/EC)
	• Maize by-products	• 12 000 (EU – EC Recommendations 2006/576/EC)
	• Complementary and complete feedstuffs for cattle and poultry, etc. with exception of:	• 5 000 (EU – EC Recommendations 2006/576/EC)
	✓ Complementary and complete feedstuffs for pigs	• 900 (EU – EC Recommendations 2006/576/EC)
	✓ Complementary and complete feedstuffs for calves, lambs and kids	• 2000 (EU – EC Recommendations 2006/576/EC)
	✓ Swine feed (advisory level)	• 1000 (USA – FDA Guidance for Industry #186, 2010)
	• Poultry feed (advisory level)	• 5000 (USA – FDA Guidance for Industry #186, 2010)
	• Ruminants feed (for beef and dairy)	• 10 000 (USA – FDA Guidance for Industry #186, 2010)
ZEA	• Cereals and cereal products without maize by-products	• 2000 (EU – EC Recommendations 2006/576/EC)
	• Maize by-products	• 3000 (EU – EC Recommendations 2006/576/EC)
	• Complementary and complete feedstuffs for dairy cattle, calves, goats (e.g. kids) and sheep (e.g. lambs)	• 500 (EU – EC Recommendations 2006/576/EC)
	• Complementary and complete feedstuffs for piglets, gilts (young sows)	• 100 (EU – EC Recommendations 2006/576/EC)
	• Complementary and complete feedstuffs for fattening pigs and sows	• 250 (EU – EC Recommendations 2006/576/EC)

(Continued)

Table 2 *Continued*

Mycotoxins	Cereals or cereal products and animal products	Maximum permitted levels or guidance values µg/kg (country)
Fumonisins FB1 + FB2	• Maize and maize by-products	• 60 000 (EU – EC Recommendations 2006/576/EC)
	• Complementary and complete feedstuffs for pigs, horses, rabbits, pets	• 5000 (EU – EC Recommendations 2006/576/EC)
	• Complementary and complete feedstuffs for poultry, calves, lambs, kids	• 20 000 (EU – EC Recommendations 2006/576/EC)
	• Complementary and complete feedstuffs for adult ruminants and mink	• 50 000 (EU – EC Recommendations 2006/576/EC)
	• Complementary and complete feedstuffs for fish	• 10 000 (EU – EC Recommendations 2006/576/EC)
	• Swine feed	• 10 000 (USA – FDA Guidance for Industry #221, 2011)
	• Poultry feed	• 30 000 (USA – FDA Guidance for Industry #221, 2011)
	• Cattle feed	• 100 000 (USA – FDA Guidance for Industry #221, 2011)
T-2 + HT-2	• Unprocessed barley (e.g. malting barley) and maize	• 200 (EU – EC Recommendations 2013/165/EU)
	• Unprocessed rye, wheat, and other cereals	• 100 (EU – EC Recommendations 2013/165/EU)
	• Unprocessed oats (with husk)	• 1000 (EU – EC Recommendations 2013/165/EU)
	• Oat milling products (husks) for feed and compound feed	• 2000 (EU – EC Recommendations 2013/165/EU)
	• Other cereal products for feed and compound feed	• 500 (EU – EC Recommendations 2013/165/EU)
	• Compound feed, excluding feed for cats	• 250 (EU – EC Recommendations 2013/165/EU)
	• General feed (guideline based on species sensitivity)	• 100–1000 (Canada – CFIA Feed Contaminants Guidelines)

limits do not take into consideration joint mycotoxin exposure and mycotoxin interaction, synergistic or additive effects of some mycotoxins, and the increased toxicity of such mycotoxin combinations, even at lower contamination levels than those permitted. The effectiveness of the current regulatory measures is, therefore, questionable as the regulations are based only on the toxicity of individual mycotoxins, without taking into account potential mycotoxin interactions. In order to ensure adequate food safety and effective control, various additional regulatory measures must be introduced, which should take into consideration synergistic or additive mycotoxin interactions. The expertise of qualified specialists in food science and technology, veterinary and human medicine, and agriculture is crucial for introducing such regulatory and quality control measures (Stoev, 2024a).

The purpose of this chapter is to identify the most prevalent mycotoxins in feedstuffs and agricultural commodities and to evaluate the potential harm of such mycotoxin contamination to animal health and its possible transmission to humans. The underestimated harm of joint mycotoxin exposure in animals is briefly explained before the effectiveness of current provisions for the regulation of mycotoxins in feedstuffs is discussed in relation to animal health. A brief investigation is made into potential veterinary preventive measures, hygiene control, and risk assessment in regard to various foodborne ailments and diseases caused by mycotoxins in the field in order to reduce mycotoxin content in meat and other animal products, and to prevent these mycotoxins from entering commercial channels.

2 Prevalence of mycotoxins in animal feed and animal-derived products

Seasonal weather conditions within each geographical area during critical growth stages of each plant species are of particular importance in terms of levels of mycotoxin contamination. The variation in results is generally a consequence of individual circumstances, such as the type of analyzed samples, the periods of the surveys, and the weather fluctuations in each particular year. In addition, various environmental conditions, such as increased humidity, temperature, excessive rainfall, drought conditions, damage by insects, and the agronomic practices used could provoke stress and further contribute to the development of mold in plants in the field prior to harvesting (Munkvold, 2003; Teller et al., 2012). Weather conditions known to lead to extensive AF contamination include high temperatures, low rainfall, and subsequent drought stress, while *Fusarium spp*. producing DON and ZEA are mainly associated with cool weather conditions and an excessively wet growing season (Munkvold, 2003; Reyneri, 2006). Some mycotoxins, such as OTA and AFs, are produced primarily when feed is stored for a prolonged period and are, therefore, known as storage mycotoxins. Others, such as FUMs, trichothecenes, DON, ZEA, and other *Fusarium* mycotoxins, are mostly produced in field conditions before harvesting. In addition, the international grain trade may further distribute mycotoxin-contaminated materials outside of their natural geographical areas of occurrence, which further complicates the potential prediction of mycotoxin contamination of feedstuffs or food.

EU regulatory limits and guidance values in animal feeds are given in Table 2 . However, with regard to milk and milk-based products designed for humans, the limit for AFM1 is 0.05 µg/kg (EC Regulation No 2023/915).

AFs have most often been found as natural contaminants in South Asia (78% positive samples with average contamination levels of 128 µg/kg), followed by Southeast Asia (55% positive samples with average contamination levels of 61 µg/kg). ZEA was the most prevalent contaminant in North Asia (56% positive samples with average contamination levels of 386 µg/kg). DON was found as a natural contaminant in North Asia (78% positive samples with average contamination levels of 1060 µg/kg). However, the highest average DON contamination has been found in North America (68% positive samples with average contamination levels of 1418 µg/kg) (Schatzmayr and Streit, 2013). DON was also equally prevalent in European and Canadian food, with about 57% of the European (SCOOP, 2003) and about 59% of the Canadian food (CFIA, 2012). In Austrian feeds and feed raw materials, DON was found in around 60% of investigated cereal samples other than maize and in around 95% of maize samples (Streit et al., 2012). High DON contamination levels were also found in liquid pig feed samples in the Netherlands, and 10% of the same sample exceeded the maximum permitted values (Adamse et al., 2012). FUMs were the most frequent contaminant in South America (77% positive samples with average contamination levels of 2691 µg/kg). OTA was the most prevalent contaminant in South Asia (55% positive samples with average contamination levels of 20 µg/kg). OTA was also observed as a frequent contaminant in Eastern European samples (49% positive) assessing OTA exposure of the EU population, but the average contamination level was much lower – 4 µg/kg (Schatzmayr and Streit, 2013; SCOOP, 2002). A very high prevalence of OTA was also found in wheat products (94%) in Canada (CFIA, 2012), but the highest concentrations of OTA have been identified in southern Europe (Paterson et al., 2018) (Table 3, Table 4).

OTA, in addition to AFs, has been reported to contaminate animal products, including dried meat and other meat products such as sausages and salami, as well as eggs, presenting a global public health problem (Völkel, 2011; Pizzolato Montanha et al., 2018). ZEA and DON were also reported to contaminate meat but to a lower degree (Marc, 2022). AFs have been reported to be the most important contaminants of milk and dairy products such as cheese or yogurt (Marc, 2022). The most prevalent contaminants of eggs are AFB1, OTA, ZEA,

Table 3 A brief overview of prevalence (%) and median (μg/kg) values of natural mycotoxin contamination of feed worldwide (Gruber-Dorninger et al., 2019).

Continent/Region (n)	AFB1 (%; μg/kg)	OTA (%; μg/kg)	FUMs (%; μg/kg)	DON (%; μg/kg)	ZEA (%; μg/kg)	T-2 (%; μg/kg)
Central Europe (21036)	13; 1.6	12; 2.8	43; 187	70; 428	45; 40	31; 11
North Europe (1958)	6; 3.1	8; 1.9	22; 186	74; 504	29; 35	30; 34
South Europe (3527)	29; 2.1	21; 2.6	75; 607	53; 324	36; 44	12; 25
Central America (367)	9; 3.9	4; 2.5	82; 929	70; 316	38; 60	4; 3.1
North America (5471)	10%; 8.7	4%; 2.4	48%; 652	64; 505	32; 102	4; 29
South America (17332)	24; 3.2	5; 17	75; 1390	27; 344	47; 51	22; 31
North Africa/Middle East (1075)	22; 2.4	20; 3.1	67; 347	48; 236	45; 31	8; 14
Sub-Saharan Africa (208)	76; 23	32; 7.2	73; 789	50; 352	52; 38	3; 3.0
South Africa (1077)	9; 2.2	6; 2.2	63; 266	63; 363	42; 30	1; 4.4
Oceania (1695)	11; 2.0	8; 3.6	22; 106	34; 158	22; 105	2; 16
Central Asia (15)	8; 1.4	13; 22	25; 13	27; 28	23; 1.5	7; 25
South Asia (1136)	82; 20	60; 4.6	69; 288	23; 96	20; 37	1; 13
East Asia (13232)	17; 10	14; 2.9	61; 810	85; 418	58; 90	11; 16

and DON (Iqbal et al., 2014), which appears to be a potential public health problem.

Contamination peaks of mycotoxins are often traceable to key regions and generally linked to extreme or uncommon weather conditions. Taking into account climate change in Europe and throughout the world, an increase in magnitude or frequency of human/animal exposure to mycotoxins is expected to occur, which could further increase animal and human health concerns. In this regard, mycotoxins that have not previously been reported in foods/feedstuffs from EU countries might occur as a result of the change in distribution of various key fungal species in regions experiencing such climate changes, e.g. wider dissemination of *Fusarium* fungi and mycotoxins is expected to be seen in EU countries (Miraglia et al., 2009; Battilani et al.,

Table 4 A brief overview of ranges of natural mycotoxin contamination of feed or feed ingredients in Europe (Stoev, 2024b).

Mycotoxin	Feeds or ingredients	Range (μg/kg)	Number of samples	Positive %	Continent (region)	Reference
OTA	Wheat, maize	22–33	82	2	Europe	(Monbaliu et al., 2010)
FB1	Wheat, maize	36–5114	82	44	Europe	
ZEA	Wheat, maize	58–387	82	15	Europe	
DON	Wheat, maize	74–9528	82	63	Europe	
HT-2	Wheat, maize	22–116	82	9	Europe	
AFs	Various	0.5–66	127	25	Southern Europe	(Griessler et al., 2010)
OTA	Various	1–54	46	22	Southern Europe	
FUMs	Various	25–36 390	89	66	Southern Europe	
DON	Various	52–4827	348	66	Southern Europe	
ZEA	Various	10–2939	303	28	Southern Europe	
T-2/HT-2	Various	35–137	65	8	Southern Europe	

2016; Assunção et al., 2018; Alvito and Assunção, 2021). Therefore, a different exposure pattern to mycotoxins is expected to occur in EU countries, now or in the near future. For example, a strong *Aspergillus flavus* infection was observed in 2003, due to the hot and dry weather that year, which led to high AFB1 contamination of maize in northern Italy (Piva et al., 2006). A study of 110 samples revealed an AFB1 incidence of 75% with a mean level of 4.4 µg/kg. Due to the use of maize as a feed source for dairy cattle, a large AFM1 contamination was subsequently seen in milk, leading to thousands of tons of milk exceeding the EU maximum permitted levels of 0.05 µg/kg to be destroyed (Streit et al., 2012; EC Regulation No 165/2010). In addition, various species such as *Fusarium verticillioides*, commonly associated with warmer and drier regions such as Italy and Spain, were found to be the predominant *Fusarium* species isolated from maize grown in Germany in 2006, which subsequently increased FUMs contamination of maize to 34% of the studied samples (Goertz et al., 2010). Such high contamination levels of AFB1 in the EU demonstrate that climate change will lead to differences in mycotoxin distribution patterns in the future.

The development of predictive models for mycotoxin contamination in cereals, foods, and feedstuffs, based on regional climate data, would be a valuable tool for assessing the risk of mycotoxin contamination in each season. Such predictive models typically combine weather forecasts and mycotoxin contamination risks and are developed to help manage agricultural risks. These models use factors such as temperature, rainfall, and humidity, as well as crop phenology and historic contamination data, to predict mycotoxin levels in crops such as wheat and corn (Platzer et al., 2025; Cui et al., 2022). They provide early warning signs for farmers, grain harvesters, and policymakers to support decisions regarding treatments or harvesting and manage food safety. Although climate is one of the most influential parameters in terms of the extent of mycotoxin contamination, other measures such as crop rotation (avoiding maize as a pre-crop for wheat), tillage, or planting time (earlier planting of maize is important) are also crucial in order to reduce mycotoxin contamination of cereals (Streit et al., 2012).

3 Exposure to multiple mycotoxins as a cause of foodborne disease in animals

There is sufficient evidence to link mycotoxins to various diseases in humans and farm animals, including equine leukoencephalomalacia, porcine pulmonary edema, porcine vulvovaginitis, or rectal prolapse in pigs, stachybotryotoxicosis, mycotoxic porcine/chicken nephropathy, ergotism, and various others (Marasas et al., 1976; Stoev et al., 2010a, 2010b; Stoev, 2015, 2024a, 2024b). Different animals and poultry have different sensitivities to mycotoxins, with poultry being more resistant to the toxicity of DON, FUMs, and ZEA, while pigs are highly sensitive to T-2 and DON (Pinotti et al., 2016).

Feedstuffs contaminated by mycotoxins contribute to mycotoxin contamination of various food products originating from animals, such as milk, dairy products, meat, and eggs, due to the carryover of some mycotoxins from forages to defined food products. This situation may further contribute to the increase in human exposure to mycotoxins (Vlachou et al., 2022; Meizhen and Ping, 2023), as might thermal and chemical stability of many mycotoxins, and minimal loss during thermal treatments or production processing.

Multiple mycotoxin contamination of forages and food commodities has been proven to be responsible for many foodborne ailments or diseases in animals (Fig. 1; Stoev et al., 2010a, 2010b; Stoev, 2015). In addition, mycotoxins are reported to be responsible for various secondary bacterial diseases due to their immunosuppressive effects (Stoev et al., 2000a, 2000b).

Fusarium mycotoxins, DON, and ZEA are primarily responsible for multiple mycotoxin contamination of maize and, less often, oats, barley, and wheat. ZEA is reported to mainly contaminate cereal products, including feedstuffs (Alvito et al., 2022) as well as animal products, such as milk, meat, and eggs

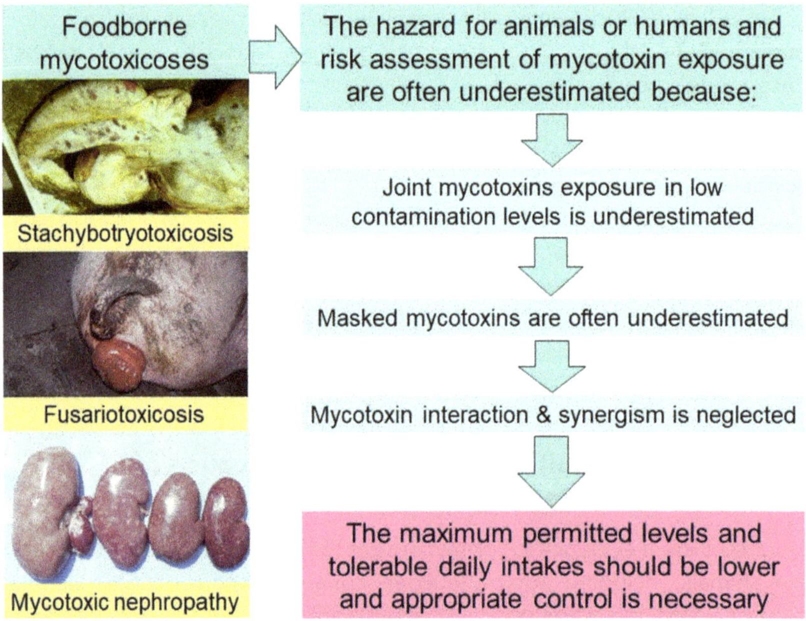

Figure 1 Spontaneous foodborne mycotoxicoses and the harm to animals or humans due to underestimated risk assessment from joint mycotoxin exposure in low contamination levels, e.g. masked mycotoxins, synergistic interaction between target mycotoxins and neglected hygiene control (Stoev, 2024b).

(Marc, 2022). DON is reported to mainly contaminate wheat, rye, corn, oats, barley, rice, and sorghum (Marc, 2022). ZEA was found to be involved in the appearance of vulvovaginitis and rectal or vaginal prolapses in female pigs (Fig. 2A). Various other estrogenic symptoms, including swelling of the mammary glands, infertility, embryo lethal resorptions, abortions in dairy cattle, decreased milk production, and reduced litter size (Zinedine et al., 2007; Stoev, 2015) have also been seen in female animals. In male animals, feminization and decrease in testosterone levels and/or spermatogenesis as well as decreased libido have been observed (Zinedine et al., 2007). ZEA has genotoxic action and could be partially responsible for breast and esophageal carcinomas (Awuchi et al., 2022). The International Agency for Research on Cancer (IARC), however, has classified ZEA in group 3 (mycotoxins that do not exert carcinogenic effects on humans) (IARC, 2002, 2012). Hyperestrogenism in female pigs (vulvovaginitis and rectal or vaginal prolapses), which is known to be the main indicator of ZEA toxicosis, is typically observed only after prolonged ZEA intake for nearly a month, whereas the first clinical symptoms of exposure to feedstuffs contaminated with *F. graminearum* and/or *F. culmorum* are usually manifested by the toxic effects of DON, including cytotoxic and emetic effects (degenerative changes in the gastrointestinal tract), neurotropic effect (paresis of the limbs), and immunosuppressive effect (secondary bacterial diseases) (Chaytor et al., 2011; Maresca, 2013; Stoev, 2015, 2023). DON can also reduce milk production in dairy cattle, while low levels of exposure to DON can induce nausea, diarrhea, and lesions in the gastrointestinal tract, growth retardation, damage to the hematopoietic systems, and immune dysregulation (Chaytor et al., 2011; Maresca, 2013). Poultry was found to be less sensitive to DON, and feed refusal has only been observed at very high contamination levels in feedstuffs (16–20 mg/kg feed) (Eriksen and Pettersson, 2004). Ruminants were found to be the least sensitive animal species to DON, which is due to the ability

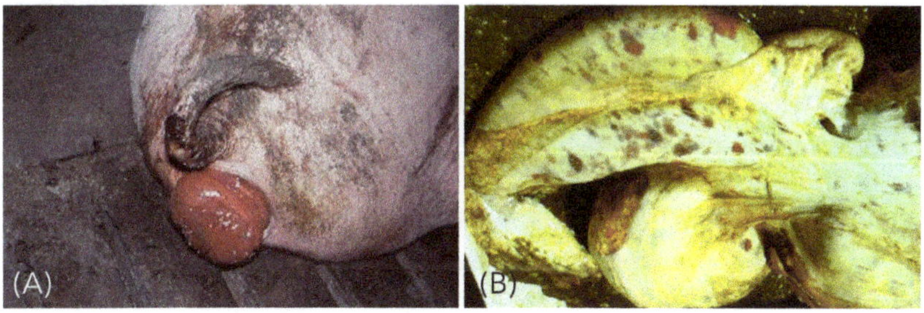

Figure 2 (A) rectal prolapse in a spontaneous case of fusariotoxicosis in a pig from Bulgaria due to the estrogenic effect of mycotoxin ZEA. (B) Spontaneous case of stachybotryotoxicosis in cattle. Strong hemorrhagic diathesis on the serosa of the abomasum and rumen and ulcerative-necrotic damages, which are translucent under the serosa (Stoev, 2016)

of rumen microflora to detoxify it (Kamle et al., 2022). This mycotoxin has been reported to mainly contaminate wheat, corn, rye, oats, and barley and rarely rice, triticale, and sorghum. Cereals are typically contaminated by DON in the field but also during storage (Marc, 2022).

Other *Fusarium* mycotoxins, such as T-2 toxin, HT-2 toxin, and DAS, which are produced mainly by *Fusarium* spp. from the section *Sporotrichiella* or *F. poae* species were found to possess strong cytotoxic, genotoxic, and immunosuppressive effects (Richard, 2007). Barley and oats were the main cultures frequently contaminated with T-2 and HT-2 (Streit et al., 2012). Damage to the hematopoietic system, eggshells and egg production, feed refusal, and growth retardation are the main symptoms of T-2 toxicity (Akande et al., 2006). Damage to the cardiovascular system, growth retardation, and lung damage are the main symptoms of DAS toxicity, which could add to the clinical finding caused by T-2 and HT-2 (Stoev, 2015). The most sensitive species to this fusariotoxicosis were found to be pigs and poultry (Streit et al., 2012), and symptoms are observed mainly following ingestion of hay, grain, or straw that has overwintered outdoors. This fusariotoxicosis in animals and humans is known as alimentary toxic aleukia (Lutsky and Mor, 1981). The clinical signs of this fusariotoxicosis are catarrhal or hemorrhagic gastroenteritis, accompanied by ulcers and necroses in the gastrointestinal system, damage to the kidney, liver, heart, brain, and peripheral ganglia, which are responsible for muscular spasms, paresis of the limbs, and tremors (Stoev, 2015). Abdominal pain, diarrhea, nausea, tremors, and weight loss are seen in the initial stages of fusariotoxicosis (Pleadin et al., 2019).

Maize contaminated by various *Fusarium* fungi has historically been suspected to be the cause of horse and pig diseases, including the death of many horses more than 150 years ago, which was subsequently attributed to exposure to the toxic fungus *F. moniliforme (F. verticillioides)*. However, the mycotoxins causing the same diseases (FUMs) were not discovered until the 1980s (Bezuidenhout et al., 1988; Gelderblom et al., 1988). The significant health hazard of these mycotoxins was recognized after a number of outbreaks of equine leukoencephalomalacia (Marasas et al., 1976; Conkova et al., 2003) and porcine pulmonary edema caused the death of many horses and pigs fed on diets containing moldy maize contaminated with FUMs in the US during 1989 and 1990 (Marasas, 2001). At that time, it was found that equine leukoencephalomalacia was caused by FUMs, among which FB1 was the most toxic. These mycotoxins were reported to contaminate mainly the maize in developing countries (Marasas et al., 1976; Conkova et al., 2003). Porcine pulmonary edema is another foodborne mycotoxicosis caused by moldy maize containing FUMs (Marasas, 2001). Heart failure and subsequent edema in the lungs of pigs can be explained by the dysfunction in myocardial contractility in pigs, which is induced by the increase of sphingosine and subsequent inhibition of L-type calcium channels in the myocardium (Haschek et al., 2009). A similar disturbance in contractility of myocardium in humans is known as Idiopathic Congestive Cardiomyopathy (ICC) and was first recognized in 1980 in rural hospitals in South Africa. The disease was apparently due to the intake of various *Fusarium* mycotoxins such as FB1, MON, and others (Stoev, 2015).

AFs are potent mycotoxins, often seen in feedstuffs or food commodities, together with other mycotoxins, which can complicate clinical and pathological findings. AFs are reported to mainly contaminate cereals and their levels increase significantly during storage. AFs were discovered due to their devastating effect on turkey poults (turkey "X" disease) and other chicks in 1960 in England, when more than 100 000 turkey poults died (Allcroft and Carnaghan, 1962). The so-called turkey "X" disease was finally related to high levels of AFs in Brazilian peanut meal imported as a feed ingredient. These mycotoxins, among which AFB1 is the most dangerous, affect mostly the liver (enlarged and yellow liver), with young animals, ducks, and turkeys being the most sensitive. Pathological signs in the early stages of intoxication include fatty degeneration and/or necroses in the liver, accompanied by proliferation of connective tissue in the interstitium, enlargement of the gall bladder, and intestinal damage. Typical signs in the later stages include icterus and cirrhosis of the liver, hydrothorax, and ascites, accompanied by a thickening of the skin around the mouth or neck and papillomas on the mucosa of the abomasus, which are characteristic symptoms in cattle. A decrease in body weight gain, immunosuppression, and anemia (Stoev, 2015; Meizhen and Ping, 2023) are additional chronic signs of aflatoxicosis in animals, poultry, or humans, which can be complicated by other co-contaminating mycotoxins with similar toxic effects on farm animals (Stoev, 2015). AFB1, in combination with other mycotoxins, can also induce edema and is often associated with Kwashiorkor disease (Misihairabgwi et al., 2019; IARC, 2012). IARC classified AFB1 as carcinogenic for humans (group 1 mycotoxin) and the cause of nearly 28% of all liver cancers (IARC, 2012). AFs was also reported to be teratogenic to embryos and to be able to cross the placental barrier (Abdulrazzaq et al., 2002).

Mycotoxic porcine nephropathy (MPN) is also a mycotoxicosis, which is caused by several mycotoxins. The main clinical symptoms of MPN are major kidney damage (Fig. 3A) (Stoev et al., 2010a,2010b), but a decrease in weight gain, nervous symptoms (Fig. 3B-C), hepatocellular damage, and a decrease in the weight of eggs (Fig. 3D) can also be seen in chicks or laying hens exposed to the same mycotoxins (Stoev et al., 2000a, 2002a,2002b; Stoev, 2010b).

This nephropathy was proven to have more complicated pathology and etiology in some countries such as Bulgaria and South Africa (Stoev, 2008). In these countries, spontaneous nephropathy in pigs (Stoev et al., 2010a, 2010b)

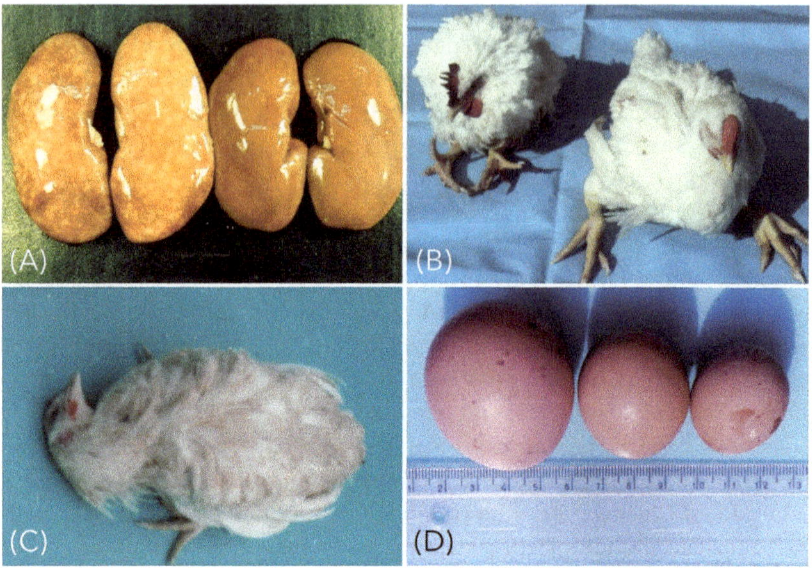

Figure 3 (A) Spontaneous case of mycotoxic porcine nephropathy identified at the slaughterhouse in Bulgaria. Enlarged and mottled appearance of kidneys from pigs aged 6–8 months (left) and normal kidneys in pigs of the same age (right) (Stoev, 2016); (B) Nervous symptoms as flexion of legs and sitting posture in chicks fed on a moldy diet containing 790 ppb OTA and 2000–5000 ppb penicillic acid (contamination levels naturally found in poultry feeds) for 10 months (Stoev, 2016); (C) Nervous symptoms such as torticollis in a chick fed on moldy diet containing 790 ppb OTA and 2000–5000 ppb penicillic acid for 70 days (Stoev, 2023); (D) Small size of eggs from laying hens exposed to 5 ppm OTA (contamination levels slightly above those in the field) in a diet with the eggs weighing 15.8 g (center) and 25.8 g (right). Normal size of an egg from a laying hen without OTA treatment (left) (Stoev, 2024b).

or chicks (Stoev et al., 2002a) described above was found to have a much more complicated etiology, being caused by the combined action of several mycotoxins such as OTA, penicillic acid (PA), and FB1, in addition to a as yet unidentified metabolite (UM), having a strong synergistic interaction (Stoev et al., 2010a, 2010b). A number of the same mycotoxins produced by storage fungi such as OTA and PA are secondary fungal metabolites encountered in feeds made from cereals or fibrous plants, stored for long periods and/or in increased humidity. Another part of these mycotoxins such as FB1 are, however, produced by field fungi, which contaminate cereals (mainly corn) before harvesting. Farms experiencing this particular nephropathy are known to have various problems with feed storage. Sometimes, however, the problem stems from certain feed enterprises where the grains, collected during moist and rainy days, have not been properly dried. Subsequently, all farms using these feed enterprises as their main feed source had nephropathy problems and growth depression among their pigs, but after changing the source of feed supply, the problems with poor growth disappeared (Stoev et al., 2010a). The same mycotoxins have synergistic (OTA and PA) (Stoev et al., 2001, 2004) or additive (OTA and FB1) interaction (Stoev et al., 2012).

Ergotism, caused by the fungus *Claviceps purpurea*, is also provoked by many different mycotoxins, e.g. ergocristine, ergocryptine, peptide alkaloids such as ergosine, ergotamine, and ergosecaline, lysergin amide derivatives (e.g. ergometrine and ergine), biogenic amines (e.g. histamine and acetylcholine) and various others, and represents another dangerous mycotoxicosis known as ergotism (Stoev, 2015). These mycotoxins are found in the fungal sclerotia, a compact mass of hyphae similar to a large, dark wheat grain. This fungus mainly contaminates wheat, rye, barley, oats, and millet. In humans, ergotism is known as St Anthony's fire and has been known to cause hallucinations. St Anthony's fire killed many people in France in the Middle Ages (Betina, 1989) and is still found in some developing countries (IPCS, 1990; Schneider et al., 1996). The main pathological/clinical symptoms of ergotism are ischemic necroses of peripheral parts of the body, e.g. tail, ears, or crown of hooves, induced by contraction of vessels, gangrene of extremities, and peripheral parts of the limbs, in addition to various gastrointestinal signs, enhanced contractions of the uterus and subsequent prolapse of the uterus and abortions (IPCS, 1990; Schneider et al., 1996).

Stachybotryotoxicosis is another mycotoxicosis commonly seen in farm animals. It is provoked by the fungal species *Stachybotrys alternans (Stachybotrys altra),* which contaminates moist substrates (straw/hay/oats) containing plenty of cellulose. This fungus can produce highly toxic compounds, such as satratoxins, verrucarins, and roridins, and is defined by its black coating. These fungal mycotoxins irritate the mucosa of the oral cavity and gastrointestinal tract, causing strong hyperemia, inflammation, swelling of the lips, and fissures in the mouth. Following mycotoxin accumulation, damage to the vessels and deep neurotrophic symmetrical necroses and ulcers on the gastrointestinal mucosa, edema, and hemorrhages are observed (Lefebvre et al., 1994; Stoev, 2015) (Fig. 2B). This mycotoxicosis shows a stationarity because this fungus can survive for long periods in the soil outside the substrate, which explains the repeated contamination of animal feedstuffs (Stoev, 2015).

Another dangerous mycotoxin is PAT, which is often seen together with other mycotoxins. PAT is reported in apples, grapes, and pears damaged by brown rot, as well as in the juice of these fruits. It is also reported to contaminate vegetables, cereals, and various types of cheese (Marc, 2022). The rotten part of the fruits should be removed before consumption in order to reduce the PAT content ingested by consumers. PAT is reported to have many toxic effects, including neurotoxic, cytotoxic, and carcinogenic effects, in addition to reproductive disturbances (Misihairabgwi et al., 2019; Sohrabi et al., 2022). The cytotoxic effect is manifested by damage to the gastrointestinal tract, liver, kidneys, and disturbances in endocrine and immune systems (Pal et al., 2017). PAT is classified by IARC as a suspected carcinogen (from group 3) (Ramalingam et al., 2019).

It is well known that most mycotoxins have strong immunosuppressive properties and can increase susceptibility to secondary bacterial infection in concentrations similar to those in the field (Boonchuvit et al., 1975; Ziprin et al., 1987; Tai and Pestka, 1988; Cooray and Jonsson, 1990; Oswald and Comera, 1998; Stoev et al., 2000b), including susceptibility to salmonellosis (Elissalde et al., 1994; Fukata et al., 1996; Gupta et al., 2008) and colibacillosis (Kumar et al., 2003). They can provoke a heavy progression of various infections such as *Pasteurella multocida* infection (Halloy et al., 2005) and porcine reproductive and respiratory syndrome (PRRS) (Ramos et al., 2010; Pósa et al., 2011, 2013, 2016) (Fig. 4), as well as worsening of various parasitic invasions, such as coccidiosis (Stoev et al., 2002b; Koynarski et al., 2007). For example, repetitive exposure to T-2 toxins has been found to increase susceptibility to various microorganisms such as *Staphylococcus, Mycobacterium bovis, Salmonella typhimurium* and *Listeria monocytogenes* infections in animals and/or chickens (Boonchuvit et al., 1975; Tai and Pestka, 1988; Ziprin et al., 1987; Cooray and Jonsson, 1990; Oswald and Comera, 1998).

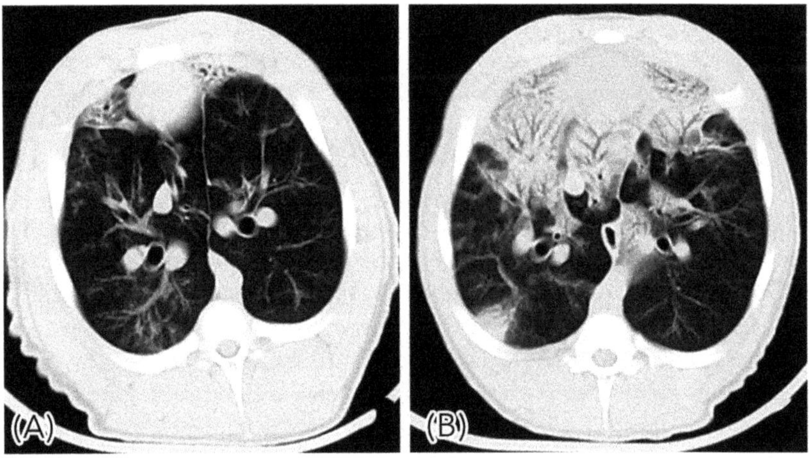

Figure 4 (A) Computed tomography (CT) image of lung lesions in *Mycoplasma hyopneumoniae* infected pig at a definite sectioning plane on day 58 of the experiment, demonstrating small foci of patchy ground glass opacification. (B) CT image of lung lesions in *M. hyopneumoniae* infected pig exposed additionally to 20 ppm FB1 via the diet (contamination levels that can be found in spontaneously contaminated animal feed) at the same sectioning plane on day 58 of the experiment, showing progressive development of pneumonic process as seen from the more enlarged size of the same damage (Stoev, 2024b).

It is important to emphasize that some experiments focused on immunosuppression as the first expressed toxic effect of OTA, which may become evident clinically before nephropathy and other biochemical changes. For the first time, a high susceptibility to natural infectious disease and/or secondary bacterial infections has been demonstrated in pigs fed on OTA-contaminated diet in levels often encountered in the field (Stoev et al., 2000b). The reported suppression of humoral and cellular immunity caused by this mycotoxin, defined in principle (NNT, 1991), has been demonstrated in practice via the natural development of secondary bacterial infections in pigs at only 1 ppm OTA in diet (Stoev et al., 2000b). It would appear that humoral immunity has been affected to the extent of allowing the development of clinical disease in pigs at only 1 ppm OTA in the diet (Stoev et al., 2000b) (Fig. 5).

Bearing this in mind, it can be assumed that combined mycotoxin exposure could compromise the immune system of animals in very low concentrations, given the synergistic or additive interaction between some key mycotoxins (Stoev et al., 2004, 2012). Therefore, it can be concluded that increased morbidity and mortality in livestock and poultry exposed to various mycotoxin combinations via feedstuffs is likely due to their increased susceptibility to secondary bacterial infections or a heavy progression of parasitic diseases or microbial infections (Stoev, 2023). Oxidative stress, which can be caused by many mycotoxins, could additionally threaten animal and human health (Karamalakova et al., 2018; Benkerroum, 2020).

It is worrying that many mycotoxins have carcinogenic (Fig. 6), genotoxic, and teratogenic (Fig. 7) effects, in addition to their immunosuppressive and toxic effects (Stoev, 2010a, 2020, 2021, 2022a, 2022b; Pfohl-Leszkowicz and Manderville, 2012; Claeys et al., 2020).

For example, FB1 has been reported to induce liver carcinomas in rats (Gelderblom et al., 1992), in addition to its nephrotoxic effect and involvement

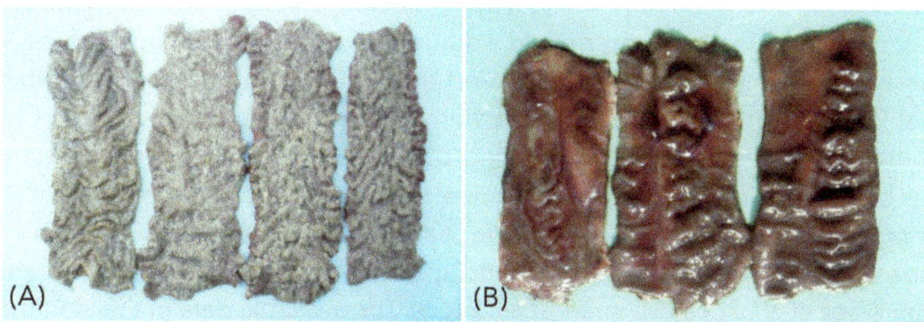

Figure 5 (A) Diphtheroid accretions on colon mucosa of a pig given 3 ppm OTA (contamination levels that can be found in spontaneously contaminated animal feed) for 17 days and sick with secondary salmonellosis. (B) Hemorrhagic dysentery associated with *Serpulina hyodysenteriae* and *Campylobacter coli* in pigs immunized against *S. choleraesuis* 47 days after commencing a diet containing 1 ppm OTA (Stoev, 2023).

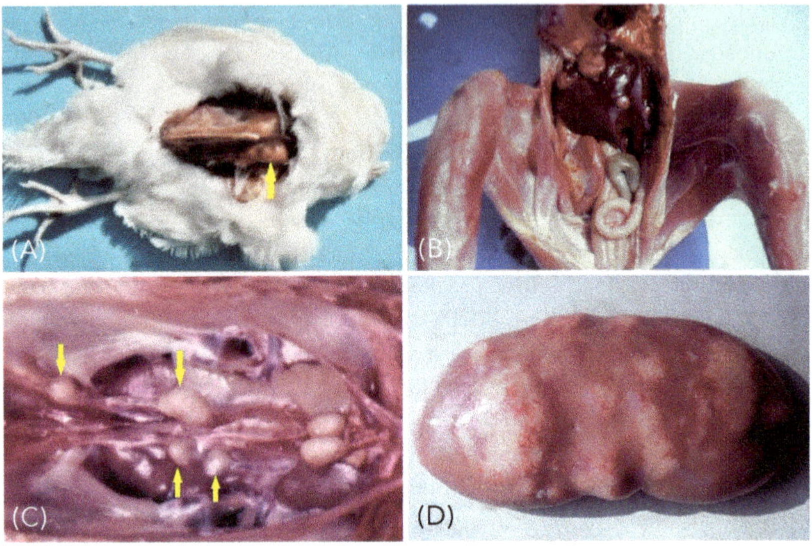

Figure 6 (A) Rhabdomyoma in the breast muscle (yellow arrow) of a female chick exposed to 5 ppm OTA (contamination levels found in spontaneously contaminated poultry feed) and 25 ppm phenylalanine via the feed, which was slaughtered at the end of the 24th month of the experiment. Large neoplasia in the region of the breast muscle, which protruded significantly above the surface. (B) Adenocarcinoma in the liver of a male chick exposed to 5 ppm OTA via the feed, which died at the end of the 10th month of the experiment. Large gray-white neoplastic foci in the diaphragmatic surface of the liver protruded significantly above the surface. (C) Carcinoma in the region of ureters (yellow arrows) of a male chick exposed to 5 ppm OTA via the feed, which died at the end of the 20th month of the experiment. Large gray-white neoplastic foci are seen along the ureters and protrude significantly above their surface. (D) Neoplastic tissue proliferation (fibroma and fibroadenoma) in kidney with spontaneous mycotoxic porcine nephropathy (Stoev, 2010a, 2023).

in animal nephropathies in Bulgaria and South Africa, together with OTA and PA (Stoev et al., 2010a, 2010b). FB1 is also suspected to cause human esophageal cancer in developing countries. In addition to animals, many individuals in these countries are chronically exposed to high levels of mycotoxins in their diet (Marasas et al., 1988). Pregnant women exposed to high levels of FUMs in their diet have a high risk of giving birth to a child with a birth defect of the brain or spinal cord. Recent studies clearly suggest that maternal intake of high levels of FB1 during early pregnancy may increase the risk of neural tube defects (NTDs), such as brain and spinal cord defects, in their newborns (van Waes et al., 2005; Missmer et al., 2006). The available data in the literature have primarily investigated the carcinogenic effect of single and (rarely) double mycotoxin exposure via in vitro studies (Bensassi et al., 2014), but in vivo studies investigating the chronic effects of multiple mycotoxins on the induction of neoplasia are scarce (De Ruyck et al., 2015; Speijers and Speijers, 2004).

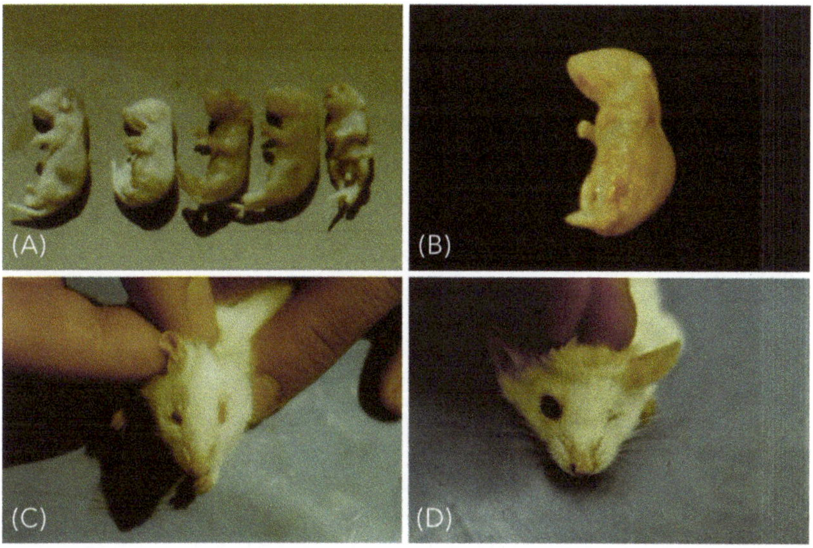

Figure 7 (A) Malformations in newborn mice whose mothers were exposed to 20 ppm OTA and 6 ppm OTB in their feed (contamination levels above these found in spontaneously contaminated animal feed) given from day 7 up to day 12 of the pregnancy – astomia and anophthalmia (both fetuses on left), 2 normal fetuses in the center and spina bifida, e.g. facial cleft and maxillary hypoplasia (fetus on right). (B) Malformations in newborn mice whose mothers were exposed to 20 ppm OTA and 6 ppm OTB in their feed given from day 7 up to day 12 of the pregnancy – peromelia in both right legs, astomia, and anophthalmia. (C) Malformation (micromonoftalmia) in a mouse whose mother was exposed to 10 ppm OTA via their feed up to day 8 of the pregnancy. (D) Malformation (monoftalmia) in a mouse whose mother was exposed to 10 ppm OTA via their feed up to day 8 of the pregnancy (Stoev, 2023).

4 Assessing risks from exposure to multiple mycotoxins

Multi-mycotoxin investigations reveal that around 75–100% of feed and food samples contain more than one mycotoxin at low contamination levels, which could additionally impact animal or human health at low multi-mycotoxin doses (Streit et al., 2012). Current regulations for mycotoxins were established using toxicological data from studies taking into account only one mycotoxin exposure at a time, and, therefore, do not address the combined toxic effects of several mycotoxins. However, the natural co-occurrence of mycotoxins in food and feed is well-known and explained by the multiple mycotoxin production of each fungus or by simultaneous contamination of food/feed by several fungi. Meanwhile, animal diets are typically prepared using multiple grain sources, which contributes significantly to multiple mycotoxin exposure (Smith et al., 2016). It is difficult to predict the toxicity of various mycotoxin combinations based on the individual toxicities of each mycotoxin because multiple mycotoxin exposure often has additive, synergistic, or antagonistic toxic effects (Speijers and Speijers, 2004).

Current data on the combined toxic effects of mycotoxins are scarce, and, therefore, the health risk from such multiple exposures is not well-known (Table 5). Grenier and Oswald (2011) reviewed more than 100 cases of mycotoxin interactions and found that most studies reported additive or synergistic interactions of simultaneously co-occurring mycotoxins when addressing the adverse effects on animal performance (Grenier and Oswald, 2011).

Various mathematical models classifying interaction effects as antagonisms, additive effects, or synergisms (based on a comparison of the observed effect) have been reported. However, the same models incorrectly assume that mycotoxin dose-effect curves are linear (simple addition of effects, factorial analysis of variance) (Kifer et al., 2020). It has been suggested that more appropriate mathematical models for assessing mycotoxin interactions include Bliss independence criterion, Loewe's additivity law, Response surface, Highest single agent (HSA) model, Combination index and isobologram analysis, Chou-Talalay's median effect approach, Code for the identification of synergism numerically efficient (CISNE), MixLow method, and various others. According to Kifer et al (2020), the only appropriate way to assess the nature of an interaction is to correctly estimate the dose-effect curves of each mycotoxin and combination and apply a well-defined model (based on Bliss or Loewe's theory) with respect to the model's assumptions and fitting the model by a direct estimation of all model parameters from a nonlinear least squares fitting. The authors also conclude that none of the proposed models is ideal because they all have their own advantages and disadvantages (Kifer et al., 2020). In addition, the same mathematical models are based on "in vitro" studies, which can often confuse the final result, because some of the synergistic effects – including the synergism between OTA and PA – can only be observed "in vivo" (Micco et al., 1991; Stoev et al., 2001).

The combined toxic effect of mycotoxins may be different depending on their physiological effects or mechanisms of action, and could exacerbate oxidative stress, immune suppression, impaired reproduction, and organ damage across species, leading to reduced growth performance, decreased milk and egg production, compromised carcass quality, and increased mortality rates (Gimeno and Martins, 2011; Akinmoladun et al., 2025). Understanding the cellular mechanism of such interactions is critical for improving risk assessment models, formulating integrated mitigation strategies, and protecting both livestock productivity and human food security.

Although toxicological interactions between mycotoxins pose significant risks to animal health and production, there are no international regulations (minimal regulations exist only with respect to the combination of AFB1, AFB2,

Table 5 The impact of joint mycotoxin exposure on body weight gain, productivity, organ damage, and immune system of different livestock species.

Mycotoxin combination	Type of interaction	Livestock species	Effect on health/productivity	Reference
AFB1+DON	Synergistic	Pigs	Immunosuppression	Chaytor et al., 2011
AFB1+DON	Synergistic	Pigs	Weight loss	Harvey et al., 1989
AFB1+DON	Synergistic	Dairy cattle	Milk yield reduction	Kiyothong et al., 2012
AFB1+DON+ZEA	Synergistic	Dairy cattle	Milk yield reduction	Kiyothong et al., 2012
AFB1+DON+ZEA	Synergistic	Dairy cows	Milk yield reduction	Wu et al., 2022
AFB1+DON+ZEA	Additive	Dairy cows	Immunosuppression	Jovaisiene et al., 2016
DON+ZEA	Additive	Dairy cows	Immunosuppression	Danicke et al., 2017
AFB1+OTA+ZEA	Synergistic	Dairy cows	Milk yield reduction	Wu et al., 2022
AFB1+OTA+ZEA	Synergistic	Dairy goats	Milk yield reduction	Huang et al., 2017
AFB1+DON	Additive	Poultry	Weight loss/rejection of feed	Huff et al., 1986
AFB1+T-2	Synergistic	Poultry	Weight loss	Huff et al., 1988
AFB1+OTA	Synergistic	Sheep	Fertility rate, litter number	Mohamed et al., 2019
AFs+OTA	Additive	Poultry	Egg production, weight loss	Verma et al., 2003
AFB1+OTA	Synergistic	Broilers	Liver/kidney damages	Qing et al., 2022
AFB1+OTA	Synergistic	Rabbits	Weight loss/increased mortality	Prabu et al., 2013
AFB1+FB1	Synergistic/additive	Broilers	Immunosuppression/liver damage	Tessari et al., 2006
AFs+FB1	Additive	Poultry	Weight loss	Kubena et al., 1995
AFs+DAS	Synergistic	Poultry	Weight loss	Kubena et al., 1993
AFB1+DAS	Synergistic	Pigs	Weight loss	Harvey et al., 1991
OTA+PA	Synergistic	Pigs	Kidney damage	Stoev et al., 2001
OTA+PA	Synergistic	Poultry	Weight loss/organs damage	Stoev et al., 2004

(Continued)

Table 5 *Continued*

Mycotoxin combination	Type of interaction	Livestock species	Effect on health/productivity	Reference
OTA+T-2	Synergistic	Broilers	Weight loss/organs damage	Kubena et al., 1989a
OTA+T-2	Synergistic	Pigs	Weight loss/organs damage	Harvey et al., 1994
OTA+DAS	Additive	Chicks	Weight loss/organs damage	Kubena et al., 1994
DON+ZEA	Synergistic	Pigs	Weight loss/feed intake reduction	Jia et al., 2020
DON+ZEA	Synergistic	Pigs	Ovarian/embryo damage	Malekinejad et al., 2007
DON+ZEA	Synergistic	Sows	Damage of reproductive system	Doll et al., 2003
DON+ZEA	Synergistic	Hens	Decrease in egg production	Gimeno and Martins , 2011
DON+T-2	Synergistic	Chickens	Weight loss/oral lesions	Kubena et al., 1989b
FB1+DAS	Additive	Turkeys	Weight loss	Kubena et al., 1997a
FB1+OTA	Additive	Turkeys	Weight loss	Kubena et al., 1997a
FB1+OTA	Additive	Pigs	Kidney/lung/brain damages	Stoev et al., 2012
FB1+T-2	Additive	Chickens	Weight loss, mortality	Kubena et al., 1997b
FB1+DON	Less than additive	Chickens	Weight loss	Kubena et al., 1997b

AFG1 and AFG2, or FB1 and FB2, or T-2 and HT-2) that address the combined effects of multiple mycotoxins. Although scientific interest in the biological effects of multiple mycotoxin exposure has increased, research in this area is still in its infancy. Comprehensive monitoring of mycotoxin co-occurrence is, therefore, crucial not only to identify the most prevalent combinations of mycotoxins but also to inform regulatory priorities and mitigation strategies aimed at ensuring feed safety (Akinmoladun et al., 2025).

Globally, contamination with multiple mycotoxins remains a serious problem, as it has been reported that 64% of feed samples worldwide contain at least two different mycotoxins (Gruber-Dorninger et al., 2019).

Despite all these studies, the extent of mycotoxin co-occurrence may still be underestimated, as most studies do not screen for the full spectrum of known mycotoxins, and limitations in analytical techniques may hinder the detection of all contaminants present in the samples studied (Freire and Sant'Ana, 2018).

Stoev and collaborators (Stoev et al., 2010a, 2010b) arrived at a similar conclusion as all analyzed feed samples from Bulgaria and South Africa taken from farms experiencing nephropathy issues were found to have multiple mycotoxin contamination, but the contamination levels of OTA were significantly lower than the levels shown to induce nephropathy symptoms. The conclusion was that this nephropathy was mainly caused by the combined synergistic or additive effects of OTA, FB1, and penicillic acid (PA).

In a UK study, screening of maize products intended for animal feed for 22 mycotoxins revealed that all 67 samples were co-contaminated with up to 12 different *Fusarium* mycotoxins (Scudamore et al., 1998). FUMs and DON were found together in 75% of the samples, but moniliformin and ZEA were also frequent co-contaminants. Another study of 330 samples by the same authors revealed maize to be the most affected by co-contamination, and 60% of the studied maize samples tested positive for more than one mycotoxin, with AFs and FUMs being the most common mycotoxin combination of about 28% of samples (Scudamore et al., 1997).

A similar investigation in Germany revealed the occurrence of 14 different *Fusarium* mycotoxins in 84 German maize samples studied following the 2006 and 2007 harvests (Goertz et al., 2010). Likewise, co-occurrence of trichothecenes with ZEA and OTA was observed in 760 Hungarian maize samples (Rafai et al., 2000).

Other studies on the occurrence of AFs and FUMs in Brazilian maize revealed that 54% of investigated samples were contaminated by both mycotoxins (Camargos et al., 2001). Such co-contamination with both mycotoxins is particularly concerning as evidence suggests that FB1 synergistically promotes liver tumors initiated by AFB1 (Grenier and Oswald, 2011). Considering that Brazil is the third most important producer of maize globally, the high levels of multiple mycotoxin contamination in Brazilian maize are of international significance.

Another study of feed and feed ingredients in 416 samples from Southern Europe carried out in 2005 and 2009 revealed that 22% of compound feed samples contained more than one mycotoxin (Griessler et al., 2010). In the same study, 23% of samples from Spain and 32% of samples from Italy contained at least two mycotoxins, with trichothecenes, ZEA, and FUMs being the major contaminants and most often co-occurring mycotoxins. A similar study of 400 feed samples across multiple livestock species in northern Spain (Muñoz-Solano and González-Peñas, 2023) found that 63.5% contained at least two mycotoxins. The highest co-contamination rate was seen in poultry feed (70%), followed by sheep feed (63%), cattle feed (62%), and pig feed (59%).

A similar survey in 277 feed samples for fattening pigs in Portugal for OTA, ZEA, and DON revealed that 10% of samples were found to contain any two-toxin combination, with ZEA and DON co-occurring most often (Almeida et al., 2011).

A similar investigation analyzing 82 samples of sow feed, wheat, and maize from different EU countries in 2008 for the presence of 23 different mycotoxins revealed that 75% of the investigated samples were co-contaminated by two or more mycotoxins (Monbaliu et al., 2010). Most of the samples were co-contaminated with DON and FUMs in combination with various other mycotoxins.

A similar investigation of 50 poultry feed samples in Slovakia investigated for trichothecenes and ZEA revealed that 84% were found to contain two or more mycotoxins. A combination of four mycotoxins (e.g. ZEA, DON, T-2, and HT-2) was found in 32% of the studied samples (Labuda et al., 2005a). Another study by the same authors on the occurrence of FB1, FB2, and moniliformin in poultry feed from Slovakia revealed that 25% of samples were co-contaminated with three mycotoxins (Labuda et al., 2005b).

In a study of dairy cattle diets in the Netherlands, DON and ZEA were found as co-occurring mycotoxins in 44% of investigated diets, with silage and compound feed being the major sources of mycotoxin exposure (Driehuis et al., 2008).

A study in Romania during 2021–2024 revealed that six mycotoxins (total AFs, FUMs, DON, ZEA, T2/HT2, and OTA) were the most frequently encountered in the southern area of the country in poultry, piglet and pig feed. The maximum highest concentrations were reported to be 5.8 ppb for AFs, 4.7 ppm for FUMs, 1.9 ppm for DON, 62.8 ppb for ZEA, 32.1 ppb for T2/HT2, and 19.7 ppb for OTA, irrespective of the type of feed. AFs and ZEA were present in all samples during the entire monitored period, and the only mycotoxin that exceeded the guidance value was DON (the recommended level of 0.9 ppm for pig feed was exceeded). It was concluded that sub-chronic and chronic simultaneous exposure to low mycotoxin levels and co-contamination were more common than acute exposure, and such low simultaneous exposure to several mycotoxins could affect animal health over time by lowering defense capacity, inducing inflammatory reactions, and damaging intestinal health, which could have major economic consequences (Bulgaru et al., 2025).

Other studies in Europe have focused on a single commodity. Among 123 Spanish barley samples from harvests studied for AFs, ZEA, and OTA in 2007 and 2008, 31% of samples revealed co-occurrence of AFB1 and OTA (Ibáñez-Vea et al., 2012a). The co-occurrence of AFB1 and ZEA was found in 12% of samples, while AFB1, OTA, and ZEA co-occurrence was seen in 27% of samples. In a subsequent study by the same authors on the occurrence of type A and B trichothecenes in the same samples, 43% revealed co-contamination with three or more trichothecenes (Ibáñez-Vea et al., 2012b). A further survey combining the data from both investigations showed that 96% of the samples were co-contaminated by three or more mycotoxins (Ibáñez-Vea et al., 2012c), with combinations of AFB1, OTA, and DON as well as AFB1, OTA, DON, and ZEA being the most frequent and most often observed in 29% and 26% of the samples, respectively.

The ultimate effect of prolonged intake of different combinations of mycotoxins, as occurs in practice, is not clear. It is assumed that the simultaneous intake of such combinations of mycotoxins, even in low feed concentrations, could be a crucial risk factor for the development of chronic kidney nephropathy in animals and humans, especially following continuous mycotoxin exposure. This is because the same mycotoxin combinations have been found in high (FB1 and PA), low, or moderate concentrations (OTA and CIT) in all feeds originating from farms with nephropathy issues in Bulgaria and South Africa (Stoev et al., 2010a, 2010b).

The frequent detection of mycotoxin co-contamination and the evidence suggesting potential additive or synergistic interactions between the most commonly co-occurring mycotoxins suggest that guidelines or maximum permitted levels should not only be set for each individual mycotoxin but also for target mycotoxin combinations, found often in feedstuffs and food. In order to define such guidelines and rules, more data and further studies are required on the impact of different mycotoxin combinations on various animal species, poultry, and humans. In addition, future studies should be conducted to improve the available knowledge on the toxicological effects of co-occurring mycotoxins, which are critical for revising the current guidelines and legislative limits for mycotoxins in feedstuffs and food to ensure better health protection for animals and humans. In this regard, potential additive or synergistic interactions with emerging mycotoxins such as enniatins, moniliformin, beauvericin, fusaproliferin, fusaric acid, culmorin, butenolide, sterigmatocystin, emodin, mycophenolic acid, alternariol, alternariol monomethyl ether, and tenuazonic acid should also be taken into account (Streit et al., 2012). Clearly, not all of the above-mentioned mycotoxins (compounds) are toxicologically dangerous at their naturally occurring levels and are, therefore, of little health concern to consumers.

5 Risks of masked mycotoxins to animal health

Another major risk and potential hazard to animal health is masked mycotoxins, which contaminate various feeds and animal-derived products (Stoev, 2015). Masked mycotoxins – or mycotoxin derivatives – cannot generally be detected or are underestimated by conventional analytical methods, which are designed primarily for parent (native) mycotoxins. It is almost impossible to carry out a thorough risk assessment of contamination levels of masked mycotoxins in feeds, due to the limited data available on their levels of exposure in different animals and the lack of sufficient data on their actual toxic properties. In this regard, actual contamination levels and the hazard of masked/modified mycotoxins are often underestimated. These plant metabolites (masked mycotoxins) have been observed for many mycotoxins, including DON, T-2 toxin, HT-2 toxin, ZEA, OTA, and FUMs and can be seen mainly in cereal commodities, such as wheat, barley or corn (Berthiller et al., 2005; 2009, 2013). Some, such as glucoside conjugates of trichothecenes, may present a potential safety threat due to their hydrolyzation to toxic compounds during the digestive process (Dall'Erta et al., 2013). The major masked mycotoxins that raise safety concerns are DON-3-glucoside, ZEA-14-glucoside, ZEA-14-sulfate, T2-glucoside. and HT2-glucoside.

Masked mycotoxins that remain undetected by conventional analytical methods could explain why naturally DON-contaminated feedstuffs generally exert a stronger toxic effect compared to feedstuffs artificially contaminated with pure DON (Dersjant-Li et al., 2003; Trenholm et al., 1994). The same conjugates (masked mycotoxins) could be produced by the fungi themselves (3-ADON, 15-ADON) or elaborated as part of the defense mechanism of infected plants (ZEA-4-Glucoside, DON-3-Glucoside) (Berthiller et al., 2009). Masked mycotoxins may exert toxic effects themselves or undergo conversion into their parent forms during the digestive process and further increase the toxicity of mycotoxin-contaminated feeds (Garais et al., 1990; Vendl et al., 2009). Such conjugates are typically not detectable when tested for parent toxins.

The European Food Safety Authority (EFSA) considers it appropriate to assume that many modified mycotoxins exhibit the same toxicity as their parent compounds and may substantially contribute to overall exposure (EFSA, 2014, 2017a, 2017b, 2018). In this regard, EFSA has extended the tolerable daily intake (TDI) for ZEA of 0.25 μg/kg bw to a group-TDI covering ZEA and its modified forms (EFSA, 2017a) (Table 6). Unfortunately, realistic estimates of

Table 6 Tolerable daily intake of mycotoxins according to EU regulations and recommendations (EC Regulations No 2023/915 and No 165/2010; JECFA, 2022; EC Recommendations 2013/165/EU; EFSA CONTAM Panel, 2016; EFSA CONTAM Panel, 2017).

Mycotoxins	TDI per kg bw (μg/kg bw)
Aflatoxins (AFs) (sum of AFB1, AFB2, AFG1, AFG2, AFM1)	No exact value – as low as reasonably achievable (ALARA principle) – less than 0.001–0.01 (the lower dose concerns carcinogenic effect)
Ochratoxin A (OTA)	0.0002–0.017 (the lower dose concerns carcinogenic effect)
Patulin (PAT)	0.4
Deoxynivalenol (DON)	1
Nivalenol (NIV)	1.2
Zearalenone (ZEA)	0.25
T-2 + HT-2 toxins + DAS	0.025
Fumonisins (FUMs)	2

animal or human dietary exposure to ZEA or DON and their modified forms, based on feed/food consumption and occurrence data, are hampered by a lack of commercially available analytical standards and certified reference materials, in addition to a lack of validated analytical methods for the simultaneous quantification of mycotoxins and their modified forms (Lorenz et al., 2019).

Masked mycotoxins can represent a large proportion of total mycotoxins and, following ingestion, can be converted to the native (parent) mycotoxins after microbial and/or animal metabolism (Berthiller et al., 2013). It can, therefore, be supposed that the actual levels of target mycotoxins in feedstuffs are often higher because masked mycotoxins have not been measured.

The group of masked mycotoxins comprises both extractable conjugated and bound (non-extractable) varieties. Unfortunately, extractable conjugated mycotoxins can only be detected by appropriate analytical methods when their structure is known and analytical standards are available. Bound mycotoxins, however, are not directly accessible and must be liberated or separated from the matrix by chemical or enzymatic treatment prior to chemical analysis (Berthiller et al., 2013).

Transformation of masked mycotoxins into parent molecules can also be undertaken in order to use conventional analytical methods, which typically involve hydrolysis (Dall'Asta et al., 2009), the type of which should be carefully selected. Unfortunately, there is no single hydrolysis method that can be applied to all masked mycotoxins. Masked mycotoxins might be less toxic than their parent compounds, but they might also be more toxic, e.g. when they are more bioavailable (Berthiller et al., 2013). It should be emphasized that there are currently no sufficient toxicokinetic and toxicodynamic studies available to evaluate the possible hazard or risk of masked mycotoxins in comparison with their parent forms. It is, therefore, impossible to perform a proper risk assessment for these mycotoxins in food, due to the lack of data on exposure and toxic properties. Recognition of the toxicological relevance of masked mycotoxins, in addition to evaluating the hazard of co-occurrence of target mycotoxins in feeds or animal-derived products, is a recent challenge that needs to be addressed by the respective regulatory bodies, food manufacturers, and monitoring authorities to protect the health of consumers and to evaluate animal and human health hazards (Stoev, 2015).

With regard to different kinds of animals and poultry, we should take into account the fact that the guidance mycotoxin values for cereals and feeds have been determined for the most tolerant animal species and should, therefore, be considered as upper guidance values. For feeds intended for more sensitive animals, the lower guidance values should be applied by manufacturers (EC Recommendations 2006/576/EC).

6 Improving risk assessment, control and regulation of mycotoxins

Safety and quality control activities along the food/feed supply chain are essential to enhance the safety of various foods and feedstuffs and to prevent foodborne diseases in animals and humans, but these activities require integration into a well-coordinated and integrated system. The following components of risk analysis, including risk assessment, risk communication, and risk management, should be considered to ensure adequate food/feed safety. In order to provide an adequate risk assessment, the following issues, such as hazard identification, hazard characterization, exposure assessment, and risk characterization, must also be assessed. In order to ensure adequate risk communication, the regular exchange of knowledge and opinions among risk evaluators, consumers, and risk managers is necessary during the process of risk analysis. However, an adequate risk management decision should not only be based on the risk assessment but also take into account some other issues, such as environmental, economic, and ethical factors, as well as effective control feasibility (EC Regulation No 178/2002; Stoev, 2015, 2024b).

To eliminate or minimize mycotoxin content in food and feeds, various key agricultural practices and storage protocols should be introduced that can effectively prevent their formation. The occurrence of mold and mycotoxins can be decreased by the application of various preventative measures before and/or after harvest, including timely harvesting, cleanup, appropriate drying and storage practices, creation of plant cultures resistant to fungal infestation, management of insect infestation, and crop rotation (Bata et al., 2001; Doko et al., 1995). The fate of each mycotoxin at each stage of the process should also be traced in order to create a reliable control and reduce the quantity of mycotoxins that can reach the consumer (be that animals or humans).

Unfortunately, complete removal of mycotoxins is not always achievable. It is essential to protect the consumer from the effects of a particular mycotoxin in the food supply when a risk assessment of the contaminant indicates that its level of exposure is unacceptable. When the risk assessment of a particular mycotoxin is considered significant, legislation should be introduced, together with measures aimed at minimizing exposure of the consumer to that mycotoxin. Setting a maximum permissible limit is a powerful tool for such effective control (Rosner, 1998) (Table 2). Two main compounds should be taken into account in each risk assessment – the toxicological effect of a particular mycotoxin and an assessment of the consumer's mycotoxin exposure. It is essential that all limits are supported by reliable and extensive research to avoid unnecessary restrictions and subsequent economic loss.

The elaboration of internationally recognized regulatory measures for mycotoxin content in food and feedstuffs is a difficult task, and, therefore, the preliminary elaboration and introduction of temporary guideline limits in cases posing a significant human/animal health hazard would be useful and easier to achieve.

The maximum levels of mycotoxins and other chemical contaminants in foods and various commodities are generally considered by the Codex Committee on Food Additives and Contaminants (CCFAC) in consultation with the Codex Commodity Committee (CCC), but the final decision is usually taken by the Codex Alimentarius Commission (CAC). It is important to emphasize that all final conclusions are also based on scientific evaluations made by the Joint FAO/WHO Expert Committee on Food Additives (JECFA) and other relevant expert organizations (Boutrif and Canet, 1998). The CAC, supported by FAO and WHO, aims to facilitate world trade and protect the health of consumers through the development of international standards for foods and feeds.

Following evaluation of the most dangerous mycotoxins, JECFA has provided a mechanism for assessing the toxic impact of each mycotoxin (JECFA, 2001), which includes determining the no-observed-effect-level (NOEL) via experimental studies and subsequent application of a safety factor, in order to estimate the provisional tolerable weekly intake (PTWI) or daily intake (PTDI). This approach evaluates the maximum tolerated levels of mycotoxins in foods/feeds but cannot be applied in regard to the carcinogenic effect of some mycotoxins such as AFs, where the "as low as reasonably achievable" (ALARA) approach or "as low as possible but technologically feasible and analytically detectable in the food/feed ready for consumption" approach should be applied (Rosner, 1998). The International Agency for Research on Cancer (IARC) provides such a mechanism for evaluating the carcinogenic properties of mycotoxins and the following groups of mycotoxins have been created: group 1 (carcinogenic to humans, incl. AFB1), group 2A (probably carcinogenic to humans), group 2B (possibly carcinogenic to humans, incl. FB1 and OTA), and group 3 (not classifiable as to its carcinogenicity to humans) (Castegnaro and McGregor, 1998; IARC, 2002). Revisions of the accepted criteria for this classification and regular updates with regard to certain mycotoxins are undertaken periodically (IARC, 2012; Stoev, 2024a, 2024b). The European Food Safety Authority (EFSA) also provides independent scientific advice on food-related risks in relation to AFs via a margin of exposure (MOE) approach. More than 200,000 analytical results on the occurrence of AFs were used in the evaluation. A benchmark dose lower confidence limit (BMDL) for a benchmark response of 10% of 0.4 μg/kg bw per day for the incidence of hepatocellular carcinomas (HCC) in male rats following AFB1 exposure was set by an MOE approach for risk characterization. However, the calculation of a BMDL from the human data was found to be inappropriate. For AFM1, a potency factor of 0.1 relative to AFB1 was used in this assessment. MOE values for AFB1 exposure ranged from 5000 to 29, and from 100 000 to 508 for AFM1 exposure. The calculated MOEs were below 10 000 for AFB1 and for AFM1, which raises a health concern (EFSA CONTAM Panel, 2020).

The group TDI of 25 ng/kg bw for T-2, HT-2, and DAS, alone or in combination, was recently decided by JECFA to be the most appropriate TDI (JECFA, 2022) and the previous group provisional maximum tolerable daily intake (PMTDI) of 60 ng/kg bw for T-2 and HT-2, accepted at the 56th meeting and updated at the 83rd meeting to include DAS, has now been changed. A PMTDI of 0.4 μg/kg bw was also set for PAT (Table 6).

Similarly, a group TDI of 1 μg/kg bw was designed for DON and its derivatives, e.g. 3-acetyldeoxynivalenol (3-ADON), 15-acetyldeoxynivalenol (15-ADON), and plant metabolite DON-3G, which is based on experiments with mice in chronic mycotoxin exposure (EFSA CONTAM Panel, 2017; Table 6). In acute cases, however, 8 μg/kg bw per eating occasion was accepted as a reference dose due to gastrointestinal damage reported in DON-exposed humans in China (EFSA CONTAM Panel, 2017; Wang et al., 2021). This TDI was based on the fact that the same three derivatives can be biotransformed into DON in humans (Vidal et al., 2018). Evaluation of the risk of DON exposure in the EU revealed that a proportion of this population is exposed to concentrations that represent a significant health hazard (Alvito et al., 2021, 2022).

The TDI of FUMs was set at 2 μg/kg bw, as FB1 is thought to cause neural tube defects in the embryo (Knutsen et al., 2018) and little data is available for the mechanism of renal excretion of FB1 in humans. The mycotoxin is poorly absorbed by the gastrointestinal system, rapidly eliminated from the blood via the hepatobiliary route, and primarily excreted through the faeces (Alvito et al., 2022), which explains the lack of necessary research from the scientific community. Additional efforts to clarify the toxicokinetic profile of this mycotoxin following oral ingestion are necessary to contribute further to a more thorough risk assessment (Alvito et al., 2022).

The TDI of AFs (the total sum of all forms of AFs) is the lowest, similar to the maximum permitted levels of AFs in feeds or food commodities (Table 6), because AFB1 is classified by IARC as a group 1 human carcinogen (IARC, 2002, 2012). AFB1 and its metabolites are excreted via the feces, urine, and milk, but these mycotoxins can also be found in animal organs, meat, or chicken eggs, representing a significant health hazard. Moreover, the typical contamination of cereal-based products, in addition to corn gluten meal, pork products, milk, and eggs, increases the potential for human exposure to AFs (Streit et al., 2012; Alvito et al., 2022). It is a worrying fact that above 80% of AFs are absorbed in the gastrointestinal tract of non-ruminant animals, mostly via passive transport, in comparison to the low rate of absorption of other dangerous mycotoxins such as OTA or FUMs (from 1% up to 60%) (Grenier and Applegate, 2013).

The TDI of OTA, similar to the TDI of AFs, has the lowest value (0.0002–0.017 μg/kg bw) as the lower dose reflects its carcinogenic effect (Table 6). OTA contamination mainly occurs in cereals such as wheat, corn, and rye flour, but also in eggs, kidneys, and sometimes in meat products (Stoev, 2015). The enterohepatic circulation of OTA contributes to its retention for a longer time in the gastrointestinal system, which can increase the health hazard. The JECFA-proposed initial value of 112 ng/kg bw PTWI for OTA corresponds to about 16 ng/kg bw PTDI (WHO, 1991). This PTWI was then decreased to 100 ng/kg bw or 14 ng/kg bw PTDI (JECFA, 1997) but was subsequently increased to 120 ng/kg bw, which is equivalent to about 17 ng/kg bw PTDI (Table 6). This calculation is based purely on the nephrotoxic properties of OTA, and not its carcinogenicity. Another calculation of TDI has been made by Kuiper-Goodman and Scott, which takes into consideration the carcinogenic effects of OTA, which range from 0.2 to 4.2 ng/kg bw, depending on the calculation methodology (Kuiper-Goodman and Scott, 1989). Taking both TDI calculations into account, the average daily intake of OTA in people and animals living in endemic areas for Balkan endemic nephropathy (BEN) in Bulgaria totaled 26.8 ng/kg bw in 1988, 36.4 ng/kg bw in 1989, and 34.2 ng/kg bw in 1990 (Stoev, 2015). These values are way above the TDI of 17 ng/kg bw (in regard to the nephrotoxic effects of OTA) and even higher when taking into consideration the carcinogenic properties of OTA (0.2 to 4.2 ng/kg bw) (Kuiper-Goodman and Scott, 1989).

With regard to veterinary hygiene controls for OTA contamination in animal products, the measures applied in some EU countries, such as Denmark, do not ensure adequate food safety as they are inadequate (Stoev, 2015). The accepted regulation in Denmark requires all 'mottled and/or enlarged kidneys' to be tested for OTA content during the slaughter of pigs, with subsequent rejection of the carcasses if OTA content is greater than 10 μg/kg (Boutrif and Canet, 1998). Such regulations, however, are inappropriate, because the mottled appearance of kidneys occurs only after prolonged OTA exposure of nearly one to three months (Krogh et al., 1973; Stoev et al., 2001). This regulation does not, therefore, guarantee OTA-free pork and cannot prevent OTA-contaminated meat from entering commercial channels, representing a potential risk to human health (Stoev, 2015). An easily implemented preventive measure would be to analyze only a few blood samples from pigs and poultry in farms experiencing nephropathy problems a few weeks (in pigs) or a few days (in poultry) prior to slaughter. If OTA is detected in the blood, the feed supplied by the producer could be replaced with a safer, mycotoxin-free alternative for one week in pigs or for several days in poultry. A potential approach to prevent OTA contamination in meat and derived products could be to extend the fasting period (feed deprivation) just before slaughter (Stoev, 2008). Such a measure is easy to perform and highly effective, due to the short half-life of OTA in pigs (72–120 hours) and especially in poultry (four hours) (NNT, 1991). In such cases, the OTA levels in blood or tissues of poultry and pigs, respectively, will decrease significantly, and any loss of pig/chicken production will be avoided. Such control measures could ensure a more effective procedure for preventing subsequent OTA exposure of humans via meat products, as compared to the toxicological studies of 'mottled kidneys' as required by Danish regulations. If these control measures cannot be performed, the removal of kidneys and livers in already slaughtered poultry, and only kidneys in already slaughtered pigs, where the largest quantity of OTA is accumulated, would be sufficient (Stoev, 2008; Stoev et al., 2002a).

It is essential that the HACCP (Hazard Analysis and Critical Control Point) system is introduced worldwide to ensure regular identification and assessment of potential hazards at different stages of food/feed production. It is also necessary to undertake adequate measures to ensure regular control of food and feedstuff safety, including prevention strategies, good manufacturing practices, and regular quality control at various stages of production, from the harvest of raw ingredients to the final consumer (Stoev, 2015). Determining the content of key mycotoxins such as DON, ZEA, and FUMs during each step of maize and cereal processing is crucial due to the high contamination levels of these mycotoxins in raw ingredients. Enforcement of regular surveillance and adequate food safety regulations is also critical to provide safe food/feed and decrease incidences of foodborne ailments. Cleaning the cereals before milling is carried out to remove fungal spores, debris, and broken grains containing high concentrations of mycotoxins (Dowell et al., 1990; FAO, 2002). The removal of bran from flour intended for bread might also decrease human or animal exposure to mycotoxins, and more lenient limits for raw ingredients should be accepted (Scudamore and Banks, 2004).

The monitoring of food and feed quality and the application of mycotoxin regulations is primarily carried out in developed countries, with such standards unavailable or ineffective in developing countries. The introduction of such regulations in developing countries is highly complicated, due to problems with food supply. Moreover, such regulations often encourage the export of the best quality crops that comply with the regulations, which increases the risk of mycotoxin exposure and subsequent health ailments in the local population (both people and animals), as the low-quality foods/feeds or food ingredients typically remain for local consumption.

Currently, many countries have elaborated their own limits and regulations, designed to ensure effective control of mycotoxin contamination in foods and feeds and the safety of their human and animal populations (Stoev, 2015, 2023). Such statutory limits for mycotoxins in various commodities are generally dependent on various important factors such as the existence of sufficient data on the toxicological effect of mycotoxins, the existence of data on the occurrence of mycotoxins in various commodities, the availability of sampling and analysis methods, as well as intercountry trade implications and the existence of sufficient food supply (Boutrif and Canet, 1998) (Table 2). The first mycotoxin limits were set for AFs in the 1960s. At the end of 2003, nearly 100 different countries, representing nearly 87% of the world's population, had set limits for mycotoxins in foods and feedstuffs, and this number continues to increase (Van Egmond and Jonker, 2004).

Introducing worldwide legislation and internationally recognized regulations is of crucial importance for minimizing human and animal exposure to each mycotoxin or mycotoxin combination when the risk assessment of such exposure is significant. However, when evaluating each particular risk assessment, the toxic effect of each mycotoxin and mycotoxin combination should be taken into account, in addition to the estimated mycotoxin exposure of various species of animals, birds, and humans, due to their different sensitivity to different mycotoxins. Simultaneous exposure to several mycotoxins via food commodities or feedstuffs, even at very low concentrations for a long time (such as exposure to OTA and PA simultaneously) is of crucial importance to the appearance of foodborne ailments and could be a significant risk for animal/human health. Therefore, the actual toxic and carcinogenic effects of various key mycotoxin combinations, which are often seen in the field, should be carefully evaluated, and new limit values must be introduced in such cases. Unfortunately, the current maximum permitted levels (MPLs) and TDI values for mycotoxins (Table 2, Table 6) as accepted in the EU for animal feed (EC Recommendations 2006/576/EC and 2013/165/EU; EC Directives 2002/32/EC and 2003/100/EC) or human food (EC Regulation No 2023/915 and EC Recommendation 2013/165/EU) are unreliable. These values consider only the known toxic effects of individual mycotoxins or groups of similar mycotoxins (such as combinations of AFs or T-2+HT-2+DAS or FB1+FB2) in different animal species while ignoring the actual mycotoxin interactions (synergistic or additive) that occur under field conditions. The United States Department of Agriculture has also accepted similar limits in the US but has ignored the actual mycotoxin interactions that occur in practice (USDA, 2017). Additional regulations and control measures should, therefore, be introduced in such cases, based on known synergistic or additive interactions of various mycotoxins and their stronger toxic effects on animals and humans in such cases, e.g. OTA and PA (Micco et al., 1991; Stoev et al., 2001, 2004) or OTA and FB1 (Stoev et al., 2012), in order to provide adequate risk assessment and food safety. International harmonization of such regulations and control measures should also be implemented to facilitate global food trade and food safety. These regulations and risk assessments should be carefully designed and based on the toxic effects of combined mycotoxin exposure, as occurs under field conditions (Stoev, 2024a, 2024b). Such MPLs and TDIs should be based on extensive studies in order to prevent excessive restrictions and unnecessary economic loss (Stoev, 2015).

In order to ensure the increased safety of the food chain and compliance with EU legislation for food and feed, the Rapid Alert System for Food and Feed (RASFF) was created in 1979. RASFF serves as a useful tool for exchanging knowledge between competent authorities on consignments of food or feed in cases where a high risk for human/animal health has been identified and appropriate preventive measures have been undertaken (Pinotti et al., 2016). Considering the increased information related to the changes in mycotoxin content at the time of food processing, more lenient limits for raw ingredients could be set if a significant decrease is seen in certain stages of the food processing chain. Therefore, in-depth knowledge of how mycotoxins are destroyed or eliminated when using key food/feed processing techniques could further justify the use of these animal feeds, despite containing high levels of key mycotoxins. Such feed ingredients or feedstuffs should be the subject of more frequent monitoring compared to other feed products. For example, the most common mycotoxin in barley is DON, whereas maize is generally a good substrate for the production of mycotoxins, such as AFB1, OTA, ZEA, and DON and, therefore, requires continuous monitoring (Marc, 2022; Stoev, 2023). Such knowledge is especially important for DON, ZEA, and FUMs in maize and cereals because these mycotoxins are often found in very high levels in raw ingredients but are significantly reduced during food processing. In addition, it has been found that 17% of the AFs tested samples did not comply with the EU maximum permitted level of 5 μg/kg applicable to feedstuffs for dairy animals (Stoev, 2023). Such insights should be taken into account in order to ensure an appropriate balance between economic issues and risk to the consumer, because the absence of such a balance could lead to unnecessary penalization of agricultural producers without significantly improving food safety for animals or humans. It is important to emphasize that certain food processing techniques, such as bran removal, might significantly decrease human/animal mycotoxin intake (FAO, 1997; Stoev, 2023).

The development of a suitable networking system for dissemination of important knowledge should be introduced and sustained at the international level, including staff training at regional and international levels. Such international regulations and control measures should be scientifically based and elaborated with the cooperation of all interested parties, including consumers, manufacturers, traders, and policymakers, to ensure their broad distribution and compliance. In the process of creating such international regulations and standards, various factors should be taken into consideration, including scientific validity, adequate risk assessment, analytical accuracy, and establishing the toxicity of mycotoxin combinations that occur most frequently in the field. In addition, key economic and political factors – including the commercial and economic interests of countries and the need to ensure an adequate food supply while avoiding unjustified rejection of raw food ingredients and potential economic difficulties for producers – should also be taken into consideration (Stoev, 2015).

7 Conclusion and future trends in research

It is well known in the scientific community that human and animal exposure to mycotoxins via food and feedstuffs cannot be fully prevented, since mycotoxins are natural contaminants of food and feed ingredients. Certain mycotoxins, such as AFs, OTA, ZEA, and, in some cases, DON are much more hazardous due to their transmission in milk of lactating cows (parent toxins or their metabolites, e.g. AFM1, ZEA, αZEA) or eggs and meat (e.g. OTA) (Marc, 2022).

Current national rules and regulations for the monitoring and control of mycotoxins in foods and feedstuffs are primarily based on evaluating the threat of each individual mycotoxin for each individual country. It is crucial to introduce carefully designed surveillance control systems and modern, internationally recognized, biomonitoring measures that evaluate animal/human exposure to mycotoxins. Such measures should be introduced worldwide for reliable control of the factors that compromise the quality of food/feed ingredients and the commodity system. It is important to highlight that introducing more restrictive food safety regulations could lead to unnecessary rejections of various raw food ingredients and related commodities, which could have serious consequences for small producers and traders and create unjustified barriers to international trade (Stoev, 2015). Therefore, a synchronization of current national regulations and elaboration of international regulations and standards for acceptable content of mycotoxins or key mycotoxin combinations in foods/feedstuffs and raw ingredients should be undertaken. Such international regulations would significantly improve consumer protection worldwide and facilitate international trade and food safety on the basis of recent scientific advances and adequate risk assessment.

The key recommendation of this review is the evaluation of risk assessment of combined mycotoxin exposure to various key mycotoxin combinations, along with a proposal to elaborate new regulatory measures in such cases, which must take into account additive and synergistic interaction between the same mycotoxins as occur in the field. Current regulations and standards in the EU and US do not take into consideration mycotoxin interactions and the combined toxicity or carcinogenicity of mycotoxins, and are based solely on their individual toxic effects. Therefore, the combined toxicity of key mycotoxin combinations and the synergistic or additive interactions between mycotoxins should be thoroughly evaluated for regulatory purposes. The potential interaction between emerging mycotoxins should also be taken into account. In this regard, knowledge gaps concerning various emerging compounds have been identified, and these should be addressed by the scientific community in the coming years to enable proper risk assessment. (Gruber-Dorninger et al., 2017). Additional experimental studies in animals or humans designed to clarify relationships between key mycotoxins and respective health outcomes (e.g. neural tube defects, idiopathic congestive cardiomyopathy and esophageal cancers in humans) are also of crucial importance. Human and animal exposure to mycotoxins in some countries (such as the Balkan countries) is often considered to be within the TDI for each individual mycotoxin; however, the combined toxic effect frequently exceeds the toxicity of the individual mycotoxins considered separately (Stoev, 2023).

A suitable networking system for worldwide dissemination of key knowledge should be introduced and sustained at the international level. Cooperation among research institutions, consumers, manufacturers, traders, and policymakers is crucial for addressing current food safety issues and facilitating broad implementation of and compliance with regulations and standards, while avoiding unnecessary rejections of raw food ingredients and subsequent economic difficulties for producers. Political and economic factors, including commercial interests and means of ensuring sufficient food supply should also be taken into consideration in the decision-making process. Efforts to improve the quality of food/feed commodities should also involve the cooperation of affected populations, as such improvements may entail increased costs for feeds or foods.

8 Where to look for further information

8.1 Further reading (EC directives, recommendations and regulations)

- EC Directive 2002/32/EC of the European Parliament and the Council on undesirable substances in animal feed (*Off. J. Eur. Union L.* 2002, 140,

10–21) and its amendment EC Directive 0203/100/EC (Off. J. Eur. Union L., 2002 and 2003).

- EC Recommendation 2006/576/EC and EC Recommendation 2013/165/EU on the presence of mycotoxins in products intended for animal feeding and cereal products (*Off. J. Eur. Union L.* 2006).
- EC Regulation No 178/2002, EC Regulation No 165/2010, and EC Regulation No 2023/915 of the European Parliament and of the Council laying down the general principles and requirements of food law, establishing the European Food Safety Authority and laying down procedures in matters of food safety and setting maximum levels for certain contaminants in foodstuffs (*Off. J. Eur. Union* 2010).
- Worldwide Regulations for Mycotoxins in Food and Feed in 2003 (Food and Agriculture Organization of the United Nations: New York, NY, USA, 2004).
- United States Department of Agriculture (USDA), GB 2761–2017: limit of mycotoxin in food of China National Food Safety Standard (Washington, USDA, 2017).

8.2 Key journals, conferences, meetings, papers and books (issued by target organizations, agencies, expert committees and programs)

- CAST (Council for Agricultural Science and Technology) In: *Mycotoxins: Risks in plant, animal, and human systems*. Task Force Report N 139. Richard, J.L., Payne, G.A. (Eds.). (Ames, IA, USA, 2003).
- EFSA CONTAM Panel (EFSA Panel on Contaminants in the Food Chain). Appropriateness to set a group health-based guidance value for some mycotoxins and risks to human and animal health related to some mycotoxins and their modified forms in foods and feeds (*EFSA J.*, 2016, 2017, 2020, https://doi.org/10.2903/j.efsa.2020.6040).
- FAO. Food and Nutrition Paper No. 64, *Worldwide Regulations for Mycotoxins* (FAO: Rome, Italy, 1997).
- *Manual on the application of the HACCP system in mycotoxin prevention and control* (Joint FAO/WHO Food Standards Programme FAO, Rome, Italy, 2002).
- Food Security and Foodborne Mycotoxicoses, Risk Assessment, Preventive Measures, and Underestimated Hazard of Masked Mycotoxins or Joint Mycotoxin Interaction, In: *Food Toxicology, Chapter 9*, Debasis Bagchi, Anand Swaroop (Eds), (CRC Press, Taylor & Francis Group, 2016, https://doi.org/10.1201/9781315371443).
- Scientific Cooperation on Questions Relating to Food (SCOOP). Assessment of dietary intake of ochratoxin A by the population of EU

Member States – Report Task 3.2.7. (European Commission, Directorate-General Health and Consumer Protection, 2002, Available at: http://ec.europa.eu/food/fs/scoop/indexen.html); Collection of occurrence data of *Fusarium* toxins in food and assessment of dietary intake by the population of EU Member States (European Commission, Directorate-General Health and Consumer Protection, 2003, Available at: http://ec.europa.eu/food/fs/scoop/indexen.html.

- Nordic Working Group on Food Toxicology and Risk Evaluation (NNT). *Nordiske Seminar-og Arbeidsrapporter* (1991:545).
- Health Evaluation of Ochratoxin A in Food Products (Nordic Council of Ministers, Copenhagen, 1991).
- International Program of Chemical Safety (IPCS). Selected mycotoxins: ochratoxins, trichothecenes, ergot (IPCS Environmental Health Criteria No. 105, WHO, Geneva, 1990).
- International Agency for Research on Cancer (IARC). Some Traditional Herbal Medicines, Some Mycotoxins, Naphthalene and Styrene (IARC, Lyon, France, Vol. 82, 2002); Mycotoxins and human health (IARC Press, Lyon, France, Vol. 158, 2012).
- Joint FAO/WHO Expert Committee on Food Additives (JECFA). Evaluation of certain food additives and contaminants (Thirtyseventh report of the JECFA, WHO technical report series 806, Geneva, 1991); Toxicological evaluation of certain food additives (WHO Food Additives Series (Forty-ninth meeting of JECFA, Geneva, 1997); evaluation of certain mycotoxins that may contaminate food (fifty-sixth meeting of JECFA, Geneva, 2001); SUMMARY AND CONCLUSIONS (Ninety-third virtual meeting, 24–25, 29–30 March and April 1, 2022).

9 Acknowledgments

This research was funded by the Faculty of Veterinary Medicine of Trakia University in Bulgaria and partially supported by the European Community under the Marie Curie Outgoing International Fellowship under the 6th framework, the Marie Curie International Research Staff Exchange Scheme under the 7th framework (PIRSES-GA-2012-316067), and the Department of Science and Technology in South Africa.

Competing interest: The author declares no competing interests.

10 References

Abdallah, M.F., et al. (2015), Occurrence, prevention and limitation of mycotoxins in feeds, *Anim. Nutr. Technology*, 15(3), 471–490, https://doi.org/10.5958/0974-181x.2015.00048.7.

Abdulrazzaq, Y.M., et al. (2002), Fetal exposure to aflatoxins in the United Arab Emirates, *Ann. Trop. Paediatr.*, 22(1), 3–9.

Adamse, P., et al. (2012), *Trend Analysis of Mycotoxins in Animal Feed*. RIKILT-Institute of Food Safety: Wageningen, The Netherlands, pp. 1–52.

Akande, K.E., et al. (2006), Nutritional and health implications of mycotoxins in animal feeds: A review, *Pak. J. Nutr.*, 5, 398–403. https://doi.org/10.3923/PJN.2006.398.403.

Akinmoladun, O.F., et al. (2025), Multiple Mycotoxin Contamination in Livestock Feed: Implications for Animal Health, Productivity, and Food Safety, *Toxins*, 17(8), 365, https://doi.org/10.3390/toxins17080365.

Allcroft, R. and Carnaghan, R.B.A. (1962), Groundnut toxicity. *Aspergillus flavus* toxin (aflatoxin) in animal products: Preliminary communication, *Vet. Rec.*, 74, 863–864.

Almeida, I., et al. (2011), Co-occurrence of mycotoxins in swine feed produced in Portugal, *Mycotoxin Res.*, 27, 177–181.

Alvito, P. and Assunção, R. (2021), Climate Change and the Impact on Aflatoxin Contamination in Foods: Where Are We and What Should Be Expected? In: *Aflatoxins in Food*, Springer: Cham, Switzerland, ISBN 9783030857615, pp. 275–288.

Alvito, P., et al. (2021), Mitigation of mycotoxins during food processing: sharing experience among Europe and South East Asia, *Sci. Technol. Cereal. Oils Foods*, 29, 59–70.

Alvito, P., et al. (2022), Current advances, research needs and gaps in mycotoxins biomonitoring under the HBM4EU–lessons learned and future trends, *Toxins*, 14, 826, https://doi.org/10.3390/toxins14120826.

Assunção, R., et al. (2018), Climate Change and the Health Impact of Aflatoxins Exposure in Portugal–an Overview, *Food Addit. Contam. Part A Chem. Anal. Control. Expo. Risk Assess.*, 35, 1610–1621.

Awuchi, C.G., et al. (2022), Mycotoxins' toxicological mechanisms involving humans, livestock and their associated health concerns: a review, *Toxins*, 14(3), 167. http://dx.doi.org/10.3390/toxins14030167.

Bata, A., et al. (2001), Investigation and a new evaluation method of the resistance of maize hybrids grown in Hungary to *Fusarium* moulds, *Phytopathology*, 149, 107–111.

Battilani, P., et al. (2016), Aflatoxin B1 Contamination in Maize in Europe Increases Due to Climate Change, *Sci. Rep.*, 6, 1–7.

Benkerroum, N. (2020), Chronic and acute toxicities of aflatoxins: mechanisms of action, *Int. J. Environ. Res. Public Health*, 17(2), 423. http://dx.doi.org/10.3390/ijerph17020423.

Bensassi, F., et al. (2014), In vitro investigation of toxicological interactions between the fusariotoxins deoxynivalenol and zearalenone, *Toxicon*, 84, 1–6, https://doi.org/10.1016/j.toxicon.2014.03.005.

Berthiller, F., et al. (2005), Masked mycotoxins: Determination of a deoxynivalenol glucoside in artificially and naturally contaminated wheat by liquid chromatography-tandem mass spectrometry, *J. Agric. Food Chem.*, 53, 3421–3425.

Berthiller, F., et al. (2009), Formation, determination and significance of masked and other conjugated mycotoxins, *Anal. Bioanal. Chem.*, 39, 1243–1252.

Berthiller, F., et al. (2013), Masked mycotoxins: A review, *Mol. Nutr. Food Res.*, 57, 165–186.

Betina, V. (1989), Biological effects of mycotoxins. In: Betina, V. (Ed.), *Mycotoxins: Chemical, Biological and Environmental Aspects*. Elsevier Science Publishers: Amsterdam, pp. 42–58.

Bezuidenhout, S.C., et al. (1988), Structure elucidation of the fumonisins, mycotoxins from *Fusarium moniliforme, J. Chem. Soc. Chem. Comm.*, 11, 743–745.

Boonchuvit, B., et al. (1975), Interaction of T-2 toxin with *Salmonella* infection in chickens, *Poultry Sci.*, 54, 1693–1696.

Boutrif, E. and Canet, C. (1998), Mycotoxin prevention and control: FAO programmes, *Revue Med. Vet.*, 149, 681–694.

Bulgaru, V.C., et al. (2025), Trends in mycotoxins co-occurrence in the complete feed for farm animals in Southern Romania during 2021–2024 period, *Toxins*, 17(4), 201, https://doi.org/10.3390/toxins17040201.

Camargos, S.M., et al. (2001), Co-occurrence of fumonisins and aflatoxins in freshly harvested Brazilian maize, *Trop. Sci.*, 41, 182–184.

Canadian Food Inspection Agency (CFIA). (2012). Report 2010–2011 targeted surveys – ochratoxin A and deoxynivalenol in selected foods. Canadian Food Inspection Agency, Ontario, Canada.

CAST (Council for Agricultural Science and Technology) (2003). In: *Mycotoxins: Risks in plant, animal, and human systems*. Task Force Report N 139. Richard, J.L., Payne, G.A. (Eds.).: Ames, IA, USA.

Castegnaro, M. and McGregor, D. (1998), Carcinogenic risk assessment of mycotoxins, *Revue Med. Vet.*, 149, 671–678.

Chaytor, A.C., et al. (2011), Effects of chronic exposure of diets with reduced concentrations of aflatoxin and deoxynivalenol on growth and immune status of pigs, *J. Anim. Sci.*, 89, 124–135. https://doi.org/10.2527/JAS.2010-3005.

Claeys, L.,et al. (2020), Mycotoxin exposure and human cancer risk: a systematic review of epidemiological studies, *Compr. Rev. Food Sci. Food Safety*, 1–16, https://doi.org/10.1111/1541-4337.12567.

Conkova, E.,et al. (2003), *Fusarial* toxins and their role in animal diseases, *Vet. J.*, 165, 214–220.

Cooray, R. and Jonsson, P. (1990), Modulation of resistance to mastitis pathogens by pre-treatment of mice with T-2 toxin, *Food Chem. Toxicol.*, 28, 687–692.

Cui, H., et al. (2022), Predictive models for assessing the risk of Fusarium pseudograminearum mycotoxin contamination in post-harvest wheat with multi-parameter integrated sensors, *Food Chemistry: X*, 16, 100472, https://doi.org/10.1016/j.fochx.2022.100472.

Dall'Asta, C., et al. (2009), Difficulties in fumonisin determination: the issue of hidden fumonisins, *Anal. Bioanal. Chem.*, 395, 1335–1345.

Dall'Erta, A., et al. (2013), Masked mycotoxins are efficiently hydrolyzed by human colonic microbiota releasing their aglycones, *Chem. Res. Toxicol.*, 26, 305–312.

Danicke, S., et al. (2017), Haematological, clinical–chemical and immunological consequences of feeding Fusarium toxin contaminated diets to early lactating dairy cows, *Mycotoxin Res.*, 33, 1–13.

De Ruyck, K., et al. (2015), Dietary mycotoxins, co-exposure, and carcinogenesis in humans: short review, *Mutat. Res. Rev. Mutat. Res.*, 766, 32–41, https://doi.org/10.1016/j.mrrev.2015.07.003.

Dersjant-Li, Y., et al. (2003), The impact of low concentrations of aflatoxin, deoxynivalenol or fumonisin in diets on growing pigs and poultry, *Nutr. Res. Rev.*, 16, 223–239.

Doko, M., et al. (1995), Incidence of levels of fumonisin contamination in maize by genotypes grown in Europe and Africa, *J. Agric. Food Chem.*, 43, 429–434.

Doll, S., et al. (2003), Effects of graded levels of Fusarium toxin contaminated maize in diets for female weaned piglets, *Arch. Anim. Nutr.*, 57, 311–334.

Dowell, F.E., et al. (1990), Aflatoxin reduction by screening farmers stock peanuts, *Peanut Sci.*, 17, 6–8.

Driehuis, F., et al. (2008), Occurrence of Mycotoxins in Feedstuffs of Dairy Cows and Estimation of Total Dietary Intakes, *J. Dairy Sci.*, 91, 4261–4271.

EC Directive 2002/32/EC of the European Parliament and of the Council on undesirable substances in animal feed, 7 May 2002, pp. 1–27 (*Off. J. Eur. Union L.* 2002, 140, 10–21).

EC Directive 2003/100/EC of amending Annex I to Directive 2002/32/EC of the European Parliament and of the Council on undesirable substances in animal feed. 31 October 2003. (*Off. J. Eur. Union L.* 2003, 285, 33–37).

EC Recommendation 2006/576/EC on the presence of deoxynivalenol, zearalenone, ochratoxin A, T-2 and HT-2 and fumonisins in products intended for animal feeding, 17 August 2006, pp 1–3 (*Off. J. Eur. Union L.* 2006, 229, 7–9).

EC Recommendations 2013/165/EU on the presence of T-2 and HT-2 toxin in cereals and cereal products, 27 March 2013, pp 1–4.

EC Regulation No 178/2002 of the European Parliament and of the Council laying down the general principles and requirements of food law, establishing the European Food Safety Authority and laying down procedures in matters of food safety, 28 January 2002, pp. 1–40.

EC Regulation No 165/2010 amending Regulation No 1881/2006 setting maximum levels for certain contaminants in foodstuffs as regards aflatoxins, 26 February 2010, *Off. J. Eur. Union* 2010, L50/8–L50/12.

EC Regulation No 2023/915 of 25 April 2023 setting maximum levels for certain contaminants in foodstuffs and repealing Regulation (EC) No 1881/2006 of 19 December 2006.

EFSA (European Food Safety Authority). (2014), Scientific Opinion on the risks for human and animal health related to the presence of modified forms of certain mycotoxins in food and feed, *EFSA J.*, 12, 3916.

EFSA (European Food Safety Authority). (2017a), Risks for animal health related to the presence of zearalenone and its modified forms in feed, *EFSA J.*, 15, 4851.

EFSA (European Food Safety Authority). (2017b), Risks to human and animal health related to the presence of DON and its acetylated and modified forms in food and feed, *EFSA J.*, 15, 4718.

EFSA (European Food Safety Authority). (2018), Appropriateness to set a group health-based guidance value for fumonisins and their modified forms, *EFSA J.*, 16, 5172.

EFSA CONTAM Panel (EFSA Panel on Contaminants in the Food Chain). (2016), Appropriateness to set a group health-based guidance value for zearalenone and its modified forms. *EFSA J.*, 14(4), 4425.

EFSA CONTAM Panel (EFSA Panel on Contaminants in the Food Chain). (2017), Knutsen, H.K., Alexander, J., Barregård, L., et al. Risks to Human and Animal Health Related to the Presence of Deoxynivalenol and Its Acetylated and Modified Forms in Food and Feed. *EFSA J.*, 15, 4718.

EFSA CONTAM Panel (EFSA Panel on Contaminants in the Food Chain). (2020), Scientific opinion – Risk assessment of aflatoxins in food. *EFSA J.*, 18(3), 6040, 1–112, https://doi.org/10.2903/j.efsa.2020.6040.

Elissalde, M.H., et al. (1994), Effect of ochratoxin A on *Salmonella*-challenged broiler chicks, *Poultry Sci.*, 73, 1241–1248.

Eriksen, G.S. and Pettersson, H. (2004), Toxicological evaluation of trichothecenes in animal feed, *Anim. Feed Sci. Technol.*, 114, 205–239.

FAO. (1997), Food and Nutrition Paper N 64, *Worldwide Regulations for Mycotoxins*; FAO: Rome, Italy.

FAO. (2002), Manual on the application of the HACCP system in mycotoxin prevention and control. Joint FAO/WHO Food Standards Programme FAO, Rome, Italy.

Freire, L. and Sant'Ana, A.S. (2018), Modified mycotoxins: An updated review on their formation, detection, occurrence, and toxic effects, *Food Chem. Toxicol.*, 111, 189–205

Fukata, T., et al. (1996), Effect of ochratoxin A on *Salmonella typhimurium*-challenged layer chickens, *Avian Dis.*, 40(4), 924–926.

Garais, M., et al. (1990), Cleavage of zearalenone-glycoside, a "masked" mycotoxin, during digestion in Swine, *J. Vet. Med.*, 37, 236–240.

Gashaw, M. (2015), Review on mycotoxins in feeds: implications to livestock and human health, *E3 J. Agric. Res. Dev.*, 5, 137-0144.

Gelderblom, W.C.A., et al. (1988), Fumonisins -novel mycotoxins with cancer-promoting activity produced by *Fusarium moniliforme, Appl. Environ. Microb.*, 54, 1806–1811.

Gelderblom, W., et al. (1992), The cancer initiating potential of the fumonisin B mycotoxins, *Carcinogenesis*, 13, 433–437.

Gimeno, A. and Martins, M.L. (2011), Mycotoxins and Mycotoxicosis in Animals and Humans. 3rd Edition, Special Nutrients Inc., Miami, pp. 1–128.

Goertz, A., Zuehlke, S., Spiteller, M., et al. (2010), Fusarium species and mycotoxin profiles on commercial maize hybrids in Germany, *Eur. J. Plant Pathol.*, 128, 101–111.

Grenier, B. and Oswald, I. (2011), Mycotoxin co-contamination of food and feed: Meta-analysis of publications describing toxicological interactions, *World Mycotoxin J.*, 4, 285–313

Grenier, B. and Applegate, T.J. (2013), Modulation of intestinal functions following mycotoxin ingestion: Meta-analysis of published experiments in animals, *Toxins*, 5, 396–430.

Griessler, K., et al. (2010), Occurrence of mycotoxins in Southern Europe, *World Mycotoxin J.*, 3, 301–309.

Gruber-Dorninger, C., et al. (2017), Emerging Mycotoxins: Beyond Traditionally Determined Food Contaminants, *J. Agric. Food Chem.*, 65(33), 7052–7070, https://doi.org/10.1021/acs.jafc.6b03413.

Gruber-Dorninger, C., et al. (2019), Global mycotoxin occurrence in feed: a ten-year survey, *Toxins*, 11(7), 375. https://doi.org/10.3390/toxins11070375.

Gupta, S., et al. (2008), Individual and combined effects of ochratoxin A and *Salmonella enterica* serovar *Gallinarum* infection on pathological changes in broiler chickens, *Avian Pathol.*, 37, 265–272.

Halloy, D.J., et al. (2005), Oral exposure to culture material extract containing fumonisins predisposes swine to the development of pneumonitis caused by *Pasteurella multocida*, *Toxicology*, 213, 34–44.

Harvey, R.B., et al. (1989), Effects of aflatoxin, deoxynivalenol, and their combinations in the diets of growing pigs, *Am. J. Vet Res.*, 50(4), 602–607.

Harvey, R.B., et al. (1991), Cocontamination of Swine diets by aflatoxin and diacetoxyscirpenol, *J. Vet. Diagn. Invest.*, 3(2), 155–160. https://doi.org/10.1177/104063879100300210.

Harvey, R.B., et al. (1994), Administration of ochratoxin A and T-2 toxin to growing swine, *Am. J. Vet. Res.*, 55(12), 1757–1761.

Haschek, W.M., et al. (2009), *Fundamentals of Toxicologic Pathology*. Academic Press, Elsevier, San Diego, California, pp 1–691.

Huang, S., et al. (2017), Effects of aflatoxin B1 combined with ochratoxin A and/or zearalenone on metabolism, immune function, and antioxidant status in lactating dairy goats, *Asian-Australas. J. Anim. Sci.*, 31, 505–513.

Huff, W.E., et al. (1986), Individual and combined effects of aflatoxin and deoxynivalenol (DON, vomitoxin) in broiler chickens, *Poult. Sci.*, 65, 1291–1298.

Huff, W.E., et al. (1988), Toxic synergism between aflatoxin and T-2 toxin in broiler chickens, *Poult.* Sci., 67, 1418–1423.

Ibáñez-Vea, M., et al. (2012a), Co-occurrence of aflatoxins, ochratoxin A and zearalenone in barley from a northern region of Spain, *Food Chem.*, 132, 35–42.

Ibáñez-Vea, M., et al. (2012b), Co-occurrence of type-A and type-B trichothecenes in barley from a northern region of Spain, *Food Control*, *25*, 81–88.

Ibáñez-Vea, M., et al. (2012c), Co-occurrence of mycotoxins in Spanish barley: A statistical overview, *Food Control*, *28*, 295–298.

International Agency for Research on Cancer (IARC). (2002), Some Traditional Herbal Medicines, Some Mycotoxins, Naphthalene and Styrene, IARC: Lyon, France, Vol. 82, ISBN 9789283215875.

International Agency for Research on Cancer (IARC). (2012), Mycotoxins and human health, IARC Press, Lyon, France, Vol. 158, pp. 87–104.

International Program of Chemical Safety (IPCS). (1990), Selected mycotoxins: ochratoxins, trichothecenes, ergot. IPCS Environmental Health Criteria No. 105. WHO, Geneva.

Iqbal, S.Z., et al. (2014), Natural incidence of aflatoxins, ochratoxin A and zearalenone in chicken meat and eggs. *Food Control*, Vol. 43, 98–103, https://doi.org/10.1016/j.foodcont.2014.02.046.

Jia R., et al. (2020), Low doses of individual and combined deoxynivalenol and zearalenone in naturally moldy diets impair intestinal functions via inducing inflammation and disrupting epithelial barrier in the intestine of piglets, *Toxicol. Lett.*, 333, 159–169.

Joint FAO/WHO Expert Committee on Food Additives (JECFA). (1997), Toxicological evaluation of certain food additives. WHO Food Additives Series, Forty-nineth meeting of JECFA, Geneva.

Joint FAO/WHO Expert Committee on Food Additives (JECFA). (2001), Evaluation of certain mycotoxins that may contaminate food. 56th meeting of JECFA, Geneva.

Joint FAO/WHO Expert Committee on Food Additives (JECFA). (2022), Ninety-third Virtual meeting, 24–25, 29–30 March and 1 April 2022. SUMMARY AND CONCLUSIONS, Issued on 12 April 2022.

Jovaisiene, J., et al. (2016), Fusarium and Aspergillus mycotoxins effects on dairy cow health, performance and the efficacy of Anti-Mycotoxin Additive, *Pol. J. Vet. Sci.*, 19, 79–87.

Kamle, M., et al. (2022), Deoxynivalenol: an overview on occurrence, chemistry, biosynthesis, health effects and its detection, management, and control strategies in food and feed, *Microbiol. Res.*, 13, 292–314.

Karamalakova, Y., et al. (2018), Oxidative-protective effects of Tinospora cordifolia extract on plasma and spleen cells after experimental ochratoxicosis, *Comp. Clin. Path.*, 27 (6), 1487–1495, doi.org/10.1007/s00580-018-2761-y.

Kifer, D., et al. (2020), Assessing the effect of mycotoxin combinations: which mathematical model is (the most) appropriate?, *Toxins*, 12(3), 153, https://doi.org/10.3390/toxins12030153.

Kiyothong, K., et al. (2012), Effect of mycotoxin deactivator product supplementation on dairy cows, *Anim. Prod. Sci.*, 52, 832–841.

Knutsen, H.K., et al. (2018), Appropriateness to set a group health-based guidance value for fumonisins and their modified forms, *EFSA J.*, 16, 5172.

Koynarski, V., et al. (2007), Experimental coccidiosis provoked by *Eimeria acervulina* in chicks simultaneously fed on ochratoxin A contaminated diet, *Res. Vet. Sci.*, 82, 225–231.

Krogh, P., et al. (1973), Occurrence of ochratoxin A and citrinin in cereals associated with mycotoxic porcine nephropathy, *Acta Path. Mcrobiol. Scand. Sect. B.*, 81, 689–695.

Kubena, L.F., et al. (1989a), Influence of ochratoxin A and T-2 toxin singly and in combination on broiler chickens1, *Poult. Sci.*, 68(7), 867–872.

Kubena, L.F., et al. (1989b), Individual and combined toxicity of deoxynivalenol and T-2 toxin in broiler chicks, *Poult. Sci.*, 68(5), 622–626.

Kubena, L.R., et al. (1993), Rottinghaus, G.E. Efficacy of a hydrated sodium calcium aluminosilicate to reduce the toxicity of aflatoxin and diacetoxyscirpenol, *Poult. Sci.*, 72, 51–59.

Kubena, L.F., et al. (1994), Influence of ochratoxin A and diacetoxyscirpenol singly and in combination on broiler chickens, *Poult. Sci.*, 73(3), 408–415.

Kubena, L.F., et al. (1995), Effects of feeding fumonisin B1 present in Fusarium moniliforme culture material and aflatoxin singly and in combination to turkey poults, *Poult. Sci.*, 74, 1295–1303.

Kubena, L.F., et al. (1997a), Individual and combined effects of fumonisin B1 present in Fusarium moniliforme culture material and diacetoxyscirpenol or ochratoxin A in turkey poults, *Poult. Sci.*, 76, 256–264.

Kubena, L.F., et al. (1997b), Individual and combined effects of fumonisin B1 present in Fusarium moniliforme culture material and T-2 toxin or deoxynivalenol in broiler chicks, *Poul. Sci.*, 76(9), 1239–1247, https://doi.org/10.1093/ps/76.9.1239.

Kuiper-Goodman, T. and Scott, P.M. (1989), Risk assessment of the mycotoxin ochratoxin A, *Biomed. Environ. Sci.*, 2, 179–248.

Kumar, A., et al. (2003). Effect of ochratoxin A on *Escherichia coli*-challenged broiler chicks, *Avian Dis.*, 47, 415–424.

Labuda, R., et al. (2005a), Incidence of trichothecenes and zearalenone in poultry feed mixtures from Slovakia, *Int. J. Food Microbiol.*, 105, 19–25.

Labuda, R., et al. (2005b), Incidence of fumonisins, moniliformin and fusarium species in poultry feed from Slovakia, *Ann. Agric. Environ. Med.*, 12, 81–86.

Lefebvre, H.P., et al. (1994), Three cases of equine stachybotryotoxicosis, *Revue Med. Vet.*, 145(4), 267–269.

Lorenz, N., et al. (2019), A critical evaluation of health risk assessment of modified mycotoxins with a special focus on zearalenone, *Mycotoxin Res.*, 35, 27–46.

Lutsky, I.I. and Mor, N. (1981), Alimentary toxic aleukia (septic angina, endemic panmyelotoxicosis, alimentary hemorrhagic aleukia): t-2 toxin-induced intoxication of cats, *Am. J. Pathol.*, 104(2), 189–191.

Malekinejad, H., et al. (2007), Exposure of oocytes to the Fusarium toxins zearalenone and deoxynivalenol causes aneuploidy and abnormal embryo development in pigs, *Biol. Reprod.*, 77, 840–847.

Marasas, W.F.O., et al. (1976), Leukoencephalomalacia: a mycotoxicosis of Equidae caused by *Fusarium moniliforme* Sheldon, *Onderstepoort J. Vet. Res.*, 43, 113–122.

Marasas, W.F.O., et al. (1988), *Fusarium moniliforme* contamination of maize in oesophageal cancer areas in Transkei, *S. Afr. Med. J.*, 74, 110–114.

Marasas, W.F.O. (2001), Discovery and occurrence of the fumonisins: a historical perspective, *Environ. Health. Persp.*, 109, 239–243.

Marc, R.A. (2022), Implications of Mycotoxins in Food Safety, Chapter 1, In: *Mycotoxins and Food Safety - Recent Advances*, London: IntechOpen, pp 1–146, https://www.intechopen.com/books/11023.

Maresca, M. (2013), From the gut to the brain: journey and pathophysiological effects of the food-associated trichothecene mycotoxin deoxynivalenol, *Toxins.*, 5, 784–820. https://doi.org/10.3390/TOXINS5040784.

Meizhen, Y.U. and Ping, L.I.U. (2023), Discussion on emergency management of food safety from the perspective of foodborne diseases caused by mycotoxins, *Food Sci. Technol.*, 43, https://doi.org/10.1590/fst.114622.

Micco, C., et al. (1991), Effect of combined exposure to ochratoxin A and penicillic acid on residues and toxicity in broilers, *La Ravista della Societa Italiana di Scienza dell'Allimentazione*, 20, 101–108.

Miraglia, M., et al. (2009), Climate change and food safety: an emerging issue with special focus on Europe, *Food Chem. Toxicol.*, 47, 1009–1021.

Misihairabgwi, J.M., et al. (2019). Mycotoxin contamination of foods in Southern Africa: a 10-year review (2007–2016), *Crit. Rev. Food Sci. Nutr.*, 59(1), 43–58. http://dx.doi.org/10.1080/10408398.2017.1357003.

Missmer, S.A., et al. (2006), Exposure to fumonisins and the occurrence of neural tube defects along the Texas-Mexico border, *Environ. Health Persp.*, 114, 237–241.

Mohamed, M.Y., et al. (2019), Ameliorating effects of organic and inorganic mycotoxin binders on the performance of Ossimi sheep, *Egypt. J. Sheep Goats Sci.*, 14, 33–48.

Monbaliu, S., et al. (2010), Occurrence ofmycotoxins in feed as analyzed by a multi-mycotoxin LC-MS/MS method, *J. Agric. Food Chem.*, 58, 66–71.

Munkvold, G.P. (2003), Cultural and genetic approaches to managing mycotoxins in maize, *Annu. Rev. Phytopathol.*, 41, 99–116.

Muñoz-Solano, B. and González-Peñas, E. (2023), Co-occurrence of mycotoxins in feed for cattle, pigs, poultry, and sheep in Navarra, a Region of Northern Spain, *Toxins*, 15(3), 172, https://doi.org/10.3390/toxins15030172.

Nordic Working Group on Food Toxicology and Risk Evaluation (NNT). (1991), *Nordiske Seminar-og Arbeidsrapporter* (1991:545). Health Evaluation of Ochratoxin A in Food Products, Nordic Council of Ministers, Copenhagen. 1991, pp. 1–29.

Oswald, I.P. and Comera, C. (1998), Immunotoxicity of mycotoxins, *Revue Med. Vet.*, 149, 585–590.

Pal, S., et al. (2017), Toxicological effects of patulin mycotoxin on the mammalian system: an overview, *Toxicol. Res.*, 6(6), 764–771, http://dx.doi.org/10.1039/c7tx00138j.

Paterson, R.R.M., et al. (2018), Predominant mycotoxins, mycotoxigenic fungi and climate change related to wine, *Food Res. Int.*, 103, 478–491, https://doi.org/10.1016/j.foodres.2017.09.080.

Pfohl-Leszkowicz, A. and Manderville, R. (2012), An update on direct genotoxicity as a molecular mechanism of ochratoxin A carcinogenicity, *Chem. Res. Toxicol.*, 25, 252–262.

Pinotti, L., et al. (2016), Mycotoxin contamination in the EU feed supply chain: a focus on cereal byproducts, *Toxins*, 8, 45.

Piva, G., et al. (2006), Emerging issues in Southern Europe: aflatoxins in Italy. In: Barug, D. and Bhatnagar, D. (Eds.), *The Mycotoxin Factbook, Food and Feed Topics*, Wageningen Academic Publishers: Wageningen, The Netherlands, pp. 139–153.

Pizzolato Montanha, F., et al. (2018), Mycotoxins in dry-cured meats: a review, *Food Chem Toxicol.*, 111, 494–502, https://doi.org/10.1016/j.fct.2017.12.008.

Platzer, A., et al. (2025), Investigating the correlations between weather factors and mycotoxin contamination in Corn: evidence from long-term data, *Toxins*, 17(2), 77, https://doi.org/10.3390/toxins17020077.

Pleadin, J., et al. (2019), Mycotoxins in food and feed (Chapter 8). In: Toldrá, F. (Ed.), *Advances in Food and Nutrition Research*. Academic Press: Cambridge, pp. 297–345.

Pósa, R., et al. (2011), Interaction of *Bordetella bronchiseptica*, *Pasteurella multocida* and fumonisin B_1 in the porcine respiratory tract followed up by computed tomography, *Can. J. Vet. Res.*, 75, 176–183.

Pósa, R., et al. (2013), Use of computed tomography and histopathologic review for lung lesions produced by the interaction between mycoplasma hyopneumoniae and fumonisin mycotoxins in Pigs, *Vet. Pathol.*, 50(6), 971–979.

Pósa, R., et al. (2016), A comparative pathological finding in pigs exposed to fumonisin B1 and/or Mycoplasma hyopneumoniae, *Tox. Ind. Health*, 32(6), 998–1012.

Prabu, P., et al. (2013), Toxicopathological studies on the effects of aflatoxin B1, ocharatoxin A and their inter action in New Zealand White rabbits, *Exp. Toxicol. Pathol.*, 65, 277–286.

Qing, H., et al. (2022), Combined toxicity evaluation of ochratoxin A and aflatoxin B1 on kidney and liver injury, immune inflammation, and gut microbiota alteration through pair-feeding pullet model, *Front. Immunol.*, 13, 920147, https://doi.org/10.3389/fimmu.2022.920147.

Rafai, P., et al. (2000), Evaluation of mycotoxin-contaminated cereals for their use in animal feeds in Hungary, *Food Addit. Contam.*, 17, 799–808.

Ramalingam, S., et al. (2019), The effects of mycotoxin patulin on cells and cellular components, *Trends Food Sci. Technol.*, 83, 99–113. http://dx.doi.org/10.1016/j.tifs.2018.10.010.

Ramos, C.M., et al. (2010), Experimental trial of the effect of fumonisin B_1 and the PRRS virus in swine, *J. Anim. Vet. Adv.*, 9, 1301–1310.

Reyneri, A. (2006), The role of climatic condition on micotoxin production in cereal, *Vet. Res. Comm.*, 30, 87–89.

Richard, J.L. (2007), Some major mycotoxins and their mycotoxicoses–an overview, *Int. J. Food Microbiol.*, 119(1–2), 3–10

Rosner, H. (1998), Mycotoxin regulations: an update, *Revue de Medecine Veterinaire*, 149, 679–680.
Schatzmayr, G. and Streit, E. (2013), Global occurrence of mycotoxins in the food and feed chain: facts and figures, *World Mycotox. J.*, 6, 213–222.
Schneider, D.J., et al. (1996), First report of ergot-alkaloid toxicity in South Africa, *Onderstepoort J. Vet. Res.*, 63, 97–108.
Scientific Cooperation on Questions Relating to Food (SCOOP). (2002), Assessment of dietary intake of ochratoxin A by the population of EU Member States – Report Task 3.2.7. European Commission, Directorate-General Health and Consumer Protection, Available at: http://ec.europa.eu/food/fs/scoop/index_en.html.
Scientific Cooperation on Questions Relating to Food (SCOOP). (2003), Collection of occurrence data of Fusarium toxins in food and assessment of dietary intake by the population of EU Member States. European Commission, Directorate-General Health and Consumer Protection, Available at: http://ec.europa.eu/food/fs/scoop/index_en.html.
Scudamore, K.A., et al. (1997), Occurrence of mycotoxins in raw ingredients used for animal feeding stuffs in the United Kingdom in 1992, *Food Addit. Contam.*, 14, 157–173.
Scudamore, K.A., et al. (1998), Mycotoxins in ingredients of animal feeding stuffs: II. Determination of mycotoxins in maize and maize products, *Food Addit. Contam.*, 15, 30–55.
Scudamore, K.A. and Banks, J.N. (2004), The fate of mycotoxins during cereal processing. In: Barug, D., et al. (Eds.), *Meeting the Mycotoxin Menace*. Wageningen Academic Publishers, pp. 165–181 (Proceedings of the 2nd World Mycotoxin Forum, Nordwijk, the Netherlands, 17–18 February 2003).
Smith, M.-C., et al. (2016), Natural co-occurrence of mycotoxins in foods and feeds and their in vitro combined toxicological effects, *Toxins.*, 8, 94, https://doi.org/10.3390/toxins8040094.
Sohrabi, H., et al. (2022), Patulin and Trichothecene: characteristics, occurrence, toxic effects and detection capabilities via clinical, analytical and nanostructured electrochemical sensing/biosensing assays in foodstuffs, *Crit. Rev. Food Sci. Nutr.*, 62(20), 5540–5568, http://dx.doi.org/10.1080/10408398.2021.1887077.
Speijers, G.J.A. and Speijers, M.H.M. (2004), Combined toxic effects of mycotoxins, *Toxicol. Lett.*, 153, 91–98.
Stoev, S.D., et al. (2000a), Influence of ochratoxin A and an extract of artichoke on the vaccinal immunity and health in broiler chicks, *Exp. Toxicol. Pathol.*, 52, 43–55.
Stoev, S.D., et al. (2000b), Susceptibility to secondary bacterial infections in growing pigs as an early response in ochratoxicosis, *Exp. Toxicol. Pathol.*, 52, 287–296.
Stoev, S.D., et al. (2001), Experimental mycotoxic nephropathy in pigs provoked by a mouldy diet containing ochratoxin A and penicillic acid, *Vet. Res. Commun.*, 25, 205–223.
Stoev, S.D., et al. (2002a), Spontaneous mycotoxic nephropathy in Bulgarian chickens with unclarified mycotoxin aetiology, *Vet. Res.*, 33, 83–94.
Stoev, S.D., et al. (2002b), Clinicomorphological studies in chicks fed ochratoxin A while simultaneously developing coccidiosis, *Vet. Res. Commun.*, 26, 189–204.
Stoev, S.D., et al. (2004), Experimental mycotoxicosis in chickens induced by ochratoxin A and penicillic acid and intervention by natural plant extracts, *Vet. Res. Commun.*, 28, 727–746.

Stoev, S.D. (2008), Complex etiology, prophylaxis and hygiene control in mycotoxic nephropathies in farm animals and humans, special issue "mycotoxins: mechanisms of toxicological activity - Treatment and prevention", section "molecular pathology", *Int. J. Mol. Sci*, 9, 578–605.

Stoev, S.D. (2010a), Studies on carcinogenic and toxic effects of ochratoxin A in chicks. Special issue "Ochratoxins", *Toxins*, 2, 649–664.

Stoev, S.D. (2010b), Studies on some feed additives and materials giving partial protection against the suppressive effect of ochratoxin A on egg production of laying hens, *Res. Vet. Sci.*, 88, 486–491.

Stoev, S.D., et al. (2010a), Mycotoxic nephropathy in Bulgarian pigs and chickens: complex aetiology and similarity to Balkan Endemic Nephropathy, *Food Addit. Contam. A.*, 27, 72–88.

Stoev, S.D., et al. (2010b), Complex etiology and pathology of mycotoxic nephropathy in South African pigs, *Mycotox. Res.*, 26, 31–46.

Stoev, S.D., et al. (2012), Experimental mycotoxic nephropathy in pigs provoked by a mouldy diet containing ochratoxin A and fumonisin B_1, *Exp. Toxicol. Pathol.*, 64, 733–741.

Stoev, S.D. (2015), Foodborne mycotoxicoses, risk assessment and underestimated hazard of masked mycotoxins and joint mycotoxin effects or interaction, *Environ. Toxicol. Pharmacol.*, 9, 794–809.

Stoev, S.D. (2016), Food Security and Foodborne Mycotoxicoses, Risk Assessment, Preventive Measures, and Underestimated Hazard of Masked Mycotoxins or Joint Mycotoxin Interaction, *In: Food Toxicology, Chapter 9,* Debasis Bagchi, Anand Swaroop (Eds), CRC Press, Taylor & Francis Group, 2016, ISBN 9781498708746, pp 169–199, https://doi.org/10.1201/9781315371443.

Stoev, S.D. (2020), Long term preliminary studies on toxic and carcinogenic effect of individual or simultaneous exposure to ochratoxin A and penicillic acid in mice, *Toxicon*, 184, 192–201.

Stoev, S.D. (2021), Follow up long term preliminary studies on carcinogenic and toxic effects of ochratoxin A in rats and the putative protection of phenylalanine, *Toxicon*, 190, 41–49, https://doi.org/10.1016/j.toxicon.2020.11.010.

Stoev, S.D. (2022a), New Evidences about the Carcinogenic Effects of Ochratoxin A and Possible Prevention by Target Feed Additives, *Toxins*, 14 (6), 380, https://www.mdpi.com/2072-6651/14/6/380/pdf.

Stoev, S.D. (2022b), Studies on teratogenic effect of ochratoxin A given via moldy diet in mice in various sensitive periods of the pregnancy and the putative protection of phenylalanine, *Toxicon*, 210, 32–38, https://doi.org/10.1016/j.toxicon.2022.02.012.

Stoev, S.D. (2023), Foodborne diseases due to underestimated hazard of joint mycotoxin exposure at low levels and possible risk assessment, *Toxins*, 15, 464, https://doi.org/10.3390/toxins15070464.

Stoev, S.D. (2024a), Food security, underestimated hazard of joint mycotoxin exposure and management of the risk of mycotoxin contamination, *Food Control,* 159, 110235, https://doi.org/10.1016/j.foodcont.2023.110235.

Stoev, S.D. (2024b), Food Security and Foodborne Mycotoxicoses–What Should Be the Adequate Risk Assessment and Regulation?, *Microoraganisms,* 12, 580, https://doi.org/10.3390/microorganisms12030580.

Streit, E., et al. (2012), Current situation of mycotoxin contamination and co-occurrence in animal feed–Focus on Europe, *Toxins*, 4, 788–809.
Tai, J.H. and Pestka, J.J. (1988), Impaired murine resistance to *Salmonella typhimurium* following oral exposure to the trichothecene T-2 toxin, *Food Chem. Toxicol.*, 26, 691–698.
Teller, R.S., et al. (2012), Effect of physical damage to ears of corn before harvest and treatment with various additives on the concentration of mycotoxins, silage fermentation, and aerobic stability of corn silage, *J. Dairy Sci.*, 95, 1428–1436.
Tessari, E.N.C., et al. (2006), Effects of aflatoxin B1 and fumonisin B1 on body weight, antibody titres and histology of broiler chicks, *Br. Poult. Sci.*, 47, 357–364.
Trenholm, H.L., et al. (1994), Effects of feeding diets containing Fusarium (naturally) contaminated wheat or pure deoxynivalenol (DON) in growing pigs, *Can. J. Anim. Sci.*, 74, 361–369.
Van Egmond, H.P. and Jonker, M.A. (2004), Worldwide Regulations for Mycotoxins in Food and Feed in 2003; Food and Agriculture Organization of the United Nations: New York, NY, USA, 2004.
van Waes, J.G., et al. (2005), Maternal fumonisin exposure and risk for neural tube defects: Mechanisms in an in vivo mouse model, *Birth Defects Res A. Clin. Mol. Teratol.*, 73, 487–497.
Vendl, O., et al. (2009), Simultaneous determination of deoxynivalenol, zearalenone, and their major masked metabolites in cereal-based food by LC-MS-MS, *Anal. Bioanal. Chem.*, 395, 1347–1354.
Verma, J., et al. (2003), Effect of varying levels of aflatoxin, ochratoxin and their combinations on the performance and egg quality characteristics in laying hens, *Asian-Australas. J. Anim. Sci.*, 16, 1015–1019.
Vidal, A., et al. (2018), Humans Significantly Metabolize and Excrete the Mycotoxin Deoxynivalenol and Its Modified Form Deoxynivalenol-3-Glucoside within 24 Hours, *Sci. Rep.*, 8, 1–11.
Vlachou, M., et al. (2022), Ochratoxin A in slaughtered pigs and pork products, *Toxins*, 14, 67, https://doi.org/10.3390/toxins14020067.
Völkel, I. (2011), The carry-over of mycotoxins in products of animal origin with special regard to its implications for the European food safety legislation, *Food Sci. Nutr.*, 2, 852–867.
United States Department of Agriculture (USDA). (2017), GB 2761–2017: limit of mycotoxin in food of China National Food Safety Standard. Washington, USDA.
Wang, X., et al. (2021), Comprehensive dietary and internal exposure assessment of deoxynivalenol contamination in a high-risk area in China using duplicate diet studies and urinary biomarkers, *Food Control*, 124, 107830, https://doi.org/10.1016/j.foodcont.2020.107830.
World Health Organization (WHO). (1991), Evaluation of certain food additives and contaminants. Thirty-seventh report of the joint FAO/WHO Expert Committee on Food Additives (WHO technical report series 806, Geneva, pp 29–31).
Wu, X., et al. (2022), Effects of dietary natural mycotoxins exposure on performance, biochemical parameters and milk small molecule metabolic pathways of lactating cows, *Agriculture*, 12(3), 420, https://doi.org/10.3390/agriculture12030420.

Zinedine, A., et al. (2007), Review on the toxicity, occurrence, metabolism, detoxification, regulations and intake of zearalenone: an oestrogenic mycotoxin, *Food Chem. Toxicol.*, 45, 1–18, https://doi.org/10.1016/J.FCT.2006.07.030.

Ziprin, R.I., et al. (1987), T-2 toxin effects on the serum amyloid P-component (SAP) response of *Listeria monocytogenes* and *Salmonella typhimurium* infected mice, *Toxicol. Lett.*, 39, 177–184.

Chapter 5

Techniques for controlling mycotoxin contamination of animal feed products

Anthony Pokoo-Aikins, Toxicology & Mycotoxin Research Unit, U.S. National Research Center - USDA-ARS, USA; and Joseph Opoku, Mycotoxin Prevention and Applied Microbiology Unit, National Center for Agricultural Utilization Research - USDA-ARS, USA

1 Introduction

A properly formulated feed may consist of a well-balanced mix of high-quality ingredients and optimal nutrient balance. Feed accounts for the majority of animal production costs; in the case of poultry, feed expenses can reach as high as 75% of the total costs of production (Olugbenga et al., 2015; Alqaisi et al., 2017). Feed produced globally in 2023 was 1295.42 metric tonnes (Kirkpinar et al., 2025). One of the main components of animal feed is corn, which has an annual global production of about 1 billion metric tons (García-Lara & Serna-Saldivar, 2019). The quality and safety of feed significantly influence animal performance and health. While there have been notable advancements in genetics, management practices, and health products that enhance animal production, the overall performance and economic benefits of animal products can be compromised without proper attention to feed safety.

One of the prominent concerns regarding feed safety is mycotoxin contamination. Mycotoxins are harmful secondary metabolites produced by fungi found in agricultural products. These toxins can be present in cereal

http://dx.doi.org/10.19103/AS.2026.0172.17

grains (such as corn, oats, wheat, barley, and rye), oilseeds (like soybeans and peanuts), forages, and by-products (including spent grains and dried distillers' grains with solubles), other feed ingredients, and finished feeds. Mycotoxins compromise the quality and safety of feed, posing health and production challenges for animals that consume contaminated products (Pokoo-Aikins et al., 2024). Some adverse effects of mycotoxins on animal production include reduced feed intake, diminished weight gain, decreased feed efficiency, compromised immunity, reduced fertility, damaged organs, decreased egg, meat, and milk yield, and, in severe cases, mortality. Mycotoxin contamination in animals, particularly poultry, often goes unrecognized, making it difficult to detect and treat. The most commonly tested mycotoxins in animal feed products include aflatoxins (AFs), fumonisins (FUM), deoxynivalenol (DON), zearalenone (ZEN), ochratoxin (OTA), and T-2 toxin. However, testing for emerging mycotoxins is still limited. It is nearly impossible to find animal feed ingredients, especially grains, completely free of mycotoxins. Some of the field or pre-harvest mycotoxins include FUM, DON, ZEN, and T-2 toxin, whereas post-harvest mycotoxins include AFs, OTA, and citrinin (Nada et al., 2022). Mycotoxigenic fungi can thrive wherever temperature, moisture, and humidity conditions are favorable. These fungi can infect crops in the field and during post-harvest storage, processing, and feed mixing.

Pre-harvest mycotoxin issues are often caused by poor agronomic practices, inadequate farm management, adverse weather conditions (such as drought or irregular rain), and changes in the weather and virulent mycotoxigenic fungi. Post-harvest contamination can occur under warm and moist conditions. Mitigation strategies include planting mycotoxigenic pathogen-resistant crop varieties, treating soil, providing irrigation during dry conditions, managing pest infestations, and ensuring proper planting and harvesting times. For post-harvest control, measures such as drying grains to reduce the moisture content to about 12–14% (depending on the length of storage), minimizing heat and humidity, and applying additives to inhibit mycotoxigenic fungi growth can be beneficial. However, the effectiveness of strategies to control mycotoxin contamination in animal feed can be hindered by the variability of contamination at different crop and feed production phases (pre-harvest, post-harvest: storage, processing, and formulation). This makes tracking contamination challenging unless regular and appropriate sampling and testing are conducted. Regular testing or using feed additives to mitigate the effects of toxins on animals should be encouraged. However, some of the challenges with testing are proper sampling to get a true picture of the mycotoxin levels in the samples or avoid uneven distribution and underestimation of the mycotoxins in the samples, the turnaround time of test results for methods such as HPLC, LC-MS/MS is usually longer whereas many of the rapid tests may be faster but best for quick field testing for preliminary information, the cost of testing is often expensive with high equipment and reagents costs, having skilled personnel to test the samples accurately, limited access to testing laboratories, disparity in tolerance levels based on geographical region.

Additionally, some control methods may impact the nutritional value and palatability of the feed, making it difficult for animals to consume it. For example, heat and chemical treatments can generate strong volatiles that alter flavor and feed quality. Some control methods may also be economically unfeasible or impractical on a large scale, complicating their implementation.

A holistic approach is required to manage mycotoxin contamination effectively. This approach aims to mitigate its effects on animal production and enhance overall yield, animal welfare, and the economic stability of farms. The subtle and often unnoticed impact of mycotoxin contamination in feed is particularly concerning, as it impairs performance and can be challenging to detect, ultimately leading to significant economic losses.

1.1 Emerging and masked mycotoxins

Due to current advancements in mycotoxin testing of feed, some lesser-known mycotoxins, generally referred to as emerging and masked mycotoxins, are detected during testing. Among the most concerning are found in Table 1 cereals and feed and are most often secondary metabolites of *Fusarium* and *Alternaria* species. In contrast, masked (or modified) mycotoxins are chemically or biologically modified derivatives of regulated toxins. These conjugated forms are produced mainly through plant detoxification pathways and may escape conventional analytical detection, yet they can be hydrolyzed back to their toxic parent compounds in the gastrointestinal tract, making them toxicologically relevant and a hidden risk in food and feed safety assessments (Berthiller et al., 2013; Fraeyman et al., 2017; Gruber-Dorninger et al., 2019; Rychlik et al., 2014). It has been reported that these lesser-known toxins also significantly contribute to the negative effects of mycotoxins on animal performance and health. Like the co-occurrence of some mycotoxins such as FUM and DON, a similar trend is also being seen with emerging mycotoxins, where there are two or more together or a combination of more commonly prevalent toxins with emerging mycotoxins in a single sample. The challenge with these emerging mycotoxins is that there is currently limited information about their toxicity in several animal species. These findings underscore the need for comprehensive studies to understand the mechanisms of action, altered toxicity profiles, chronic exposure outcomes, and expected co-occurrence patterns, taking into account animal species, geographic, storage conditions, plant stress factors, and grain type. Emerging mycotoxins are currently not regulated, yet they are increasingly detected in agricultural produce.

Table 1 Emerging and masked mycotoxins that contaminate crops and their toxicological effects

Category	Mycotoxin	Producer organism	Main crops affected	Toxicological concern
Emerging	Enniatins (ENNs A, A1, B, B1)	*Fusarium* spp.	Wheat, barley, oats, maize	Ionophoric, cytotoxic
	Beauvericin (BEA)	*Fusarium* spp.	Maize, wheat	Ionophoric, cytotoxic
	Fusaproliferin (FUS)	*F. proliferatum*, *F. subglutinans*	Maize	Cytotoxic
	Alternariol (AOH)	*Alternaria* spp.	Cereals, fruits	Genotoxic
	Alternariol monomethyl ether (AME)	*Alternaria* spp.	Cereals, fruits	Genotoxic
	Tenuazonic acid (TeA)	*Alternaria* spp.	Cereals, fruits	Cytotoxic
	Altertoxins (ATX I-III)	*Alternaria* spp.	Cereals, fruits	Mutagenic
Masked	DON-3-glucoside (DON-3G)	*Fusarium* spp.	Maize, wheat	Hydrolyzed to DON in gut
	15-acetyl-DON (15-ADON)	*Fusarium* spp.	Maize, wheat	Toxic metabolite of DON
	3-acetyl-DON (3-ADON)	*Fusarium* spp.	Maize, wheat	Toxic metabolite of DON
	Zearalenone-14-glucoside (ZEA-14G)	*Fusarium* spp.	Maize, wheat	Estrogenic
	Zearalenone-14-sulfate (ZEA-14S)	*Fusarium* spp.	Maize, wheat	Estrogenic
	α-Zearalenol (α-ZOL)	*Fusarium* spp.	Maize, wheat	Estrogenic
	β-Zearalenol (β-ZOL)	*Fusarium* spp.	Maize, wheat	Estrogenic
	Hydrolyzed fumonisins (HFB1, HFB2)	*Fusarium* spp.	Maize	Nephrotoxic
	N-(carboxymethyl)fumonisin B1	*Fusarium* spp.	Maize	Nephrotoxic
	Ochratoxin α (OTα)	*Aspergillus*, *Penicillium* spp.	Cereals, grapes	Nephrotoxic
	Citrinin glucosides	*Penicillium* spp.	Maize, rice	Nephrotoxic
	Patulin conjugates	*Penicillium* spp.	Apples, pears	Genotoxic

Adopted from de Oliveira et al., 2022; Gruber-Dorninger et al., 2019; Fraeyman et al., 2017; Kamle et al., 2022; Kolawole et al., 2024; Milani, 2013.

2 Pre-harvest interventions to control mycotoxins in feed

2.1 Interventions pre-harvest

Preharvest interventions are essential for ensuring host plants are healthy and protected from pathogens. These strategies aim to reduce host–pathogen interactions, limit pathogen growth and spread, and create environments conducive to healthy plant growth. Understanding the lifecycle of mycotoxigenic fungi and how they interact with host plants under various environmental conditions enables farmers to implement targeted strategies. Key factors include knowing how pathogens overwinter, their proliferation conditions, dispersal methods, and which growth stages plants are most vulnerable to infection and mycotoxin contamination. Effective monitoring and early detection can lead to timely interventions that minimize contamination risks, ultimately enhancing harvest quality, safety, and food supply.

2.1.1 Reducing host vulnerability

This is essential for developing crop varieties exhibiting genetic resistance to mycotoxin-producing fungi. The effectiveness of this resistance is influenced by a range of factors, including plant genetics and environmental conditions (Majumdar et al., 2022). Although various crops display differing resistance levels, it is essential to note that none are completely immune (Nankam & Pataky, 1996; He et al., 2013), and the strengths and weaknesses of the various approaches used in reducing host vulnerabilities are indicated in Table 2. Traditional breeding methods have produced some successes; however, they often necessitate a lengthy and cost-intensive process involving multiple generations of selection (Guo et al., 2017; Henry et al., 2009). Traditional breeding and marker-assisted selection (MAS) are widely used for complex traits like Fusarium head blight resistance, which involve multiple QTLs. MAS enables selection based on genetic markers but still requires multi-generational selection (Steiner et al., 2017).

In contrast, recent advances in genome editing, genetic engineering, and microbiome-based approaches are accelerating the development of resistant cultivars, potentially reducing yield loss and mycotoxin contamination. These methods may include the development of enzymes capable of degrading or reducing mycotoxins, such as AFs and FUM (Gomes et al., 2024; Liu et al., 1998; Schmidt et al., 2021; Schmidt et al., 2023). Some engineered plants are being developed to express enzymes like amidohydrolases or ochratoxinase that degrade mycotoxins in planta or during storage (often >70–90% degradation in lab assays) (Sánchez-Arroyo et al., 2024). Additionally, the microbiome associated with plants can enhance resistance through metabolites that

Table 2 Approaches to reduce host vulnerability

Approach	Method/Tool	Benefits	Limitations
Traditional breeding and MAS	QTL mapping, marker-assisted selection	Well-established; integrates with elite varieties	Slow, multi-year process; incomplete resistance
Genome editing and transgenic tech	CRISPR/Cas9, transgenes like AFP proteins	Precise edits; can drastically reduce mycotoxins (>50 %)	Regulatory barriers; risk of resistance evolution
Microbiome-mediated resistance	Endophytes, rhizosphere engineering, extremophiles	Enhances innate host immunity; eco-friendly	Field variability; development of effective consortia
Combining strategies	Pyramiding genes, microbiome + editing	Durable, multi-layered defense	Requires integrated, multidisciplinary programs

strengthen plant defense. Beneficial endophytes and rhizospheric microbes, including extremophilic fungi and PGPR, induce plant defense pathways or detoxify mycotoxins via microbial metabolites (Trivedi et al., 2020; Liu et al., 2024). Effective breeding programs should target specific pathogens and broaden their scope to address diverse mycotoxigenic fungi, incorporating resistance traits from cultivars and landraces used by producers (Gao et al., 2021; Rose et al., 2019). An important note with regard to host resistance is that once deployed in the field, these technologies can have a limited life span, as mycotoxigenic fungi can evolve countermeasures that allow them to defeat the genetic resistances of host plants. Therefore, deploying multiple resistance genes and rotating trait targets will be invaluable to mitigating the evolution of new virulent strains (Anderson et al., 2010; Gomes et al., 2024; Mierziak & Wojtasik, 2024).

2.1.2 Soil health amendment and management

Cultivating healthy soil is crucial for nurturing strong and resilient plants. Farmers can significantly reduce the risk of mycotoxin contamination in crops by adopting several strategic approaches to improve soil quality. The incorporation of organic matter, such as biochar, compost, and cover crops, has been shown by some studies to be effective in suppressing the growth of mycotoxigenic fungi. For example, biochar amendments can reduce fungal pathogen incidence by more than 80%, while compost and cover crops often lower inoculum potential by 30–70% under field conditions (Bonanomi et al., 2007; Pascale et al., 2020; Iacomino et al., 2022). However, excessive application of organic matter can also increase inoculum potential for certain fungi and disease risk by 20%, necessitating careful management (Bonanomi et al., 2007; Pascale et al., 2020). Maintaining optimal soil pH (6.0–7.0) with lime or gypsum amendments can further reduce the severity of mycotoxigenic fungi that thrive in acidic conditions by about 40–60% (Allmaras et al. 1987, Fang et al., 2012). Regular soil testing, combined with timely amendments, helps in achieving this delicate balance. Agronomic practices such as tillage play a significant role in managing the prevalence of mycotoxigenic fungal species. Tillage is used to manage postharvest crop residues, which often harbor plant pathogens that can overwinter and become inoculum sources in the subsequent season (Jasarevic et al., 2025). While tillage can increase the abundance of specific pathogens and saprophytes, ploughing or deep tillage before seeding can help bury infected crop residues and reduce the survival of mycotoxigenic fungi, thereby minimizing the initial inoculum load in the subsequent crop and reducing the proliferation of mycotoxigenic fungi and ensuring a healthier growing environment (Munkvold, 2003; Schmale & Munkvold, 2009).

2.1.3 Plant nutrient management

This is essential for optimal plant health, requiring a balance of macronutrients (nitrogen, phosphorus, potassium) and micronutrients (zinc, iron, manganese) to strengthen crops against diseases. Proper nutrient ratios help plants develop robust structures and defenses, reducing vulnerability to plant pathogens, including mycotoxin-producing fungal pathogens (López et al., 2023). Additionally, organic amendments enhance soil fertility and root systems, making plants more resilient to environmental stress. Although the application of fertilizer is important, overapplication of nutrients, especially nitrogen and phosphorus, can lead to excessive weed growth and increase competition and disease susceptibility (Wang et al., 2022). Previous studies have shown contrasting results regarding the impact of nitrogen (N) fertilization on mycotoxins such as FUM contamination in maize. However, it is widely recognized that maintaining balanced nitrogen levels is essential to minimize FUM accumulation. Both nitrogen deficiency and excessive nitrogen application can significantly increase FUM levels (Ono et al., 2011; Marroca et al., 2008; Blandino et al., 2008). It is also important to note that healthy-looking plants and grains can still be contaminated by mycotoxins, especially when weather conditions are favorable for mycotoxin contamination or when either biotic or abiotic stress triggers mycotoxin synthesis by the pathogen.

2.2 Optimal irrigation strategies to minimize fungal risks

Plants need sufficient water for healthy growth and to develop effective defense systems. However, natural environmental conditions and/or agronomic farming practices can lead to drought and/or flooding of farmlands. Both drought and flooding can compromise plant health, making them vulnerable to pathogens such as mycotoxigenic fungi. Drought stress during grain filling has been consistently associated with elevated FUM and AFs contamination. For example, maize grown under drought and heat stress in Argentina showed FUM and AFs levels up to 2–3 times higher than under optimal moisture conditions (Zanon et al., 2023). Similarly, U.S. field studies have reported FUM B1 concentrations exceeding 20 mg/kg in drought-stressed maize, far above the FDA advisory level of 5 mg/kg for human food (Munkvold, 2003). Conversely, flooding and waterlogging create anaerobic soil conditions that reduce root oxygenation and favor fungal proliferation. Nielsen et al. (2023) observed that 10–14 days of waterlogging increased Fusarium ear rot incidence by 40–60%, accompanied by higher deoxynivalenol (DON) levels. Abdelmoneim et al. (2024) further demonstrated that prolonged flooding stress increased AFs contamination in maize kernels by up to 70% compared to non-flooded controls. Weather modeling studies also predict that changes in the weather will exacerbate both drought and flooding risks, thereby amplifying the frequency and severity of mycotoxin contamination events in maize (Battilani et al., 2013). Therefore, it is crucial to manage water availability, avoiding both drought and prolonged flooding, to mitigate fungal outbreaks and mycotoxin contamination in maize (Nielsen et al., 2023; Abdelmoneim et al., 2024).

To ensure an adequate water supply, various methods can be employed, including precision drip irrigation, deficit irrigation (DI), and smart scheduling (Table 3). Precision drip irrigation is effective for delivering water directly to the plant's root zone, minimizing prolonged surface wetness that can facilitate fungal spore germination (Frimpong et al., 2023). This method reduces leaf wetness duration, which is vital for limiting pathogen transmission associated with contaminated irrigation water (Abdelmoneim et al., 2024). Additionally, using drip irrigation under plastic mulch has been shown to enhance water productivity and significantly reduce AFs levels in maize production.

Integrating deficit irrigation and smart scheduling facilitates strategic water application during drought-sensitive growth stages, enhancing water use efficiency and preventing excessive soil moisture. DI is significant for stabilizing yields and lowering disease risks by avoiding root zone over-saturation (Fereres & Soriano, 2007; Abdelmoneim et al., 2024). Modern smart irrigation systems utilize soil moisture sensors, leaf-wetness sensor data, and predictive modeling, including IoT solutions and machine-learning controllers, to improve water-use efficiency, effectively prevent pathogen buildup and fungal infection risks (Nguyen et al., 2023; Saikai et al., 2023; Rodríguez-Lira et al., 2024; Ali et al., 2025). These combined strategies underscore the importance of technologically informed irrigation management as a foundation for sustainable crop protection and mycotoxin mitigation.

In contrast, overhead irrigation tends to increase leaf wetness and enhances the splash dispersal of pathogens. The deployment of leaf wetness sensors, particularly biomimetic models that replicate plant surfaces, has

Table 3 Types of crop irrigation and their benefits

Strategy	Benefit	References
Drip irrigation	Water delivered directly to the root; reduced leaf wetness	Frimpong et al., 2023; Abdelmoneim et al., 2024
Deficit irrigation (DI)	Saves water while avoiding over-irrigation	Fereres & Soriano, 2007
Smart IoT/ML systems	Optimize irrigation based on real-time data	Frimpong et al., 2023; Ali et al., 2025; Abdelmoneim et al., 2025
Morning irrigation timing	Leaves dry faster, reducing fungal risk	Das et al., 2023; Nielsen et al., 2023
Leaf wetness monitoring	Improves disease forecasting	Nguyen et al., 2023; Abdelmoneim et al., 2024

become invaluable for accurately measuring the duration of leaf wetness. Such data significantly enhances disease forecasting capabilities, enabling growers to avoid high-risk irrigation periods that could lead to increased pathogen proliferation (Abdelmoneim et al., 2024; Nguyen et al., 2023). By understanding and harnessing these advanced irrigation methods, farmers can improve productivity while minimizing risks of diseases and mycotoxin contamination, ultimately leading to more sustainable agricultural practices.

2.2.1 Crop rotation and planting density

Crop rotation and optimal planting density are strategic, vital agronomic practices that influence soil health, microenvironment and disease risks (Table 4). Planting different crop types in succession can enrich the soil biome and improve overall soil structure. Adjusting planting density and spacing is essential for enhancing air circulation around crops, which can significantly impact humidity levels. Promoting optimal plant arrangement encourages healthier growth and makes it less favorable for mycotoxin-producing fungi to thrive. Long-term diversified crop rotations improve microbial network complexity, reduce pathogen-dominated fungal taxa, and bolster soil structure and nutrient cycling (Dong et al., 2023; Panneerselvam et al., 2023). Thus, crop rotation and optimal planting density can help suppress mycotoxin-producing

Table 4 Rotation and density strategies to reduce fungal risks

Practice Area	Strategy	Effect on Microclimate and Soil	Impact on Fungus/Mycotoxins
Crop rotation	Rotate cereals with non-host crops (e.g. legumes, maize-wheat)	Breaks pest/fungal cycles; increases microbial diversity	Reduced *Fusarium* inoculum, lower trichothecenes in succeeding crops (Chen et al., 2023a; Dong et al., 2023)
	Diversified rotations (e.g. potato-maize, inclusion of cover crops)	Enhances structure, suppressive microbial taxa	Lowers *Alternaria*, *Colletotrichum*, *Fusarium* incidences (Li et al., 2023)
Planting density	Moderate density (e.g. ~28 000 plants/acre; wide rows)	Improved air circulation; reduced canopy humidity	Minimizes aflatoxin and ear rot risk while maintaining yields
	Lower seeding rates or strategic row spacing (30" vs 40")	Better light penetration; lower fungal favorable microclimate	Reduced ear mold and mycotoxin levels in corn and cereals (Saleem et al., 2025)
Combined approach	Rotation + optimal spacing	Synergistic benefits: healthy soil + healthy canopy	Suppresses soilborne pathogens and infection risk effectively

fungi by altering pathogen inoculum levels and humidity around the host. Thoughtful planning in crop spacing creates an environment conducive to robust plant health and reduced disease incidence (Wang et al., 2022).

2.2.2 Weed management

Weed is any undesirable plant that grows in agricultural fields or other cultivated areas where it is not needed. It may be a remnant from a previous crop or volunteer plants. Weed management is critical in agriculture, not only for optimizing crop yield but also for minimizing disease risks associated with fungal infections. Weeds compete with crops for essential nutrients, water, and light. They also provide food, shelter, and reproductive sites for various pest organisms, including plant pathogens and insects serving as alternate hosts for these organisms (Bhowmick et al., 2018). Thus, they serve as a reservoir for fungal spores, which can lead to infection of crops and subsequent mycotoxin production.

Farmers use different methods to effectively control weeds in their farms before planting or at a certain growth stage of their crop (Table 5). They sometimes use tillage as an effective method for controlling weed growth and reducing seed banks. Depending on the type of agricultural system, complementing tillage with other weed control methods, such as consistent mowing, will prevent weeds from reaching the reproductive stage, reducing seed production and weed reestablishment. It is important to note that excessive tillage may adversely affect beneficial soil fungi and degrade the soil structure over time (Saleem et al., 2022).

Farmers have, over the years, adopted the use of cultural practices to control weeds. Incorporating organic mulches, such as straw, wood chips,

Table 5 Weed management and its influence on fungal disease risks

Control strategy	Mechanism	Effect on weeds and microenvironment	Impact on fungal risk/mycotoxins
Tillage and mowing	Disturbs weeds; reduces seed bank and biomass	Lowers weed cover but can disrupt soil fungi	Fewer fungal reservoirs; improved aeration
Mulching and cover cropping	Suppresses weeds; alters soil microclimate	Shades soil; improves structure; releases allelochemicals	Reduces viable weed-based fungal reservoirs
Selective herbicide use	Targets specific weed patches	Controls resistant weeds without major ecosystem disruption	Limits competition; removes fungal harbor plants
Biological control agents	Introduces specific fungal or bacterial weed pathogens	Lowers weed densities ecologically	Reduces potential reservoirs; avoids herbicide residues

and compost, or utilizing cover crops like rye and sunn hemp have been used effectively in suppressing weed growth while also limiting fungal accumulation on the soil surface. Notably, certain cover crops from the Brassicaceae family release allelopathic compounds, which inhibit soil-borne pathogens (Santos et al., 2021).

Spot treatments with selective herbicides minimize weed populations without broad-spectrum ecological disruption, reducing competition and potential fungal reservoirs (Chen et al., 2023b). Use of bioherbicides, fungi, or bacteria that specifically antagonize weeds can minimize weed biomass without relying on chemical herbicides (Den Breyeen et al., 2022; Harding & Raizada, 2015). However, care is needed to avoid interactions with crop pathogens and beneficial fungi. Thus, the risk of mycotoxin contamination can be reduced by implementing effective weed management strategies to maintain crop health and reduce the prevalence of mycotoxigenic fungi and subsequent mycotoxin contamination of crops.

2.2.3 Field hygiene practices

Maintaining clean and tidy fields is a significant preventative measure. Regularly removing crop debris and diseased plant materials decreases the sources of fungal inoculum and diminishes the overall pathogen presence in the field. Enhanced field sanitation practices as found in Table 6, such as disinfecting tools and equipment, can significantly reduce the likelihood of mycotoxin-related issues (Oliver et al., 2021).

2.2.4 Surveillance and monitoring

Proactive crop monitoring is imperative for early detection of fungal infections and diseases. Utilizing advanced field sensors and technology, such as drones

Table 6 Hygiene practices and their impact on inoculum and disease risks

Practice	Mechanism	Impact on inoculum and disease risk
Removal of crop debris	Eliminates residue that harbors fungal spores	Lowers primary inoculum and disease carry over
Leaf shredding/residue incorporation	Accelerates decay and reduces pathogen survival estimated	Suppresses crop residue-borne fungal inoculum
Tool and equipment sanitation	Prevents cross contamination between fields	Limits pathogen spread and introduction of resistant pathogens

Table 7 Benefits and limitations of different crop monitoring methods

Monitoring method	Technology/tool	Strengths	Limitations
Manual scouting	Visual inspection and symptom recording	Cost-effective, specific diagnosis	Labor-intensive, limited spatial coverage
Drone-based imaging	RGB, multispectral, thermal, LiDAR sensors	Large-area, high-resolution detection; early stress signals	Costs and technical expertise needed
ML/DL disease modeling	CNN, deep learning on imagery	High accuracy; automated disease classification	Needs training datasets and calibration
Soil and canopy sensors	IoT devices measuring humidity, moisture	Continuous environmental data; predictive risk insight	Sensor maintenance and data management

Table 8 IPM strategies and their impact on crops

IPM component	Approach/tool	Impact on pest damage and fungal risk
Biological control	*Beauveria*, *Metarhizium*, *Trichoderma* agents	Suppresses insect pests that damage plants and allow fungal infection; reduces chemical reliance
Cultural/mechanical practices	Crop rotation, sanitation, residue removal	Lowers both insect vector density and fungal inoculum
Field scouting and thresholds	Monitoring pest levels, using action thresholds	Limits pest damage and secondary fungal colonization
Targeted chemical application	Spot treatments instead of broad spraying	Minimizes fungicide resistance, reduces environmental exposure
Pest monitoring systems	Spatial/temporal networks for early detection	Reduces unnecessary pesticide use with maintained control

and soil moisture sensors as found in Table 7, enables farmers to keep a close watch on plant health. This timely intervention can help implement immediate management strategies to address potential threats before they escalate (Pérez et al., 2023).

2.2.5 Integrated pest management

Holistic approach that combines various methods to manage pest populations while minimizing the use of chemicals has been summarized in Table 8. Although chemicals are still used to control plant diseases, excessive use can increase production costs, fungicide resistance, and environmental pollution. By implementing practices such as natural pest predators, crop scouting, and targeted insecticides, farmers can protect crops from pest-related damage that creates openings for harmful fungi. Healthy, resilient plants are less prone to pest infestations, lowering the risk of fungal infections and associated mycotoxin contamination (García et al., 2023).

3 Interventions post harvest

It is difficult to come by feed ingredients or finished feed that are mycotoxin-free. Warm, moist environments of temperatures greater than 25°C and humidities greater than 85% are typically conducive to fungal growth; therefore, these conditions introduce a higher potential for mycotoxin production. Although pre-harvest interventions play a crucial role in reducing fungal growth and contamination when crops are growing in the field, post-harvest strategies deter mycotoxin development and accumulation during storage, processing, and distribution. Grains contaminated by mycotoxins are influenced by factors such as moisture content at harvest, holding time before drying, storage type and quality, insect damage, and infestation. Insufficient drying during harvest and high moisture of feed ingredients, such as corn, often lead to hot spots and fungal growth during storage.

New ingredients coming in should be tested either by the supplier before shipping or verified at receiving, which should then be stored properly to prevent fungal growth. Moisture and temperature should be controlled, regular cleaning of equipment, transportation carts, storage bins, and sanitation of feed mills is vital in controlling mycotoxins.

Feed mills, ideally, should also retest the finished feed for mycotoxins before the feed is delivered to the farm. Rapid mycotoxin testing kits such as lateral flow and ELISA could be used to screen for mycotoxins at the farm and the feed mill for quick results. Sampling, sample preparation (grinding), and testing often takes about 30 min per truck to get the results to determine if the grains have to be accepted, rejected or diverted. Waiting for long periods to get results often leads to congestion of truck lines, economic risk for farmers especially if their loads are rejected, downgraded, or diverted. Positive samples are sent to analytical labs for reconfirmation and accurate quantification, which often takes about 2–48 h to get results. Feed mills should have a hazard analysis and critical control point (HACCP) system of approach to be able to identify other risk factors. Feed additives such as vitamins have grain products in them as carriers; therefore, it is important to check them for mycotoxins, as they can be an unexpected source of contamination. Samples should be tested frequently to know when there is a problem, as a safety margin.

3.1 Pre-feed formulation interventions

3.1.1 Storage management

Reducing mycotoxin content in feed involves several strategies. The separation of contaminated materials from healthy feed commodities can be achieved by removing damaged, discolored, lightweight or moldy sections of grains and burning them. In some operations, the contaminated grains are fed to some insects that detoxifies the toxins and the insect meal generated is incorporated into animal feed in regions where insect meal is allowed in animal feed. Proper storage and handling to prevent fungal growth are effective measures of reducing mycotoxin contamination. Proper storage management can be done by controlling the temperature, humidity, and moisture levels in the environment (Peraica et al., 2002). Reducing the oxygen level in stored grains can also prevent some storage mycotoxins such as AFs (Magan & Aldred, 2007). Moisture content is a measure of water related to the drying of products, and it is often expressed as a percentage. Water activity (0.70) a measure of water available for microbial use. A water activity of 0.70 helps prevent spoilage and allows for long term storage. An example of environmental conditions conducive to mycotoxin-specific production in certain grains can be found in Tables 9 and 10. To minimize the risk of mycotoxin development, Magan & Aldred (2007) recommend hermetic storage systems and moisture-absorbing materials as effective strategies for extending the shelf life of stored grain. With proper conditions, grain can be stored for up to a decade.

3.1.2 Decontamination techniques

After pre-harvest mycotoxin interventions, if animal feed products test positive for mycotoxins or as a proactive measure, decontamination is the next step of action. Effective decontamination methods must inactivate or remove feed toxins, leave no harmful residues, preserve nutritional value, and prevent regrowth, all while being cost-effective. Common post-harvest methods for controlling mycotoxins in animal feed are categorized as physical, chemical, and biological, which can destroy, suppress, modify, or adsorb the toxins (Assaf et al., 2017; Karlovsky et al., 2016; Muhialdin et al., 2020).

3.1.2.1 Physical decontamination

Physical methods provide practical, cost-effective, and scalable solutions that can be applied at multiple stages of the feed production and storage process. These methods include drying, cleaning, heat treatment, extrusion, and Ultraviolet (UV) treatment, which aim to either prevent mycotoxin-producing

Table 9 Common pre-harvest grain mycotoxins and their optimum conditions for fungal growth and mycotoxin production

Mycotoxin	Typical producer fungi	Optimum temperature (°C)	Optimum water activity (aw)	Main grain/host
Fumonisins	*Fusarium verticillioides*, *F. proliferatum*	15–30	0.90–0.995	Corn, wheat, sorghum, barley, oats
Zearalenone (ZEN)	*Fusarium graminearum*, *F. culmorum*	~25	0.96	Maize, corn, wheat, barley, rye
Deoxynivalenol (DON)	*Fusarium graminearum*, *F. culmorum*	26–30	0.995	Corn, wheat, barley, oats
T-2 toxin	*Fusarium sporotrichioides*, *F. langsethiae*	12–24	0.88–0.96	Wheat, barley, corn, oats, rye, rice, beans
Ergot alkaloids	*Claviceps purpurea*	15–25	~0.80	Wheat, rye, barley

Source: Adopted from Milani (2013).

Table 10 Common post-harvest grain mycotoxins and their optimum conditions for fungal growth and mycotoxin production

Mycotoxin	Typical producer fungi	Optimum temperature (°C)	Optimum water activity (aw)	Main grains/commodities
Aflatoxins	*Aspergillus flavus, A. parasiticus*	~33	0.99	Corn, peanuts, wheat, cotton seeds, nuts
Ochratoxin A	*Aspergillus ochraceus, Penicillium verrucosum*	15–30	0.85–0.98	Wheat, barley, oats, soy, beans
Citrinin	*Penicillium citrinum, P. expansum*	20–30	0.75–0.85	Wheat, barley, corn, rice
Alternaria toxins (AOH, AME, TeA, ATX)	*Alternaria alternata, A. tenuissima*	5–30	0.90–0.98	Corn, wheat, oats
Patulin	*Penicillium expansum*	20–25	0.95–0.99	Apples, pears, fruit-derived feeds
Penicillic acid	*Penicillium aurantiogriseum*	20–25	0.95–0.99	Corn, wheat, barley
Roquefortine C	*Penicillium roqueforti, P. crustosum*	20–25	0.95–0.99	Cereals, silage, stored grains
Mycophenolic acid	*Penicillium roqueforti*	20–25	0.95–0.99	Silage, stored cereals
Gliotoxin	*Penicillium* spp., *Aspergillus fumigatus*	25–30	0.95–0.99	Stored grains, silage, compost
Penitrem A (tremorgenic)	*Penicillium crustosum, P. commune*	20–25	0.95–0.99	Moldy cereals, nuts, food waste, silage
Verruculogen (tremorgenic)	*Penicillium verrucosum, P. crustosum*	20–25	0.95–0.99	Cereals, silage
Thomitrems A & E (tremorgenic)	*Penicillium thomii*	20–25	0.95–0.99	Cereals, silage, stored feed

Source: Adopted from Milani (2013); Mannaa & Kim (2017); Pitt & Hocking (2009); Mostrom (2021); Waratuke (2017).

fungi's growth or minimize the concentration and bioavailability of existing mycotoxins in feed and ingredients.

Drying involves reducing the moisture content of newly harvested crops to reduce fungal growth, which would in turn assist in reducing the potential for mycotoxins. This should be done immediately after harvest to minimize mycotoxin contamination risks. Mycotoxigenic fungi such as *Aspergillus, Fusarium, and Penicillium* species proliferate in moist environments; hence, lowering the moisture content below critical thresholds (i.e. below 13–14%) significantly inhibits fungal growth and subsequent toxin production (Bankole & Mabekoje, 2004). Some studies suggest that quick drying using high temperatures of between 100–180 °F for lengths of 6–48 hours is reported to lower moisture content to less than 20%, thus preventing fungal growth and its accompanying mycotoxin contamination in grains. This process can harm or alter the nutrient contents of the corn (Timm et al., 2020). It is well established that cereal grains, such as corn, which form a major component of animal feed, are typically received into the mill in bulk and often mixed with dust, debris, and foreign material due to the threshing and subsequent handling methods.

Cleaning is a critical step in feed processing because it can reduce fungal growth. The removal of foreign materials in grains is one of the most effective ways to reduce the risk of mycotoxin contamination. Several studies have demonstrated that mycotoxins tend to concentrate in grain dust, and exposure to this dust can increase health risks in animals (Pietri et al., 2012; Hartmann et al., 2020). Effective dust collection and separation systems can help divert mycotoxins into fine particulate fractions, thereby preventing their inclusion in raw materials or finished products (Čolović et al., 2019). The removal of impurities such as stones, dust, pieces of wood, seeds of weeds, corn cobs, husk, etc during the cleaning phase can effectively reduce the risks of fungal growth during feed processing and storage (Kabak et al., 2006; Schaarschmidt and Fauhl-Hassek et al., 2021). It is important to note that separation systems are also effective for eliminating broken kernels, which tend to have higher mycotoxin concentrations than intact ones. Some grains undergo milling before storage, separating tissues into bran, germ, and endosperm fractions through sieving and aspiration. While milling generally reduces mycotoxin levels, results vary by cereal batch and fraction type. It should not be considered a primary mycotoxin mitigation strategy (Schaarschmidt & Fauhl-Hassek, 2021). Furthermore, the clean material from milling usually goes to human consumption. Techniques like screening and gravity separation are more frequently used than air classification for removing broken kernels and coarse impurities (Čolović et al., 2019). Other conventional sorting methods typically rely on gravity separation, centrifugal force, and air flotation (Pascale et al., 2020). For example, sorting and cleaning have been reported to lead to a 29–69% reduction of FUM in corn samples, a 23–90% reduction in DON, a 25–80% reduction in T-2 toxin (Pascale et al., 2011; Sydenham et al., 1994; Tibola et al., 2016; Trenholm et al., 1991) . Additionally, cleaning can result in a mass loss of about 1–10%, depending on the intensity of the cleaning. The cleaning involves the removal of unwanted materials such as weeds, foreign materials, cobs, husk pieces, broken kernels, damaged kernels, moldy kernels, dust, and fines (Pascale et al., 2022).

Thermal treatment during feed processing, such as extrusion and pelleting, can reduce the activity of some biological pathogens such as fungi (Huss et al., 2018; Jones, 2011). Optimized storage conditions are closely linked to drying, with temperature and humidity being two key variables that influence the survival and activity of fungal spores. Storage facilities for feed and ingredients should be well-ventilated, insulated against external moisture, and routinely monitored to maintain stable internal conditions. Heat treatment can be effective in reducing mycotoxin contamination (Murphy et al., 2006) (Table 9).

Ultraviolet-C (UV-C) radiation is another physical method employed in effectively controlling external mycotoxin contamination and levels in grains post-harvest. UV-C treatment has proven to be an effective strategy for reducing levels of mycotoxins, including AFs, patulin, ochratoxin A, and DON in grains (Jubeen et al., 2012; Tikekar et al., 2014; Popović et al., 2018). Ferreira et al. (2021) observed that, in the absence of UV-C treatment, the total fungal colony count increased by 63% in brown rice, 163% in black rice, and 4% in red rice after storage. Additionally, irradiation using Gamma and electron beams is reported to degrade mycotoxins (Pankaj et al., 2018; Mohamed et al., 2022). Ozone treatments can effectively degrade mycotoxins such as AFs, DON, ZEA, and FUM (Conte et al., 2020; Sujayasree et al., 2022). High-voltage atmospheric cold plasma producing reactive gas species can degrade AFs by over 60% through brief exposure to ozone and nitric oxide (Javed et al., 2025; Tang et al., 2024). While effective, these treatments can be costly, inefficient, and sometimes potentially hazardous to operators.

3.2 During feed formulation interventions

3.2.1 Mycotoxin decontamination

3.2.1.1 Adsorbents

These involve the use of substances to adsorb or transform mycotoxins into less toxic metabolites to reduce the risk associated with health hazards in animals and humans. Some adsorbents, such as Zeolite, bentonite, montmorillonite, hydrated sodium calcium aluminosilicate, polymers, are effective in decontamination of mycotoxins, whereas others such as kaolinite, limestone, talc, inorganic phosphate adsorbents are less effective due to low binding to the toxins, poor adsorptive ability, poor interaction, and minimal surface area charge (Elliott et al., 2020). The differences in the efficacy of these different adsorbents in animal feed to mitigate the negative impacts of mycotoxins often create differences or inconsistencies in research results. In the past few years, various adsorbent materials have been evaluated for their potential to bind mycotoxins in grains and animal feeds. The adsorbents use chemical interactions like van der Waals forces and hydrogen bonds, along with pore structure, to bind with mycotoxins (Di Gregorio et al., 2014). Activated charcoal, e.g. has non-specific binding characteristics but may also decrease nutrient availability by binding vitamins or amino acids (Kihal et al., 2022) (Table 10).

The addition of adsorbents to animal feed can help bind the toxins and help limit the uptake of toxins in the gastrointestinal tract (GIT), and facilitate the elimination of mycotoxins from the system. Inorganic adsorbents used in animal feed include phyllosilicates (clay mineral adsorbents), e.g. hydrated sodium aluminosilicate and bentonites. Other mineral-based adsorbents include zeolites. An adsorbent that is able to bind to more than a single mycotoxin, with a high adsorption rate, and does not bind to nutrients in feed is most desirable. Other products used include yeast cell wall products and synthetic polymers (e.g. polyvinylpyrrolidone and cholestyramine). The addition of feed additives in animal diets is common as they act by binding the toxins in the GIT of the animals and excreting them. Feed additives, such as clay minerals, yeast cell walls, and prebiotics (bentonite, montmorillonite, B-glucans, and mannan oligosaccharides), have been shown to bind multiple mycotoxins in the gut of animals (Kabak et al., 2006; Phillips et al., 2008; Yiannikouris & Jouany, 2002). Most mineral adsorbents tend to be most effective against AFs, and modified clays can adsorb other mycotoxins such as zearalenone and ochratoxin (Papaioannou et al., 2005; Wan et al., 2013).

3.2.2 Biological detoxification

This involves transforming mycotoxins into less harmful metabolites using biological agents such as enzymes and microorganisms. Enzyme function is largely determined by the structure and arrangement of the active site, a small, specialized region within the protein that interacts with specific substrates and facilitates their chemical transformation. This active site includes a catalytic center and a binding site, both essential for the enzyme's activity. Certain types of enzymes, bacteria, and yeast can bind or degrade mycotoxins (Adebo et al., 2022; Kabak et al., 2006). These enzymes are mostly specific to some mycotoxins and can speed up the degradation process. Enzymatic detoxification of mycotoxins is currently widely explored due to its advantages such as non-interference with nutrients and the involvement with bacteria, molds, and yeasts (Hathout & Aly, 2014; Wang et al., 2019). Microbial cell wall adsorbents, such as yeast, can bind mycotoxins by reducing their bioavailability.

Many studies have not fully identified the intermediate and final metabolites produced during the detoxification process. Evaluations of safety and efficacy must include an analysis of these by-products to ensure they are non-toxic and do not introduce new risks. Additionally, there is a significant gap in assessing these biologics under commercial production conditions. Most studies involving these biologicals are conducted in vitro using Petri dishes, 96-well plates, or small-scale feed bins and results are not readily translatable to commercial production systems (Liu et al., 2022b). When trials are performed in vivo, diets are usually prepared as mash. Although laying hens can be fed mash diets, pig and broiler chicken diets are predominantly pelleted, requiring exposure to high temperatures during feed processing (Coradi et al., 2016; Myers et al., 2013). Conversely, these high temperatures may reduce the impact of the biologics added to the feed to mitigate mycotoxins. Thus, it is important to investigate the effectiveness and safety of these biologics in real-world applications such as commercial crop production, grain storage, and animal feeding prior to widespread adoption and their stability during pelleting.

3.2.2.1 Chemical detoxification and suppression

Chemical agents can be used in animal feed to improve feed safety; however, it is important to note that the legality of the use of these products can vary widely by country. Oxidizing agents and bisulfate are commonly used in animal feed to suppress bacteria and mold/fungus growth (Cochrane, 2015; Johnson et al., 2011). Organic acids or blends used in feed to inhibit fungal growth in finished feed and feed ingredients, but are not very effective against mycotoxins that are already present in the feed (Lorán et al., 2022; Satterlee et al., 2023). The most widely used is propionic acid (Kabak et al., 2006), while others such as ammonia, copper sulfate, and benzoic, formic, and sorbic acids are also used in controlling fungal growth at varying concentrations for different mycotoxins. The use of organic acids such as propionic acid during storage can reduce or suppress the growth of fungus/mold, especially during storage (Kabak et al., 2006). Alkaline products, such as sodium carbonate, ammonia, sodium hydroxide, and potassium hydroxide, can be used to prevent fungal growth in feed and feed products. Feed additives (commercially available feed additives). Polyphenols in natural oils that are sometimes added to feed products can destroy the cell wall and cell membrane of the microorganisms due to their antifungal properties (Cai et al., 2022; Esper et al., 2014). The main issue with the use of such products is the strong odor, palatability of the products that have been treated with the chemicals and the corrosive threat to feed bins, silos, and feed lines on farms. Additionally, chemicals can also alter the nutritional properties of feed. While fungicides are commonly used, they can be harmful to beneficial fungi that are part of a healthy soil microbiome and compete with mycotoxin-producing fungi. Overuse of fungicides is discouraged because it can lead to increased resistance (Hayes et al., 2014).

3.2.2.2 Nutritional interventions

Research has demonstrated as shown in Table 11, that dietary supplementation of animal feed with antioxidants (e.g. vitamins A, B1, C and E), and some

Table 11 Categories of feed additives targeted against fungal growth and mycotoxin mitigation in farm animals

Additive	Example	Effectiveness	References
Adsorbing agents	Clay mineral	Mycotoxin reduction: AFB1-50% ↓ ZEN ↓ FUM ↓ T2-toxin ↓	Bhatti et al. (2018)
	Yeast cell walls	Aflatoxin- 80% ↓ ZEN- 41% ↓ OTA-26% ↓ T-2 Toxin-30% ↓	Murthy et al. (2002); Kolawole et al. (2019); Kolawole et al. (2019); Murthy et al. (2002)
	Polymers	DON -3.5% ↓ ZEN >70% ↓ OTA -50% ↓ AFB1 - >35% ↓ FB2 - >30% ↓	Hernández-Patian et al. (2018)
	Agric waste products: Grape pomances, Almond hulls.	AFBI-1.2-2.9 ug/mg ↓ ZEN- 1.3-2.9 ug/mg ↓	Greco et al. (2019)
Modifiers	Bacteria	AFB1 ↓ ZEN ↓	Kabak et al. (2006); Avantaggiato et al. (2014); Vila-Donat et al. (2018)
	Fungi	AFB1 ↓	Xu et al. (2022)
	Enzymes	DON ↓ OTA ↓ AFB1 ↓ ZEN ↓	Martinez et al. (2019); Zhu et al. (2016); Haque et al. (2020)
Antioxidants	L-Threonine Vitamins (eg. A, B_1, C and E), Selenium, curcumin carotenoids, Silymarin, alpha lipoic Amino acids (eg. methionine, glutamic acid, arginine, aspartate and lysine). Vitamin E, silymarin, curcumin, soybean isoflavone Retinol, ascorbic acid, alpha-tocopherol, silymarin and soybean isoflavone	AFB_1-Reduce negative effect of toxin on farm animals Reduces side effects of mycotoxins in animals DON-Helps boost immune function of farm animals FUM-Reduces oxidative stress in animals ZEN- Inhibit estrogen toxicity toxin	Mesgar et al. (2022); Khan et al. (2014); Zhao et al. (2019); Gbore and Adu (2017); Liu et al. (2022); Nagaraj et al. (1994); Gradelet et al. (1998); Liu et al. (2022); Ledur and Santurio (2020); Wang et al. (2018); Liu et al. (2022); Gao et al. (2018)

Table 11 (*Continued*)

Additive	Example	Effectiveness	References
	Grape extract	AFB1- improved broiler performance	Hassan et al. (2023)
	Essential oil	OTA-95% ↓ FUM- 51-87% ↓ AFB1-97% ↓ DON- 72% FB1	Aroyeun et al. (2009); Tian et al. (2019); Gorran et al. (2013); Perczak et al. (2019); Xing et al. (2014)
Organic acids	Propionic	FB1 ↓ T-2 toxin ↓	Dazuk et al. (2020)
	Citric	OTA-100 ↓ Citrinin-100 ↓	Awuchi et al. (2023)
	Butyric	AFB1-100% ↓	Moon et al.(2018)
Essential oils	Cinnamon	FB1 ↓ OTA ↓	Xing et al. (2014); Yin et al. (2020)
	Oregano	OTA ↓ AFLA ↓ ZEN ↓	Pattono and Gai (2020); Kocić-Tanackov et al. (2012)
	Lemongrass	OTA ↓ ↓ AFLA	Abd-El Fattah et al. (2010)

minerals help reduce the negative impact of certain toxins on farm animals (Gbore and Adu 2017; Khan et al., 2014; Nagaraj et al., 1994; Zhao et al., 2019).

4 Conclusion and future trends

The global production of animal protein primarily comes from farm animals, such as ruminants, swine, poultry, and aquaculture. Feed costs represent a significant portion of production expenses, particularly for poultry, where they can account for up to 75% or more. The quality and safety of feed are crucial for optimal animal performance. A major concern in this area is mycotoxin contamination, which can occur in various feed ingredients and pose serious health risks to animals.

Effective mitigation strategies are needed during both the pre-harvest and post-harvest periods to reduce contamination. These strategies may involve improved agronomic practices and appropriate storage conditions. However, implementing these control methods can be complex and may impact feed quality and economic feasibility. A holistic approach to managing mycotoxin contamination is essential to protect animal welfare, maintain production yields, and ensure economic stability in farming. If not addressed adequately, the subtle impacts of mycotoxin contamination can lead to significant economic losses.

To establish an effective mycotoxin management program, there is a need for a holistic approach to mitigate mycotoxin contamination in animal feed. This will require stakeholders' engagement. Stakeholders should use practices, approaches, and methods that have worked while embracing new technologies and methods such as machine learning and predictive modeling to enhance the management of feed mycotoxins. Encourage regular/routine testing throughout the supply chain of cereals and feed to make sure that mycotoxin levels in feed are within the right limits and not only when there is a problem and be proactive about the trends of and extent of mycotoxin issues. Different innovative rapid testing kits capable of testing multiple mycotoxins should be developed and made available to farmers. There has been an increase in testing of animal feed products for mycotoxins in the last few years partly due to the availability of rapid test kits for surveillance. Conventional wet chemistry/ laboratory methods are often used for confirmation. The use of advanced sensitive mycotoxin testing technologies has helped in the detection of emerging mycotoxins. Harvested grains should be properly dried to less than 15% moisture. Bins should be monitored to prevent fungal development during storage. The use of organic acids will continue to be used with new blends with phytochemicals to help reduce some of the harsh effects of organic acids. The industry needs more blends of adsorbing agents to be able to bind many of these toxins. Mitigation approach needs to be affordable, duplicatable, and straightforward.

There is an urgent need for research in the area of testing for mycotoxins in blood metabolites in farm animals because this provides the industry with quick information and be able to offer a quicker solution rather than waiting to confirm with histology information. Oftentimes, this would be too late to offer any meaningful solution.

The increase and use of predictive tools for pre-harvest and post-harvest mycotoxin contamination risk would significantly help the industry There is a need for the development of biomarker technologies for animal blood metabolite testing for multi-mycotoxin exposure. More research is needed on emerging mycotoxins and developing products to counteract their effects on farm animals. The use of advanced technology should be encouraged in mycotoxin mitigation in both pre and post-harvest interventions (e.g. artificial intelligence, predictive modeling, machine learning to predict/forecast mycotoxin incidence). There is a need for automated testing instruments/ devices for multi-mycotoxin simultaneous analysis to standardize testing/ results and improve high throughput processes to lessen turnaround time for results. Consolidation of all mycotoxin surveillance data collected by different entities would be beneficial to farmers. With some of the aforementioned research opportunities, the industry could be well on its way to proactively tackling mycotoxin issues in animal feed.

Published by Burleigh Dodds Science Publishing Limited, 2026.

5 Acknowledgements

We would like to thank Dr Frank Idan for providing valuable information and insights that greatly contributed to the development of this chapter.

6 Where to look for more information

Further information on the subject is divided into preharvest and post-harvest: Readers seeking deeper insight into mycotoxins, their impacts, and current mitigation strategies can draw on several comprehensive resources across scientific literature. Foundational texts such as *Determining Mycotoxins and Mycotoxigenic Fungi in Food and Feed* (De Saeger, 2011) and reviews by Yiannikouris and Jouany (2002) and Pettersson (2004) provide detailed overviews of toxin occurrence, detection, and control in food and feed systems. More recent syntheses, including Ng'ang'a and Niyonshuti (2022) and Magnoli et al. (2019), discuss evolving risk management approaches within the animal feed industry. For readers interested in mitigation technologies, several studies explore mineral- and biologically based detoxification agents (Bočarov Stančić et al., 2024; Satterlee et al., 2023), as well as enzymatic degradation pathways for aflatoxins and fumonisins (Liu et al., 1998; Schmidt et al., 2021; Schmidt et al., 2023; Gomes et al., 2024; Sánchez Arroyo et al., 2024). Genetic resistance remains an important area of research, with traditional breeding successes documented by Guo et al. (2017) and Henry et al. (2009), and emerging strategies emphasizing the deployment of multiple resistance genes to limit pathogen adaptation (Anderson et al., 2010; Mierziak & Wojtasik, 2024). Because environmental stress strongly influences mycotoxin risk, readers may also benefit from broader discussions on climate smart and water efficient agricultural practices (Frimpong et al., 2023; Nguyen et al., 2023; Saikai et al., 2023; Rodríguez Lira et al., 2024; Ali et al., 2025). Together, these resources offer a robust foundation for understanding the complexity of mycotoxin contamination and the diverse tools available for its management.

7 References

Abd-El Fattah, S. M., et al. (2010). The use of lemongrass extracts as antimicrobial and food additive potential in yoghurt. *Journal of American Science*, *6*, 582–594.

Abdelmoneim, A. A., et al. (2024). Towards affordable precision irrigation: an experimental comparison of weather-based and soil water potential-based irrigation using low-cost IoT-tensiometers on drip irrigated lettuce. *Sustainability*, *16*(1), 306.

Abdelmoneim, A. A., et al. (2025). IoT sensing for advanced irrigation management: a systematic review of trends, challenges, and future prospects. *Sensors (Basel, Switzerland)*, *25*(7), 2291.

Adebo, J. A., et al. (2022). Fermentation of cereals and legumes: impact on nutritional constituents and nutrient bioavailability. *Fermentation*, *8*(2), 63.

Ali, A., et al. (2025). Smart irrigation technologies and prospects for enhancing water use efficiency for sustainable agriculture. *AgriEngineering*, *7*(4), 106.

Allmaras, R. R., et al. (1987). Lime and gypsum effects on pea-root-pathogen inoculum and related factors in a wheat-peas rotation1. *Agronomy Journal*, *79*(3), 439–445.

Alqaisi, O., et al. (2017). Time series livestock diet optimization: cost-effective broiler feed substitution using the commodity price spread approach. *Agricultural and Food Economics*, *5*(1), 25.

Anderson, J. P., et al. (2010). Plants versus pathogens: an evolutionary arms race. *Functional Plant Biology*, *37*(6), 499–512.

Aroyeun, S. O., et al. (2009). Grading of fermented and dried cocoa beans using fungal contamination, ergosterol index and ochratoxin a production. *Mycobiology*, *37*, 215–217.

Assaf, J. C., et al. (2017). A comparative study of procedures for binding of afatoxin M1 to Lactobacillus rhamnosus GG. *Brazilian Journal of Microbiology*, *49*, 120–127. https://doi.org/10.1016/j.bjm.2017.05.003

Avantaggiato, G., et al. (2014). Assessment of multi-mycotoxin adsorption efficacy of grape pomace. *Journal of Agricultural and Food Chemistry*, *62*(2), 497–507.

Awuchi, C. G., et al. (2023). High-pressure acidified steaming with varied citric acid dosing can successfully detoxify mycotoxins. *Food Science & Nutrition*, *11*(6), 2677–2685.

Bankole, S. A. & Mabekoje, O. O. (2004). Mycotoxin contamination of maize grains in the humid tropical zones of South-Western Nigeria. *Journal of Food Protection*, *67*(3), 556–560.

Battilani, P., et al. (2013). AFLA-maize, a mechanistic model for *Aspergillus flavus* infection and aflatoxin B1 contamination in maize. *Computers and Electronics in Agriculture*, *94*, 38–46.

Berthiller, F., et al. (2013). Masked mycotoxins: a review. *Molecular Nutrition & Food Research*, *57*(1), 165–186.

Bhatti, S. A., et al. (2018). Comparative efficacy of Bentonite clay, activated charcoal and Trichosporon mycotoxinivorans in regulating the feed-to-tissue transfer of mycotoxins. *Journal of the Science of Food and Agriculture*, *98*, 884–890.

Bhowmick, M., et al. (2018). *Weeds as alternate hosts of crop pests and their management* (pp. 72–76). West Bengal State Seed Corporation Technologist Association.

Blandino, M., et al. (2008). Influence of nitrogen fertilization on mycotoxin contamination of Maize kernels. *Crop Protection*, *27*, 222–230.

Bočarov-Stančić, A., et al. (2024). Agents of different origins for reduction of mycotoxins' level in feed. *Annals of Animal Science, 24*(3), 707–729.

Bonanomi, G., et al. (2007). Suppression of soilborne fungal diseases with organic amendments. *Journal of Plant Pathology*, *89*, 311–324.

Cai, J., et al. (2022). Antifungal and mycotoxin detoxification ability of essential oils: a review. *Phytotherapy Research*, *36*(1), 62–72.

Chen, X., et al. (2023a). Weed management strategies in sustainable agriculture. *Journal of Crop Protection*, *145*, 123–129.

Chen, W., et al. (2023b). Soil and phytomicrobiome for plant disease suppression and management under climate change: a review. *Plants (Basel, Switzerland)*, *12*(14), 2736.

Cochrane, R. A. (2015). *Chemical mitigation of microbial pathogens in animal feed and ingredients* (Doctoral dissertation, Kansas State University).

Published by Burleigh Dodds Science Publishing Limited, 2026.

Čolović, R., et al. (2019). Decontamination of mycotoxin-contaminated feedstuffs and compound feed. *Toxins, 11*(11), 617.

Conte, G., et al. (2020). Mycotoxins in feed and food and the role of ozone in their detoxification and degradation: An update. *Toxins*, *12*(8), 486.

Coradi, P. C., et al. (2016), Mycotoxin levels in corn fractions and feeds. *Journal of Food Process Engineering*, *39*, 215–225.

Das, B., et al. (2023), Ensemble surface soil moisture estimates at farm-scale combining satellite-based optical-thermal-microwave remote sensing observations. *Agricultural and Forest Meteorology, 339,* 109567. https://doi.org/10.1016/j.agrformet.2023.109567.

Dazuk, V., et al. (2020). Laying hens fed mycotoxin-contaminated feed produced by *Fusarium* fungi (T-2 toxin and fumonisin B1) and Saccharomyces cerevisiae lysate: Impacts on poultry health, productive efficiency, and egg quality. *Microbial Pathogenesis*, *149*, 104517.

de Oliveira, R. C., et al. (2022). Evaluating the potential of Pediococcus pentosaceus as a biocontrol agent against tenuazonic acid–producing Alternaria alternata on livestock feeds. *Journal of Applied Microbiology*, *133*(5), 3020–3029.

De Saeger, S. (2011). *Determining Mycotoxins and Mycotoxigenic Fungi in Food and Feed*. Vol. 203, Woodhead.

Den Breeyen, A., et al. (2022). Plant pathogens as introduced weed biological control agents: could antagonistic fungi be important factors determining agent success or failure?. *Frontiers in Fungal Biology*, *3*, 959753.

Di Gregorio, M. C., et al. (2014). Mineral adsorbents for prevention of mycotoxins in animal feeds. *Toxin Reviews*, *33*, 125–135.

Dong, F., et al. (2023). Effect of crop rotation on *Fusarium* Mycotoxins and *Fusarium* species in cereals in Sichuan Province (China). *Plant Disease*, *107*(4), 1060–1066.

Elliott, C. T., et al. (2020). Potential adverse effects on animal health and performance caused by the addition of mineral adsorbents to feeds to reduce mycotoxin exposure. *Mycotoxin Research*, *36*, 115–126. https://doi.org/10.1007/s12550-019-00375-7.

Esper, R. H., et al. (2014). Potential of essential oils for protection of grains contaminated by aflatoxin produced by *Aspergillus flavus*. *Frontiers in Microbiology*, *5*, 269.

Fang, X., et al. (2012). Reduced severity and impact of Fusarium wilt on strawberry by manipulation of soil pH, soil organic amendments and crop rotation. *European Journal of Plant Pathology*, *134*(3), 619–629.

Fereres, E. & Soriano, M. A. (2007) Deficit irrigation for reducing agricultural water use. *Journal of Experimental Botany*, *58*, 147–159.

Ferreira, C. D., et al. (2021). Postharvest UV-C irradiation for fungal control and reduction of mycotoxins in brown, black, and red rice during long-term storage. *Food Chemistry*, *339*, 127810.

Fraeyman, S., et al. (2017). Emerging *Fusarium* and *Alternaria* mycotoxins: occurrence, toxicity and toxicokinetics. *Toxins*, *9*(7), 228.

Frimpong, F., et al. (2023). Water-smart farming: review of strategies, technologies, and practices for sustainable agricultural water management in a changing climate in West Africa [Review]. *Frontiers in Sustainable Food Systems*, 7, 1110179. https://doi.org/10.3389/fsufs.2023.1110179.

Gao, M., et al. (2021). Ca2+ sensor-mediated ROS scavenging suppresses rice immunity and is exploited by a fungal effector. *Cell*, *184*, 5391–5404.

Gao, X., et al. (2018). Dietary silymarin supplementation alleviates zearalenone-induced hepatotoxicity and reproductive toxicity in rats. *The Journal of Nutrition*, *148*(8), 1209–1216.

García, J., et al. (2023). Integrated pest management: sustainable approaches to crop health. *Entomological Research*, *98*(3), 213–220.

García-Lara, S. & Serna-Saldivar, S. O. (2019). Corn history and culture. In S. O. Serna-Saldivar (Ed.), *Corn* (pp. 1–18). Woodhead Publ Ltd.

Gbore, F. A. & Adu, O. A. (2017). Ameliorative potential of vitamin E on the impact of dietary fumonisin B1 on reproductive performance of female rabbits. *Journal of Agriculture and Rural Development in the Tropics*, *118*(2), 161–169.

Gomes, A. S., et al. (2024). Resistance of transgenic maize cultivars to mycotoxin production–systematic review and meta-analysis. *Toxins*, *16*(8), 373.

Gorran, A., et al. (2013). Aflatoxin b1-reduction of aspergillus flavus by three medicinal plants (lamiaceae). *Food Control*, *31*, 218–223.

Gradelet, S., et al. (1998). Dietary carotenoids inhibit aflatoxin B1-induced liver preneoplastic foci and DNA damage in the rat: role of the modulation of aflatoxin B1 metabolism. *Carcinogenesis*, *19*(3), 403–411.

Greco, D., et al. (2019). Comparative efficacy of agricultural by-products in sequestering mycotoxins. *Journal of the Science of Food and Agriculture*, *99*, 1623–1634.

Gruber-Dorninger, C., et al. (2019). Global mycotoxin occurrence in feed: a ten-year survey. *Toxins*, *11*(7), 375.

Guo, B., et al. (2017). Evaluation of maize inbred lines for resistance to pre-harvest aflatoxin and fumonisin contamination in the field. *The Crop Journal*, *5*(3), 259–264.

Hartmann, N., et al. (2020). Assessment of mycotoxin exposure in grain dust: Health implications for workers in grain handling facilities. *Toxins*, *12*(11), 678.

Hassan, R. A., et al. (2023). Ameliorative impacts of grape seed extract on growth performance, immune response, antioxidant capacity and biochemical constituents in broilers exposed to fumonisin B1. *Egyptian Journal of Nutrition and Feeds*, *26*(2), 179–192.

Harding, D. P. & Raizada, M. N. (2015). Controlling weeds with fungi, bacteria and viruses: a review. *Frontiers in Plant Science*, *6*, 659.

Hathout, A. S. & Aly, S. E. (2014). Biological detoxification of mycotoxins: a review. *Annals of Microbiology*, *64*, 905–919.

Hayes, B. M., et al. (2014). Activation of stress signalling pathways enhances tolerance of fungi to chemical fungicides and antifungal proteins. *Cellular and Molecular Life Sciences: CMLS*, *71*(14), 2651–2666.

Haque, M. A., et al. (2020). Mycotoxin contamination and control strategy in human, domestic animal and poultry: a review. *Microbial pathogenesis*, *142*, 104095.

He, X., et al. (2013). Identification and characterization of international Fusarium head blight screening nurseries of wheat at CIMMYT, Mexico. *European Journal of Plant Pathology*, *136*, 123–134.

Henry, W. B., et al. (2009). Evaluation of maize inbred lines for resistance to Aspergillus and Fusarium ear rot and mycotoxin accumulation. *Agronomy Journal*, *101*, 1219–1226.

Hernández-Patlan, D., et al. (2018). Chitoneous materials for control of foodborne pathogens and mycotoxins in poultry. In R. S. Dongre (Ed.), *Chitin-chitosan-myriad functionalities in science and technology* (1st ed., pp. 261–282). IntechOpen Publishers.

Huss, A., et al. (2018). Physical and chemical methods for the reduction of biological hazards in animal feeds. In S. Ricke, et al. (Eds.), *Food and feed safety systems and analysis* (pp. 83–95). Academic Press.

Iacomino, G., et al. (2022). The suppressive effects of biochar on above- and belowground plant pathogens and pests: a review. *Plants, 11*(22), 3144.

Jasarevic, M., et al. (2025). Durum wheat kernel contamination by mycotoxigenic fungi and mycotoxins as affected by soil tillage and fertilization management in a Mediterranean environment. *Journal of Plant Pathology*, *107*(3), 1363–1376.

Javed, M., et al. (2025). A review of decontamination of *Aspergillus* spp. and aflatoxin control for grains and nuts with atmospheric cold plasma. *Toxins*, *17*(3), 129.

Jones, F. T. (2011). A review of practical Salmonella control measures in animal feed. *Journal of Applied Poultry Research*, *20*(1), 102–113.

Johnson, T. M., et al. (2011). The use of sodium bisulfate as a best management practice for reducing ammonia emissions from poultry manures. In *NPWMS Proceedings*.

Jubeen, F., et al. (2012). Effect of ultraviolet radiation on aflatoxins in rice and maize. *Journal of the Chemical Society of Pakistan*, *34*(4), 931–936.

Kabak, B., et al. (2006). Strategies to prevent mycotoxin contamination of food and animal feed: a review. *Critical Reviews in Food Science and Nutrition*, *46*, 593–619.

Kamle, M., et al. (2022). Citrinin mycotoxin contamination in food and feed: impact on agriculture, human health, and detection and management strategies. *Toxins*, *14*(2), 85.

Karlovsky, P., et al. (2016). Impact of food processing and detoxification treatments on mycotoxin contamination. *Mycotoxin Research*, *32*, 179–205.

Khan, W. A., et al. (2014). Dietary vitamin E in White Leghorn layer breeder hens: a strategy to combat aflatoxin B1-induced damage. *Avian Pathology*, *43*(5), 389–395.

Kihal, A., et al. (2022). The efficacy of mycotoxin binders to control mycotoxins in feeds and the potential risk of interactions with nutrient: a review. *Journal of Animal Science*, *100*, skac328.

Kirkpinar, F., et al. (2025). Current situation, future goals, and strategies of the global feed and feed additives industries. https://doi.org/10.5772/intechopen.1011333.

Kocić-Tanackov, S. D., et al. (2012). Antifungal activity of oregano (Origanum vulgare L.) extract on the growth of Fusarium and Penicillium species isolated from food. *Hemijska industrija*, *66*(1), 33–41.

Kolawole, O., et al. (2019). Comparative in vitro assessment of a range of commercial feed additives with multiple mycotoxin binding claims. *Toxins*, *11*, 659.

Kolawole, O., et al. (2024). A systematic review of global occurrence of emerging mycotoxins in crops and animal feeds, and their toxicity in livestock. *Emerging Contaminants*, *10*, 100305.

Ledur, P. C. & Santurio, J. M. (2020). Cytoprotective effects of curcumin and silymarin on PK-15 cells exposed to ochratoxin A, fumonisin B1 and deoxynivalenol. *Toxicon*, *185*, 97–103.

Li, Q.-M., et al. (2023). Crop rotations increased soil ecosystem multifunctionality by improving keystone taxa and soil properties in potatoes. *Frontiers in Microbiology*, *14*, 1034761.

Liu, D. L., et al. (1998). Detoxification of aflatoxin B1 by enzymes isolated from Armillariella tabescens. *Food and Chemical Toxicology: An International Journal Published for the British Industrial Biological Research Association*, *36*(7), 563–574.

Liu, L., et al. (2022b). Biological detoxification of mycotoxins: current status and future advances. *International Journal of Molecular Sciences*, *23*, 1064.

Liu, M., et al. (2024). Bioenzymatic detoxification of mycotoxins. *Frontiers in Microbiology*, *15*, 1434987. https://doi.org/10.3389/fmicb.2024.1434987.

Liu, M., et al. (2022a). Invited review: remediation strategies for mycotoxin control in feed. *Journal of Animal Science and Biotechnology*, *13*(1), 19.

López, M., et al. (2023). The role of nutrient management in crop health and disease resistance. *Journal of Plant Nutrition*, *46*(5), 502–510.

Lorán, S., et al. (2022). Inhibition of *Aspergillus parasiticus* growth and aflatoxins production by natural essential oils and phenolic acids. *Toxins*, *14*(6), 384.

Magan, N. & Aldred, D. (2007). Post-harvest control strategies: minimizing mycotoxins in the food chain. *International Journal of Food Microbiology*, *119*(1–2), 131–139.

Magnoli, A. P., et al. (2019). Impact of mycotoxin contamination in the animal feed industry. *Current Opinion in Food Science, 29*, 99–108.

Majumdar, R., et al. (2022). Editorial: host plant resistance mechanisms against fungal pathogens. *Frontiers in Plant Science*, *13*, 1103046.

Mannaa, M. & Kim, K. D. (2017). Influence of temperature and water activity on deleterious fungi and mycotoxin production during grain storage. *Mycobiology*, *45*(4), 240–254.

Marroca A., et al. (2008). On Fumonisin incidence in monoculture Maize under no-till, conventional tillage and two nitrogen fertilisation levels. *Journal of the Science of Food and Agriculture*, *88*, 1217–1221.

Martínez, G., et al. (2019). Effect of fosfomycin, Cynara scolymus extract, deoxynivalenol and their combinations on intestinal health of weaned piglets. *Animal Nutrition, 5, 4,* 386–395. https://doi.org/10.1016/j.aninu.2019.08.001.

Mesgar, A., et al. (2022). Effect of dietary L-threonine and toxin binder on performance, blood parameters, and immune response of broilers exposed to aflatoxin B1. *Toxins*, *14*(3), 192.

Mierziak, J. & Wojtasik, W. (2024). Epigenetic weapons of plants against fungal pathogens. *BMC Plant Biology*, *24*, 175.

Milani, J. M. (2013). Ecological conditions affecting mycotoxin production in cereals: a review. *Veterinarni Medicina*, *58*(8), 405–411.

Mohamed, A. B., et al. (2022). Efficacy of electron beam irradiation in reduction of mycotoxin-producing fungi, aflatoxin, and fumonisin, in naturally contaminated maize slurry. *Toxicon: X*, *16*, 100141.

Moon, Y. S., et al. (2018). Organic acids suppress aflatoxin production via lowering expression of aflatoxin biosynthesis-related genes in *Aspergillus flavus*. *Food Control*, *88*, 207–216.

Mostrom, M. S. (2021). *Tremorgenic neuromycotoxicosis in dogs*. MSD Veterinary Manual.

Muhialdin, B. J., et al. (2020). Review on the biological detoxification of mycotoxins using lactic acid bacteria to enhance the sustainability of foods supply. *Molecules (Basel, Switzerland)*, *25*(11), 2655.

Munkvold, G. (2003). Epidemiology of fusarium diseases and their mycotoxins in maize ears. *European Journal of Plant Pathology*, *109*, 705–713.

Murphy, P. A., et al. (2006). Food mycotoxins: an update. *Journal of Food Science*, *71*(5), R51.

Murthy, T. N., et al. (2002). Evaluation of glucomannan for its adsorbing ability of aflatoxin B1 and T-2 toxin in the gastrointestinal tract of broiler chickens. *Mycotoxin Research, 18*, 20–23.

Myers, A. J., et al. (2013). The effects of diet form and feeder design on the growth performance of finishing pigs. *Journal of Animal Science, 91*(7), 3420–3428.

Nada, S., et al. (2022). Prevention and practical strategies to control mycotoxins in the wheat and maize chain. *Food Control, 136*, 108855.

Nagaraj, R. Y., et al. (1994). Toxicity of corn culture material of *Fusarium proliferatum* M-7176 and nutritional intervention in chicks. *Poultry Science, 73*(5), 617–626.

Nankam, C. & Pataky K. K. (1996). Resistance to kernel infection by Fusarium moniliforme in the sweet corn inbred IL 125b. *Plant Disease, 80*, 593–598.

Ng'ang'a, W. and Niyonshuti, E. (2022). Animal feeds Mycotoxins and risk management. *Mycotoxins and Food Safety - Recent Advances*. IntechOpen. doi:10.5772/intechopen.102010.

Nguyen, B. H., et al. (2023). A bio-mimetic leaf wetness sensor from replica molding of leaves. *Advanced Sensor Research, 2*, 2200033.

Nielsen, D., et al. (2023). Innovative irrigation practices for crop health and mycotoxin reduction. *Irrigation Science, 41*(2), 150–159.

Oliver, S., et al. (2021). Field hygiene practices: a key component in disease management. *Plant Protection Science, 57*(4), 267–275.

Olugbenga, S. O., et al. (2015). Optimized nutrients diet formulation of broiler poultry rations in Nigeria using linear programming. *Journal of Nutrition & Food Sciences, S14*, 1.

Ono, E., et al. (2011). Effect of cropping systems and crop successions on fumonisin levels in corn from Northern Paraná State, Brazil. *European Journal of Plant Pathology, 131*, 653–660. https://doi.org/10.1007/s10658-011-9839-6

Pankaj, S. K., et al. (2018). A review of novel physical and chemical decontamination technologies for aflatoxin in food. *Trends in Food Science & Technology, 71*, 73–83.

Panneerselvam, P., et al. (2023). Long-term impact of pulses crop rotation on soil fungal diversity in aerobic and wetland rice cultivation. *Fungal Biology, 127*(6), 1053–1066.

Papaioannou, D., et al. (2005). The role of natural and synthetic zeolites as feed additives on the prevention and/or the treatment of certain farm animal diseases: a review. *Microporous and Mesoporous Materials, 84*(1–3), 161–170.

Pascale, M., et al. (2011). Distribution of T-2 and HT-2 toxins in milling fractions of Durum wheat. *Journal of Food Protection, 74*, 1700–1707.

Pascale, M., et al. (2020). Aflatoxin reduction in maize by industrial-scale cleaning solutions. *Toxins, 12*(5), 331.

Pascale, M., et al. (2022). Industrial-scale cleaning solutions for the reduction of Fusarium toxins in maize. *Toxins, 14*(11), 728.

Pattono, D. & Gai, F. (2020). Ochratoxin a (OTA) occurrence in meat and dairy products: prevention and remediation strategies. In R. Hess (Ed.), *Ochratoxin A and aflatoxin B1* (pp. 1–30). Nova Science Publisher.

Peraica, M., et al. (2002). Prevention of exposure to mycotoxins from food and feed. *Arhiv za higijenu rada i toksikologiju, 53*(3), 229–237.

Pérez, M., et al. (2023). The use of modern technology in crop surveillance and monitoring. *Farming Technology Journal, 18*(1), 78–90.

Pettersson, H. (2004). Controlling mycotoxins in animal feed. *Mycotoxins in food: Detection and control*, 262–304.

Phillips, T. D., et al. (2008). Reducing human exposure to aflatoxin through the use of clay: a review. *Food Additives and Contaminants*, *25*(2), 134–145.

Pietri, A., et al. (2012). Occurrence of mycotoxins and ergosterol in maize harvested over 5 years in Italy. *Food Additives & Contaminants: Part B*, *5*(2), 73–78.

Pitt, J. I. & Hocking, A. D. (2009). *Fungi and food spoilage* (3rd ed.). Springer.

Pokoo-Aikins, A., et al. (2024). Mycotoxin contamination and the nutritional content of corn targeted for animal feed. *Poultry Science*, *103*(12), 104303.

Popović, V., et al. (2018). Ultraviolet light in decontamination of mycotoxins in food and feed: a review. *Acta Agriculturae Serbica*, *23*(45), 173–183.

Rodríguez-Lira, D.-C., et al. (2024). Trends in machine and deep learning for plant disease detection and irrigation optimization. *Agriculture*, *14*(12), 2188.

Rose, J. L., et al. (2019). *Preharvest management strategies and their impact on mycotoxigenic fungi and associated mycotoxins*. IntechOpen.

Rychlik, M., et al. (2014). Proposal of a comprehensive definition of modified and other forms of mycotoxins including "masked" mycotoxins. *Mycotoxin Research*, *30*(4), 197–205.

Saikai, Y., et al. (2023). Deep reinforcement learning for irrigation scheduling using high-dimensional sensor feedback. *PLOS Water*, *2*(9), e0000169.

Saleem, M. F., et al. (2022). Effect of short-term zero tillage and legume intercrops on soil quality, agronomic and physio-logical aspects of cotton under arid climate. *Land*, *11*(2), 289.

Saleem, N., et al. (2025). Accessing the effect of phyllotaxy and planting density on light use efficiency in field-grown maize using 3D reconstructions. *ArXiv, abs/2503.06887.*

Sánchez-Arroyo, A., et al. (2024). *Aspergillus niger* ochratoxinase is a highly specific, metal-dependent amidohydrolase suitable for OTA biodetoxification in food and feed. *Journal of Agricultural and Food Chemistry*, *72*, 18658–18669.

Santos, C. A., et al. (2021). Biofumigation with species of the Brassicaceae family: a review (Version 1). *SciELO Journals*.

Satterlee, T., et al. (2023). Synergistic effects of essential oils and organic acids against *Aspergillus flavus* contamination in poultry feed. *Toxins*, *15* (11), 635.

Schaarschmidt, S. & Fauhl-Hassek, C. (2021). The fate of mycotoxins during the primary food processing of maize. *Food Control*, *121*, 107651.

Schmale, D. G. & Munkvold, G. P. (2009). Mycotoxins in crops: a threat to human and domestic animal health. *The Plant Health Instructor.*, *3*, 340–353.

Schmidt, M. A., et al. (2021). Enzymatic degradation is an effective means to reduce aflatoxin contamination in maize. *BMC Biotechnology*, *21*, 70. https://doi.org/10.1186/s12896-021-00730-6.

Schmidt, M. A., et al. (2023), Enzymatic degradation of aflatoxin accumulation in Dry Maize Kernels as a promising post-harvest mitigation strategy. *Journal of Biotechnology and Bioprocessing*, *4*(3).

Steiner, B., et al. (2017). Breeding strategies and advances in line selection for Fusarium head blight resistance in wheat. *Tropical Plant Pathology*, *42*, 165–174.

Sujayasree, O. J., et al. (2022). Ozone: an advanced oxidation technology to enhance sustainable food consumption through mycotoxin degradation. *Ozone: Science & Engineering*, *44*(1), 17–37.

Sydenham, E. W., et al. (1994). Fumonisin-contaminated maize: physical treatment for the partial decontamination of bulk shipments. *Food Additives & Contaminants: Part A, 11*, 25–32.

Tang, L., et al. (2024). Effect of dielectric barrier discharge high voltage atmospheric cold plasma on Aspergillus flavus inactivation and aflatoxin B1 degradation on inoculated raw peanuts. *Innovative Food Science & Emerging Technologies, 97*, 103820.

Tian, F., et al. (2019). Comparison of the antifungal and Antiaflatoxigenic potential of liquid and vapor phase of *Thymus vulgaris* essential oil against *aspergillus flavus. Journal of Food Protection, 82*, 2044–2048.

Tibola, C. S., et al. (2016). Effect of cleaning, sorting and milling processes in wheat mycotoxin content. *Food Control, 60*, 174–179.

Tikekar, R. V., et al. (2014). Ultraviolet-C light inactivation of bacterial spores on almonds and in vitro. *Journal of Food Science, 79*(5), M895–M900.

Timm, N. D. S., et al. (2020). Effects of drying methods and temperatures on protein, pasting, and thermal properties of white floury corn. *Journal of Food Processing and Preservation, 44*(10), e14767.

Trenholm, H. L., et al. (1991). Two physical methods for the decontamination of four cereals contaminated with deoxynivalenol and zearalenone. *Journal of Agricultural and Food Chemistry, 39*, 356–360.

Trivedi, P., et al. (2020). Plant-microbiome interactions: from community assembly to plant health. Nature reviews. *Microbiology, 18*(11), 607–621.

Vila-Donat, P., et al. (2018). A review of the mycotoxin adsorbing agents, with an emphasis on their multi-binding capacity, for animal feed decontamination. *Food and Chemical Toxicology, 114*, 246–259.

Wan, X. L., et al. (2013). Toxicity of increasing aflatoxin B1 concentrations from contaminated corn with or without clay adsorbent supplementation in ducklings. *Poultry Science, 92*(5), 1244–1253.

Wang, J., et al. (2022). The impact of planting density on crop productivity and disease resistance. *Crop Science, 62*(6), 432–438.

Wang, L., et al. (2019). Aflatoxin B1 degradation and detoxification by *Escherichia coli* CG1061 isolated from chicken cecum. *Frontiers in Pharmacology, 9*, 1548.

Wang, X., et al. (2018). Protective role of selenium in immune-relevant cytokine and immunoglobulin production by piglet splenic lymphocytes exposed to deoxynivalenol. *Biological Trace Element Research, 184*(1), 83–91.

Waratuke, K. (2017). *Tremorgenic mycotoxin intoxication in dogs*. Today's Veterinary Practice.

Xing, F., et al. (2014). Degradation of fumonisin b1 by cinnamon essential oil. *Food Control, 38*, 37–40.

Xu, R., et al. (2022). Nutritional impact of mycotoxins in food animal production and strategies for mitigation. *Journal of Animal Science and Biotechnology, 13*(1), 69.

Yiannikouris, A. & Jouany, J. P. (2002). Mycotoxins in feeds and their fate in animals: a review. *Animal Research, 51*(2), 81–99.

Yin, L., et al. (2020). Study the antibacterial mechanism of cinnamaldehyde against drug-resistant Aeromonas hydrophila in vitro. *Microbial Pathogenesis, 145*, 10420.

Zanon, M. A., et al. (2023). Aflatoxins and fumonisins in maize under a climate change scenario. Biocontrol strategies at the pre-harvest stage. *European Journal of Plant Pathology, 167*(4), 551–567. https://doi.org/10.1007/s10658-023-02735-7.

Zhao, Y., et al. (2019). Bio-precipitation of calcium and magnesium ions through extracellular and intracellular processes induced by Bacillus licheniformis SRB2. *Minerals*, *9*(9), 526. https://doi.org/10.3390/min9090526.

Zhu, Y., et al. (2016). Innovative technologies for the mitigation of mycotoxins in animal feed and ingredients–A review of recent patents. *Animal Feed Science and Technology*, *216*, 19–29.

Chapter 6

Exposure, transfer and uptake of environmental contaminants in animal feed: risk assessment, decontamination and detoxification interventions

Martin Rose, Jorvik Food and Environmental Chemical Safety Ltd (JFECS), United Kingdom

1 Introduction

Safe animal feed is essential to protect animal health and to prevent human health risks from food produced using unsafe feed. Awareness of the importance of safe animal feed is gaining as an increasing number of incidents can be traced back to the use of contaminated feed.

This chapter aims to provide a brief introduction to contaminants likely to pose a significant threat to health, either directly to the animal or indirectly to humans as a result of using animal products in food for human consumption. It is important to note that natural toxins (mycotoxins, plant toxins), contaminants related to transmissible spongiform encephalopathies (TSEs), microbiological contaminants, and allergens in feed are dealt with elsewhere in this book.

http://dx.doi.org/10.19103/AS.2026.0172.20

2 Environmental contaminants: persistent organic pollutants (POPs), heavy metals and other toxins

2.1 Organic contaminants

Perhaps the most obvious example of contaminated feed posing a potential threat to human health is contamination with persistent organic pollutants (POPs) such as dioxins. POPs are organic chemical substances that, due to their physical and chemical properties, persist in the environment for extended periods after being released. These substances can disperse widely through natural processes involving soil, water, and air; accumulate in the fatty tissue of living organisms (including humans); increase in concentration at higher levels of the food chain; and exhibit toxicity to humans and wildlife.

A notable case involving POPs was the 1999 incident in Belgium, in which a polychlorinated biphenyl (PCB) mixture was used directly in animal feed. PCBs were used in transformer oil (used in electrical equipment) which had been disposed of in a municipal waste collection facility where waste fats and oils were also collected to be used in animal feed (Bernard et al., 1999, Bernard et al., 2002, Hoogenboom et al., 2015, Van Larebeke et al., 2001, De Bont et al., 2004, Traag et al., 2006). The PCBs were in turn contaminated with low levels of polychlorinated dibenzodioxins and polychlorinated dibenzofurans (PCDD/Fs; 'dioxins'). Both PCBs and PCDD/Fs are toxic and bioaccumulate. Where contaminated feeds were used, the resulting food products (eggs, chicken, and other meat) posed a significant health risk to consumers. A later incident in Ireland in 2009 also involved PCBs and dioxins, but this time PCBs were present in fuel oil burnt to dry feed as part of the production process (Hoogenboom et al., 2015, Heres et al., 2010, Tlustos et al., 2012, Marnane, 2012). Again, dioxins were present at lower levels when compared to PCBs, arising both as environmental contamination and also arising as a result of heating PCBs which resulted in dioxin formation.

Other incidents involving PCBs and dioxins in animal feed had occurred prior to these major international food emergency incidents. Although smaller in scale, there were nevertheless major implications at an international level. During 1997–1999, Brazilian citrus pellets used in animal feed designed for export were contaminated with dioxins as a result of the use of contaminated lime in the drying process used in feed production. This resulted in elevated levels of dioxins in German milk that were uncovered by routine monitoring programmes (Malisch, 2000, Carvalhaes et al., 2002, Malisch & Kotz, 2014). Other incidents involving dioxins and PCBs, including those originating from animal feed, have been previously reported (Hoogenboom et al., 2015).

The Stockholm Convention (www.pops.int) governs various classes of POPs, including dioxins, PCBs, and other POPs that also have potential to cause contamination problems in animal feed similar to those described for dioxins and PCBs. Established in 2001, the treaty initially listed 12 POPs from pesticides (e.g. aldrin, Dichlorodiphenyltrichloroethane (DDT)), industrial chemicals (e.g. hexachlorobenzene, PCBs), and by-products (e.g. dioxins). In 2009, 17 more classes were added. These included brominated flame retardants (BFRs), e.g. polybrominated diphenyl ethers (PBDEs) and perfluorooctane sulfonate (PFOS). PFOS, found in products like electronics, fire-fighting foam, and textiles, is still produced in some countries. As restrictions grow, industries are shifting to shorter-chain alternatives. These, however, demonstrate similar persistence, bioaccumulation potential, and toxicity. As concerns grow around the safety of a particular class of POPs, replacements often appear on the market with similar properties which are also often found to pose a similar threat to environmental and human health. An example is the replacement of PBDEs with other brominated flame retardants and more recently with organo-phosphorus flame retardants. This process is known as 'regrettable substitution'.

Safety issues relating to organic contaminants in animal feed are not confined to POPs and sometimes arise as a result of deliberate actions. An example is melamine, used in manufacturing plastics. In 2007 it was found added to pet food in China to give a false indication of high protein content, resulting in several pet fatalities. The same compound was later added to milk that had been diluted also to give an artificial indication of high protein content, resulting in several infant fatalities when used to produce infant formula (Gossner et al., 2009, Yang et al., 2009, EFSA, 2010). Naturally produced organic toxins are discussed in Section 2.3.

2.2 Heavy metal contamination

Incidents of heavy metal contamination in animal feed arise from a combination of environmental and anthropogenic sources, both of which can have far-reaching consequences for the safety of the food supply. Naturally occurring geological deposits can introduce elements such as lead (Pb), cadmium (Cd), arsenic (As) (arsenic is not technically a heavy metal, rather a metalloid), and mercury (Hg) into the soil and water, where they are subsequently absorbed by crops used to produce animal feed. Industrial activities such as mining, improper waste disposal, and the use of contaminated water in irrigation can all result in the presence of these toxic metals in the environment. Agricultural practices themselves, including the application of certain fertilisers and pesticides, may also unintentionally contribute to elevated levels of heavy metals in feed ingredients as well as introducing problems in their own right, i.e. through the presence of fertiliser and pesticide residues in animal feed (Hargin and Shears, 2013, Rose et al., 2010).

Once present in animal feed, heavy metals can accumulate in the tissues of livestock, poultry, and fish, resulting in chronic exposure that may result in a variety of adverse health effects. Animals consuming contaminated feed may suffer from disrupted metabolic processes, impaired growth and development, reproductive disorders, and organ damage. In severe cases, high concentrations of these metals can lead to acute poisoning episodes, further threatening animal welfare.

The bioaccumulation of heavy metals means that these toxins can be transferred from feed, through animals, to products such as meat, milk, eggs, and fish. This pathway places public health at risk, as chronic ingestion of contaminated animal products can lead to neurological, renal, and developmental disorders in humans, particularly in vulnerable populations including infants, the elderly, and pregnant individuals.

Several high-profile incidents illustrate the global nature and seriousness of the problem of contamination with heavy metals and metalloids. For example, instances of arsenic-tainted poultry feed can arise as a result of either naturally occurring arsenic from the environment or as a result of the use of feed additives containing arsenic which were once widely used in poultry and pig production. Such incidents emphasise the importance of regular monitoring, stringent quality assurance protocols, and robust regulatory frameworks to either prevent or uncover such events at the earliest possible stage and to take action to minimise the risk to livestock, companion animals, and humans.

2.3 Other examples of toxic compounds contaminating feed

Significant contamination of feed can occur from toxic compounds occurring in the environment or from direct addition during feed production. However, contamination can also occur from specific primary and secondary metabolites produced by plants used for feed (Abdisa et al., 2024). Toxic compounds resulting from either plant or microbial origin share common features. As an example, they can produce anti-nutritional effects, reduce reproductive performance, or cause other adverse health effects (Kocyigit et al., 2023).

Natural (endogenous) toxins are present in certain plants and act as defence mechanisms against predators. Examples of endogenous natural toxins in feed include solanine in potatoes, cyanogenic glycosides in cassava and linseed, lectins and enzyme inhibitors in soya beans, and phorbol esters in jatropha (Urugo et al., 2023). Pyrrolizidine alkaloids have been found in honey due to bees collecting nectar from certain toxin-producing plants. Both pyrrolizidine alkaloids and tropane alkaloids, for example, occur in some weed species growing next to crops.

When they occur, incidents involving contaminants and toxins in feed can occur either at a local level or on a global scale, but there is also a difference in impact based on the toxicity of the contaminant concerned. It is also important not just to view individual classes of compounds in isolation. There is also the possibility of contamination by multiple substance which can lead to synergistic interactions, although our understanding of the mechanisms and magnitude of these effects is still not complete.

In response to these challenges, many countries have implemented stringent legislation setting maximum allowable limits for contaminants in both feed and food products. These regulations are supported by advances in analytical techniques that enable more sensitive detection of contaminants at increasingly lower concentrations. As a result of both the potential threat to animal and human health, alongside growing general concern about use of chemicals as animal feed additives (whether from permitted or non-permitted uses), legislation setting maximum limits for chemicals in feed has increased dramatically over recent years and is expected to increase further in the future.

Ongoing collaboration among governments, industry stakeholders, and scientific researchers is essential to continually improve feed safety standards, promptly address emerging risks, and safeguard both animal and human health (Zhou et al., 2024).

Ultimately, the complex interplay between environmental contamination, industrial activity, and agricultural practices makes contamination a major concern for the animal feed industry. Proactive risk management, rigorous enforcement of standards, and investment in research and education are all critical steps towards ensuring the ongoing safety of the global food supply chain.

3 Legislation and official control of animal feed

Recognising the potential health threats posed by contaminants in animal feed, comprehensive legislation and robust official controls have become integral to modern feed safety governance. Regulatory frameworks like the European Food Safety Authority (EFSA) and the U.S. Food and Drug Administration (FDA) use risk assessment to inform policies, set maximum limits (MLs) for contaminants and maximum residue levels (MRLs) for residues, and implement preventive controls at national, regional, and international levels. These are designed to minimise risks by enforcing stringent standards and monitoring compliance throughout the feed supply chain.

In the European Union, a suite of regulations forms the backbone of official control. The General Principles of Food Law Regulation (Regulation (EC) No 178/2002) establishes the overarching principles for food and feed law, mandating that risk assessment always prioritise the health of animals and humans. This is complemented by specific legislation, such as the Feed Additives Regulation (Regulation (EC) No 1831/2003), which governs the authorisation, safety, efficacy, and labelling of feed additives, ensuring that only rigorously evaluated substances are permitted for use.

Table 1 Global legislation on contaminants and residues in animal feed

Region/country	Primary legal/normative instruments	Scope and coverage	Main contaminant/residue categories
Global (Codex)	Codex Code of Practice on Good Animal Feeding (CAC/RCP 54-2004); various Codex maximum levels (MLs) and maximum residue limits (MRLs)	Non-binding international standards for feed safety; reference under WTO SPS Agreement	Mycotoxins (e.g. aflatoxins), heavy metals, dioxins and PCBs, pesticide residues, veterinary drug residues, microbiological hazards
European Union	Directive 2002/32/EC on undesirable substances in animal feed; Regulation (EC) No 183/2005 (feed hygiene); Regulation (EC) No 767/2009 (placing on the market and use of feed); Regulation (EC) No 396/2005 (pesticide MRLs)	Legally binding safety, hygiene, and marketing requirements for feed materials, compound feed, feed additives, and feed business operators	Undesirable substances (e.g. heavy metals, dioxins, mycotoxins, botanical contaminants), pesticide residues, carry-over of coccidiostats and veterinary drugs, microbiological and physical hazards
United Kingdom	Retained EU feed law (including Directive 2002/32/EC, Regulations 183/2005, 767/2009, 396/2005) as adapted, plus domestic instruments (e.g. Animal Feed (Composition, Marketing and Use) regulations)	Largely mirrors pre-Brexit EU framework with UK-specific enforcement and risk management	Same broad scope as EU: undesirable substances, residues of pesticides and veterinary medicines, microbiological contamination, physical contaminants
United States	Federal Food, Drug, and Cosmetic Act (FFDCA); Food Safety Modernization Act (FSMA) and Preventive Controls for Animal Food rule (21 CFR Part 507); tolerances/action levels and guidance levels set by FDA	Adulteration standards and preventive controls for animal food; authorises FDA to set tolerances or action levels	Mycotoxins, heavy metals, pesticide residues, industrial contaminants (e.g. dioxins), veterinary drug residues, microbiological hazards
Canada	Feeds Act and Feeds Regulations; Health of Animals Act; Canadian Food Inspection Agency (CFIA) policies and standards	Pre-market authorisation and safety standards for feeds; controls on contaminants and residues	Mycotoxins, metals, pesticides, industrial contaminants, veterinary drug residues, microbiological hazards

Brazil	Feed regulations under the Ministry of Agriculture, Livestock and Food Supply (MAPA) (e.g. Normative Instructions on feed manufacturing, registration and quality); alignment with Mercosur resolutions	Registration, quality, and safety standards for animal feed and feed ingredients	Mycotoxins (notably aflatoxins), pesticides, heavy metals, other chemical contaminants, microbiological hazards
China	Feed and Feed Additive Regulation; associated national standards (GB and GB/T) on hygienic specifications and contaminant limits in feed	Comprehensive framework for feed production, use, and quality; mandatory standards for contaminants	Mycotoxins, heavy metals, pesticide residues, veterinary drug residues, pathogenic microorganisms, and other hazardous substances
Australia/New Zealand	Australia: state/territory stock feed legislation, APVMA regulation of agricultural and veterinary chemicals; New Zealand: Agricultural Compounds and Veterinary Medicines (ACVM) Act and related standards; both: alignment with Codex and international MLs/MRLs	Mixed model combining chemical control (ag-vet registration) with feed quality/safety requirements; strong emphasis on residue control in foods of animal origin	Pesticide and veterinary drug residues (via MRLs), mycotoxins, metals, and other contaminants as specified in national standards or guidelines

Directive 2002/32/EC, which addresses undesirable substances in animal feed, establishes maximum permissible levels for an array of chemical contaminants and naturally occurring toxins, including heavy metals, mycotoxins, and dioxins. Authorities are required to conduct routine surveillance, enforce immediate corrective actions when limits are breached, and uphold transparency through traceability requirements.

Internationally, collaboration and harmonisation are promoted through organisations such as the Food and Agriculture Organization of the United Nations (FAO). Initiatives like the Global Feed Safety Platform (https://www.fao.org/feed-safety/en/) and the Code of Practice on Good Animal Feeding (FAO/WHO Codex Alimentarius, 2004) provide science-based standards and guidance, helping countries align their practices and uphold the integrity of the global food chain. A holistic 'One Health' approach recognises the interconnectedness of human health, animal health, and the environment. This is reflected in risk assessment by considering potential health risks to both animals and humans from feed contamination.

Through these layered legislative measures and official controls, risk assessment, hygiene, traceability, and maximum contaminant thresholds are integrated into a holistic regulatory strategy. This approach aims to ensure that animal feed, as a key component of the food system, is systematically safeguarded with a focus on protecting animal welfare and securing public health in an increasingly interconnected world.

Whilst legislation globally is extensive and complex, Table 1 summarises some of the principal legislative instruments and standards for major jurisdictions.

4 Assessing exposure

Assessing exposure to contaminants arising from animal feed is a multi-faceted process that serves as a foundation for effective risk management strategies in food safety. This process begins with the systematic identification and quantification of contaminants present in various feed ingredients. Analytical methods, ranging from rapid screening tools to advanced chromatographic techniques using mass spectrometry, are used to detect and monitor levels of chemical contaminants, such as heavy metals, POPs, and naturally occurring toxins (Rose, 2012).

Once the presence and concentration (occurrence) of these contaminants are established, the next step is to evaluate the dietary exposure of animals to these substances. This involves calculating the intake of contaminated feed based on daily consumption rates for the animal, the proportion of each ingredient in the feed, and the body weight of the animal. Exposure scenarios are often constructed to encompass both average (mean) and high-level (worst-case) consumption patterns, reflecting variations in feeding practices across different production systems.

It is also important to account for the bioavailability and absorption rates of specific contaminants, as not all ingested substances are equally taken up by the animal (Goff, 2018). Factors such as the chemical form of the contaminant (e.g. whether metals are present in elemental inorganic form or they are incorporated into an organic compound), its interaction with other dietary components, and the physiological status of the animal (e.g. age, breed, metabolic rate) can significantly influence uptake.

To maintain robust food safety controls, exposure assessments are revisited regularly, especially when new contaminants are identified or when changes occur in agricultural practices, feed formulations, or legislation. The findings from these assessments inform regulatory decisions, such as the establishment of maximum residue limits, targeted monitoring programs, and guidance for safe feed sourcing and processing, all of which are essential measures designed to protect both animal and human health throughout the food supply chain.

Table 2 shows the major chemical contaminant groups, their common sources in animal feed, exposure pathways, and potential impacts. Magnitude of exposure can vary considerably depending on the circumstances of the individual exposure incident.

5 Transfer and uptake of contaminants in food products

A comprehensive exposure assessment also considers the transfer of contaminants from animal feed into edible products such as meat, milk, eggs, and offal – a process that plays a critical role in determining potential risk to human consumers (Leeman et al., 2007). This transfer is influenced by a complex interplay of factors, including the physicochemical properties of the contaminant, the metabolic capacity of the animal, and the duration and level of exposure.

To accurately estimate the likelihood and extent of contaminant residues appearing in food products, kinetic modelling based on both experimental studies and real-world monitoring data can be used. These models track the absorption, distribution, metabolism, and elimination of contaminants within the animal's body, allowing residue levels in tissues and animal-derived foods across a range of feeding scenarios to be estimated.

Additionally, certain contaminants accumulate preferentially in specific tissues or products. For example, lipophilic compounds such as dioxins and PCBs are known to concentrate in fatty tissues and milk, whereas heavy metals like cadmium may concentrate in organs such as the liver and kidneys.

Tables 2 Exposure to chemical contaminants from animal feed

Contaminant group	Typical sources in feed	Exposure pathway to animals	Key risks to animal health	Key risks to human health (via food chain)
Mycotoxins (e.g. aflatoxins, DON, fumonisins, zearalenone)	Fungal growth on cereals, silage, oilseeds; poor storage conditions	Ingestion of contaminated grains or forages	Immunosuppression, reduced productivity, hepatotoxicity, reproductive effects	Carcinogenicity (e.g. aflatoxin B1 → M1 in milk), endocrine disruption
Pesticide residues (insecticides, herbicides, fungicides)	Treated crops used as feed ingredients	Ingestion of treated crops or by-products	Neurological effects, endocrine disruption, reduced feed intake	Residues in meat, milk, eggs; chronic toxicity risks
Veterinary drug residues (antibiotics, antiparasitics, coccidiostats)	Carry-over from medicated feed; cross-contamination in feed mills	Ingestion of medicated or contaminated feed	Altered gut microbiota, toxicity at high doses	Antimicrobial resistance development; residues in edible tissues
Heavy metals (lead, cadmium, arsenic, mercury)	Contaminated soils, water, mineral supplements, industrial fallout	Ingestion of contaminated feed or water	Neurotoxicity, kidney damage, reduced growth	Bioaccumulation; carcinogenicity; nephrotoxicity
Dioxins and PCBs	Contaminated fats/oils, industrial emissions, contaminated clays	Ingestion of contaminated feed fats	Immunotoxicity, reproductive toxicity	POPs in meat, milk, eggs; carcinogenicity
PAHs (polycyclic aromatic hydrocarbons)	Drying processes using direct combustion; environmental contamination	Ingestion of contaminated feed materials	Liver damage, reduced performance	Carcinogenicity; accumulation in animal fats
Radionuclides (e.g. Cs-137, Sr-90)	Fallout, contaminated soils, uptake by forage crops	Ingestion of contaminated forages	Internal irradiation, bone and tissue accumulation	Transfer to milk/meat; long-term radiation exposure
Processing contaminants (e.g. acrylamide, furans)	Heat-treated feed ingredients (e.g. toasted soy, bakery by-products)	Ingestion of processed feed	Limited evidence; potential neurotoxicity	Potential carcinogenicity; low but possible transfer
Environmental contaminants (PFAS, microplastics)	Contaminated water, biosolids, packaging residues	Ingestion via feed or water	Emerging evidence: immune and metabolic effects	Bioaccumulation; long half-lives; potential endocrine and immune effects
Nutritional imbalances leading to toxicities (e.g. excess copper in sheep feed)	Formulation errors, species-inappropriate mineral mixes	Ingestion of incorrectly formulated feed	Acute or chronic toxicity, organ damage	Indirect – animal welfare and food safety concerns

Exposure assessments should be periodically updated to reflect changes in animal husbandry practices, feed ingredient sourcing, and emerging contaminants of concern. These assessments form the scientific backbone for regulatory decisions regarding food safety, guiding risk management actions such as feed bans, withdrawal periods before slaughter or milking, and advice given in a public health context. Ultimately, rigorous evaluation of contaminant transfer and uptake ensures that both animal health and consumer safety are protected throughout the food supply chain (Hoogenboom et al., 2006, Hoogenboom et al., 2015, Rose et al., 2012, Fernandes et al., 2023, Rigby et al., 2023).

Table 3 provides 'order-of-magnitude typical' ranges for transfer of contaminants for chronic exposure at or near steady state, rather than worst-case single-dose values. These are given as approximate percentages of the ingested dose excreted or deposited in the listed food type.

The values presented previously are deliberately broad, intended as 'typical transfer ranges' rather than precise values. The following section looks at transfer values for different contaminant classes

6 Transfer values for different classes of environmental contaminants

6.1 Lipophilic persistent organic pollutants

6.1.1 Dioxins, polychlorinated biphenyls and organochlorine pesticides

Dioxins and dioxin-like PCBs can transfer a substantial fraction into milk fat; cumulative transfer efficiencies are commonly in the 20–80% range under steady-state chronic exposure. Laying hens show similar behaviour for egg yolk fat, with 30–70% of ingested TEQ ultimately found in yolk fat over a sustained exposure in experimental feeding trials. For meat, both for dioxin-like PCBs and non-dioxin-like PCBs, 50–90% of the ingested dose can eventually be accounted for in body fat, with lower percentages in lean muscle due to lower lipid content (FESA, 2010b, 2018).

Legacy organochlorine pesticide compounds such as DDT/DDE, lindane, and HCB show broadly similar behaviour to PCBs: high deposition in fat, appreciable carry-over into milk and eggs in the 10–60% range (Fries, 1996, Lorber, 2002).

6.2 Mycotoxins

6.2.1 Aflatoxin B1/AFM1 and others

Numerous metabolic and residue studies show carry-over of Aflatoxin B1 (AFB1) in feed to AFM1 in cows' milk, typically in the range 0.3–6% of the ingested

Table 3 Typical transfer from feed to food of animal origin

Contaminant class	Representative compounds	Milk (dairy cow, goat)	Meat (muscle of cattle, pigs, poultry)	Eggs (laying hens)	Key points
Dioxins and dl-PCBs	2,3,7,8-TCDD; PCB-126; PCB-153	20–80% of ingested TEQ in milk fat over time	50–90% to body fat; lower % to lean muscle	30–70% to egg yolk fat	Strongly lipophilic, long half-lives, high bioaccumulation in fat.
Non-dioxin-like PCBs	PCB-118, PCB-180	10–60%	40–90%	20–60%	Similar behaviour to dl-PCBs; higher in high-fat tissues and yolk.
Organochlorine pesticides (legacy)	DDT/DDE, HCB, lindane	10–60%	40–90%	20–60%	POP-like fate; legacy contamination still relevant in some regions.
Aflatoxin B1 → AFM1	AFB1 in feed; AFM1 in milk	0.3–6% of ingested AFB1 excreted as AFM1 in milk	<1%	<1%	Transfer highly influenced by milk yield, metabolic status, and exposure level.
Other mycotoxins (DON, ZEN, fumonisins)	DON, ZEN, FB1	<1%	<1%	<1%	Extensive metabolism and low carry-over; residues typically trace.
Patulin, OTA, T-2/ HT-2	Patulin, ochratoxin A, T-2	Generally <1% to milk, eggs; <5% to meat (OTA in pigs can be higher)	Up to ~5–10% (OTA in porcine kidney/liver)	<1–2%	OTA accumulates notably in pigs; others mostly low transfer.
Metals (Pb, Cd)	Lead, cadmium	<1%	1–10% (liver, kidney > muscle)	1–5%	Preferential accumulation in liver/ kidney; low in milk and eggs.
Mercury (methylmercury)	MeHg from fishmeal	1–10%	10–40%	5–20%	Higher transfer for organic Hg; relevant mainly with fish-based feeds.
PFAS	PFOS, PFOA, PFNA, PFHxS	1–20%	10–50%	5–30%	Persistent, bind to proteins (esp. in liver); feed a key driver of residues.

Veterinary drugs (residues via feed)	Ionophores, antibiotics, antiparasitics	Typically 0–10% (if misused in feed or via cross-contamination)	0–20%	0–10%	Very compound-specific; withdrawals normally keep residues below MRLs.
Radionuclides	Cs-137, Sr-90, I-131	Cs-137: ~5–30%; Sr-90: higher to milk (bone/teeth); I-131: high to milk	Cs-137: 10–50% to muscle; Sr-90: to bone	Cs, Sr: 5–20% to eggs	Cs behaves like K; Sr like Ca; I like iodine – hence high transfer to milk.
Plant alkaloids	Pyrrolizidine alkaloids, tropane alkaloids	<0.1%	<0.1%	<0.1%	Generally low carry-over, though reactive metabolites may be detectable at trace levels.

AFB1 dose, often clustering around 1–3% under commercial conditions. Carry-over is higher in high-yielding cows and can be influenced by stage of lactation, rumen function, and health status.

Transfer of AFB1 or AFM1 to muscle (meat) and eggs is generally <1%, with concentrations usually below regulatory concern where feed is controlled (EFSA, 2004, JECFA, 2018, Veldman, 1992, Van Eijkeren, 2006)

6.2.2 Deoxynivalenol, zearalenone, fumonisins, OTA and T-2/HT-2

In ruminants and monogastrics, DON, ZEN, and fumonisins are extensively metabolised, with very low levels appearing unchanged in milk, meat, or eggs; carry-over generally <1%, often far below.

Pigs show notable accumulation of OTA in kidney and liver, with feed-to-tissue transfer up to around 5–10% in some studies, while transfer to milk in cows and eggs in poultry remains low (<1–2%).

T-2/HT-2 and patulin are typically rapidly metabolised, with negligible transfer to edible tissues and products under normal exposure (see links from: https://www.efsa.europa.eu/en/topics/topic/mycotoxins (accessed 12 January 2026); and EFSA, 2011b, 2013, Battacone, 2012).

6.3 Metals

Lead (Pb) is poorly absorbed from the gut, with most of that consumed being excreted. Transfer to milk and eggs is generally <1% of intake, with deposition into liver and kidney reaching typically a few percent (1–5%) under sustained exposure conditions. Cadmium (Cd) also favours accumulation in liver and kidney, with low transfer to milk and eggs (<1–2%). Around 1–10% of intake may eventually be accounted for in edible offal after prolonged exposure. Mercury (particularly methylmercury) from fishmeal can be more efficiently absorbed, with 10–40% of intake being deposited in muscle and organs, while 1–10% appears in milk and 5–20% in eggs for poultry receiving high fishmeal diets (JECFA, 2011, EFSA, 2009, 2010c, 2012).

6.4 Radionuclides (Cs-137, Sr-90, I-131)

Caesium-137 (Cs-137) is chemically similar to potassium and is relatively efficiently transferred from feed to milk and meat. Empirical transfer factors correspond roughly to 5–30% of ingested activity being excreted in milk over time and 10–50% being deposited in muscle, depending on diet and age. Strontium-90 (Sr-90) mimics calcium and is preferentially incorporated into bone and teeth. Transfer to milk and eggs is higher than for many metals (commonly 5–20% of intake), with bone accumulation being even higher. Iodine-131 (I-131) is transferred highly efficiently to milk due to active uptake and secretion. A substantial proportion of intake can appear rapidly in milk, which is why milk was so critical in post-Chernobyl exposure modelling (IAEA, 2010, Beresford et al., 2007).

6.5 Transfer of PFAS from feed to food

Although PFOS compounds are classed as POPs, their behaviour is somewhat different from the lipophilic POPs in that PFAS compounds such as PFOS, PFOA, PFNA, and PFHxS are highly persistent and tend to bind to serum proteins and accumulate in liver and blood, rather than adipose tissue. Feed (including contaminated forages, by-products, and fishmeal) is a key driver of these contaminants in livestock, and recent work has explicitly modelled the relationship between feed contamination levels and compliance with maximum levels in food of animal origin.

Typical transfer ranges for milk from experimental and modelling data suggest 1–20% of ingested PFOS/PFOA equivalents can appear in milk over time, depending on molecular properties such as chain length and functional group. For meat and offal, 10–50% can be accounted for in edible tissues, particularly liver (higher) and muscle (lower), reflecting strong protein binding. For eggs, laying hens show 5–30% transfer of intake to eggs (yolk and albumen combined), again with marked accumulation in liver (EFSA, 2020, European Commission, 2020, BfR, 2024)

6.6 Veterinary drugs

When authorised drugs are used according to the label, feed is not normally the primary route of residues; residues in meat, milk, and eggs are controlled through withdrawal periods linked to residue depletion kinetics. However, residues can occur via feed cross-contamination, medicated premixes, or misuse. Depending on the compound, oral bioavailability, and half-life, values of 0–10% of the ingested dose may appear in milk or eggs and up to 0–20% in tissues, but this is highly drug-specific (JECFA, various years, EMA (CVMP), various years).

6.7 Plant toxins

Plant toxins like pyrrolizidine alkaloids (PAs) generally show very low carry-over rates (<0.1%) from feed to milk, meat, and eggs, although reactive metabolites and DNA adducts can form in target organs. Tropane alkaloids and others typically exhibit low transfer to food products of animal origin at environmental exposure levels (EFSA, 2011, BfR, 2013).

7 Risk assessment for chemical contaminants in animal feed

A risk assessment for chemical contaminants in animal feed is a scientifically based process that evaluates the potential for adverse health effects in both animals and humans from indirect dietary exposure (Mantovani, 2012). The assessment follows a structured four-step procedure:

1. Hazard identification
2. Hazard characterisation
3. Exposure assessment
4. Risk characterisation

This process is crucial for upholding animal welfare and ensuring the safety of the human food supply, as contaminants in feed can transfer to foods of animal origin, such as meat, milk, and eggs.

1. Hazard identification
This step identifies biological, chemical, or physical agents in animal feed that have the potential to cause adverse health effects, such as natural toxins (e.g. mycotoxins), environmental pollutants (e.g. heavy metals, POPs), and pesticides.

2. Hazard characterisation
This step involves qualitatively and quantitatively evaluating the nature of adverse health effects associated with the identified hazards. For chemicals, this includes a dose-response assessment, which determines the relationship between the magnitude of exposure and the severity of the health effect.

Health-based guidance values (HBGVs) represent intake levels that are not considered to cause appreciable health risks. They are derived from toxicological data which are used to establish values like the Acceptable Daily Intake (ADI) or Tolerable Weekly Intake (TWI), which are examples of HBGVs.

It is important to consider animal-specific effects. The health effects on the animal consuming the contaminated feed may not be the same for all species, as some contaminants primarily impact animal welfare rather than human food safety. For example, the mycotoxin zearalenone is known to cause reproductive issues in pigs.

3. Exposure assessment
This step evaluates the likely intake of the hazardous agent by the exposed population. The process for animal feed is two-tiered:

(3a) Animal exposure
This involves estimating the intake of a chemical by an animal through its diet, based on the feed composition and concentration of the contaminant. Factors

such as animal species, its age, and feeding practices need to be considered (see previous section of this chapter).

(3b) Human exposure
To estimate human exposure it is necessary to use transfer factors to estimate the level of the contaminant in edible products (e.g. meat, milk) from the animal, and subsequently to estimate the human dietary intake. An estimate needs to consider data on food consumption patterns and potential exposure from other sources, like non-animal foods, water, air, and dermal contact, to get a full picture of overall exposure to a given contaminant.

4. Risk characterisation
This final step compares the hazard characterisation and exposure assessment to provide an estimate of the potential risk, i.e. by comparing the predicted human exposure level with the HBGV established during hazard characterisation.

8 Contamination issues specific to particular animal species

Dietary requirements and exposure risks associated with different classes of food-producing animals may be specific to the species. This section examines how the unique nutritional needs and feeding practices influence susceptibility to contaminants in feed for different species, and discusses the implications for animal health and food safety (Carter and Lipman, 2018). By outlining both the challenges and essential control measures across species, an overview of the critical role feed management plays in ensuring the integrity of animal-derived products within the food chain is described hereafter.

8.1 Pigs and pork products

The dietary habits and feed requirements of pigs differ significantly from those of ruminants (ahdb.org.uk). As monogastric animals, pigs rely heavily on high-energy diets rich in cereals, protein meals (such as soybean or canola meal), and supplementary vitamins and minerals. Their digestive systems are more sensitive to feed contaminants, which can be introduced through raw materials, additives, or improper storage. Mycotoxins, in particular, pose a considerable risk due to the susceptibility of grains to fungal contamination during cultivation or storage.

Exposure assessment in pigs must account for the types of feed used, the life stage of the animals (e.g. weaners, growers, finishers, or breeding stock), and the potential for cumulative effects from repeated low-level exposure to contaminants. Certain residues may accumulate in fat or organs, subsequently affecting the safety of pork products for human consumption. Consequently, regular testing of feed ingredients for contaminants and strict adherence to good agricultural and manufacturing practices are vital in reducing risk and ensuring both animal health and food safety.

8.2 Poultry, meat and eggs

Poultry, including chickens raised for meat and eggs, have distinct nutritional needs that shape the composition and monitoring of their feed (Ravindran, 2014). Their diets typically consist of cereal grains (such as corn and wheat), plant-based protein sources (like soybean meal), and a blend of vitamins and minerals to support rapid growth and egg production. Because chickens are highly efficient converters of feed to edible products, any contaminants present in their rations can be quickly transferred into eggs or poultry meat, often at concentrations that mirror those in the feed.

The quality and safety of poultry feed are therefore critically important. Mycotoxins, introduced by mouldy grains, remain among the most significant contaminants, potentially impairing bird health and reducing productivity, as well as posing a risk to consumers. Additionally, residues of pesticides, heavy metals, or unintended additives in feed ingredients can accumulate in tissues or eggs. As with other livestock, the life stage and production purpose (broilers versus layers) influence both feed formulation and susceptibility to dietary contaminants.

Routine screening of feed materials, proper storage conditions to prevent mould growth, and sourcing from reputable suppliers are essential strategies for minimising the risk of feed-borne contaminants in poultry production. By implementing comprehensive quality assurance programs and monitoring, the integrity of both poultry and egg products destined for the food chain can be effectively protected.

8.3 Ruminants

8.3.1 Beef and dairy

Cattle, including both beef and dairy animals, have unique dietary requirements that influence how contaminants in feed may affect them and, consequently, the safety of animal-derived food products (Tretola et al., 2025). Their ruminant digestive system allows them to utilise a wide variety of feedstuffs, from fresh forage and silage to cereal grains and by-products. However, the complexity of their diet introduces several potential sources of contamination – such as mycotoxins, heavy metals, or residues from agrochemicals.

Assessing exposure in cattle involves careful consideration of their feeding practices, the diversity of feed materials, and their physiological status (e.g.

lactating versus non-lactating cows). For dairy cattle, contaminants can transfer directly into milk, making food safety assessment particularly critical. In beef cattle, accumulation in tissues may affect meat safety and quality. Effective risk management, therefore, requires both monitoring feed sources and applying best practices in feed handling and storage to minimise contamination risks.

8.3.2 Sheep and goats

Sheep, as ruminants, share some similarities in their nutritional requirements with cattle, though their smaller size and specific physiological needs demand tailored feed strategies. Their diets commonly include pasture grasses, hay, silage, and supplementary grains – each presenting unique potential for exposure to contaminants. Sheep are particularly sensitive to certain toxins, such as copper, which can accumulate rapidly in their system and lead to serious health consequences. Additionally, mycotoxins from mouldy feed, residues of pesticides used on pastureland, and heavy metals from environmental sources all pose risks to sheep health and, by extension, the safety of lamb and mutton products.

Exposure assessment in sheep requires attention to seasonal feeding patterns, variations in pasture quality, and supplemental feeding during periods of low forage availability. Lambs, lactating ewes, and breeding rams may have heightened vulnerability to contaminants due to their specific metabolic demands. Effective management practices – such as regular feed testing, avoiding grazing on recently treated land, and ensuring proper storage of supplementary feeds – are essential to mitigate risks.

Maintaining high standards in feed sourcing and handling not only safeguards sheep welfare but also helps ensure consumer confidence in sheep-derived food products. As livestock production systems evolve and environmental pressures increase, ongoing vigilance and adaptation in feed management remain fundamental to protecting both animal health and food safety.

Whilst sheep are grazers, goats are browsers and prefer shrubs, leaves, weeds, and woody plants. They are more active than sheep and require more protein and have higher energy needs. Whereas copper is toxic to sheep, it is a required nutrient for goats, who also have higher needs for calcium, phosphorus, zinc, and selenium.

8.4 Aquaculture

Aquaculture production – encompassing fish, crustaceans, and molluscs – relies heavily on formulated feeds designed to meet precise nutritional requirements for optimal growth, health, and product quality (Verdegem et al., 2023, Rose,

2023). Unlike terrestrial livestock, the dietary needs of aquatic species vary widely, reflecting their diverse physiology, feeding behaviours, and habitats. Modern aquafeeds are typically composed of a blend of fishmeal and fish oil, plant-based proteins (such as soybean or pea meal), cereal grains, and tailored vitamin and mineral premixes. As global sustainability concerns have grown, alternative ingredients like insect meal, algae, and microbial proteins are increasingly supplementing or replacing traditional marine-derived components.

Feed safety in aquaculture is of paramount importance, as contaminants can be readily absorbed by aquatic organisms and, in many cases, bioaccumulate throughout the food chain. The aquatic environment itself can serve as both a source and a pathway for contaminants, including heavy metals, PCBs, and POPs. In addition, the use of medicated feeds, growth promoters, and colour enhancers demands careful regulatory oversight to prevent unintended residues in edible products.

Quality assurance for aquafeeds includes routine screening for mycotoxins, pesticides, and environmental contaminants, with storage and handling practices tailored to the moist, temperature-variable conditions often present near water bodies. Selecting high-quality, traceable feed ingredients and maintaining rigorous documentation help to minimise risks, while research into novel feed processing technologies continues to advance the safety and nutritive value of aquaculture diets.

Ultimately, the integrity of aquaculture feeds underpins both the welfare of farmed aquatic species and the safety of seafood for consumers, demanding ongoing vigilance and innovation as production expands.

9 Approaches to decontamination and detoxification of animal feed

Given the broad spectrum of potential contaminants in both terrestrial and aquatic livestock systems, strategies to remove or neutralise hazardous substances from animal feeds are under continuous development. Methods can be broadly classified into physical, chemical, and biological approaches, often applied singly or in combination depending on the contaminant profile and the nature of the feed (EFSA, 2024).

Physical techniques include sorting, cleaning, and sieving to eliminate visibly contaminated material, as well as more advanced methods such as thermal treatment or irradiation, which can reduce microbial loads and deactivate certain toxins. Adsorbent materials can be incorporated into feed to bind specific contaminants, thus reducing their bioavailability to animals. Examples include bentonite clays, which can bind toxic metals or mycotoxins, and activated charcoal, which can bind some POPs.

Chemical decontamination methods may rely on the use of approved agents – such as organic acids, ozone, or ammonia (via ammoniation) – to break down toxins or convert them into less harmful compounds. However, strict regulatory oversight is needed to ensure that such treatments do not introduce new hazards or leave undesirable residues. Careful selection and validation of chemical detoxification methods are essential, particularly for feeds intended for sensitive species or life stages.

Biological approaches harness the ability of certain microorganisms or enzymes to degrade or transform contaminants within feed matrices. For example, specific bacterial strains or fungal cultures can be deployed to break down aflatoxins, while enzymatic preparations may target pesticide residues or undesirable anti-nutritional factors. These methods, while often species- and contaminant-specific, hold promise for their targeted action and lower environmental impact.

Effective decontamination and detoxification protocols must be tailored to the unique challenges presented by different feed types and production systems. Integrating these interventions into a broader risk management framework – encompassing feed sourcing, processing, and continuous monitoring – can further enhance overall food safety and animal well-being.

Within the European Union (EU) it is possible to use acceptable detoxification processes on animal feed products in order to conform with the provisions of Annex I of Directive 2002/32/EC. Commission Regulation (EU) 2015/7862 established criteria for establishing the acceptability of detoxification processes to ensure that the detoxified feed does not endanger animal and public health and the environment and to ensure that the characteristics of the feed are not adversely altered by the detoxification process, e.g. the feed detoxification process should not adversely affect the nutritional value of the feed.

Several methods for feed detoxification have been described in the literature. Any method used should be demonstrated in terms of the efficacy of application of the process at an established plant for feed to which the process will be applied, i.e. the amount of contaminant removed by the process should be established.

Undesirable substances may occur as mixtures and may be removed by the process at differing proportions. For example, dioxins and dioxin-like PCBs may be removed by binding to activated charcoal, but this may not be the case for non-dioxin-like PCBs which are likely to co-occur in contaminated areas. As a result, decontaminated material may contain relatively high levels of these PCBs that may go unnoticed if not included in the analysis. This may, for instance, occur when processing fish oil and fishmeal from contaminated areas. Similarly, mycotoxins produced by the same moulds that are responsible for the target compound for the decontamination may be removed in different proportions. For this reason, it is important to establish efficacy based on incurred materials and not just spiked materials.

It is also important to consider if a process may be reversible, i.e. whether the levels of the undesirable substance could increase again in the feed or inside the animal. It is unlikely that a physical process of detoxification, where the undesirable substances are removed, will be reversible, but this is not the case for a chemical or enzymatic process where the degradation products are not removed; in such an example these may revert to the target compounds.

The possibility of the formation of degradation products should be considered in case these may have adverse health effects on animals or humans as a result of food consumption. For microbiologically based detoxification processes, additional biohazard considerations apply, including the possibility that the microbes used may produce toxins.

10 Where to look for further information

10.1 Websites

- The Stockholm Convention which governs various classes of POPs, www.pops.int.
- The Food and Agriculture Organization (FAO) of the United Nations, https://www.fao.org/.
- The World Health Organization (WHO) of the United Nations, https://www.who.int/.
- The Food and Agriculture Organization of the United Nations (FAO), the World Organisation for Animal Health (OIE), the United Nations Environment Programme (UNEP), the World Health Organization (WHO) (2021) Joint Tripartite (FAO, OIE, WHO), and UNEP Statement. Tripartite and UNEP support OHHLEP's definition of 'One Health'. https://openknowledge.fao.org/server/api/core/bitstreams/54a0f96f-066e-4a16-a186-1a4bbd9d9303/content.
- The European Food Safety Authority (EFSA), https://www.efsa.europa.eu/.
- European Food Safety Authority: Mycotoxins, https://www.efsa.europa.eu/en/topics/topic/mycotoxins.
- Agriculture and Horticulture Development Board (AHDB), https://ahdb.org.uk/.

10.2 Key texts

- Persistent organic pollutants and toxic metals in foods. Martin Rose and Alwyn Fernandes (Editors). 2013. ISBN 0 85709 245 6 Pub: Woodhead.

- Animal feed contamination: effects on livestock and food safety. Editor: Professor Johanna Fink-Gremmels, Utrecht University, The Netherlands. Pub: Woodhead 2012: pp 117–130.
- Chemical hazards in foods of animal origin. ECVPH Food safety assurance, Volume 7. Editors: Frans J.M. Smulders, Ivonne M.C.M. Rietjens and Martin Rose Published: 2019.
- Encyclopedia of Meat Sciences (Second Edition 2014 and Third edition 2024) Reference Module in Food Science (Published by Elsevier).

11 References

Abdisa, T. and Dilbato, T. (2024). Toxic plants and their impact on livestock health and economic losses: a comprehensive review. *Journal of Toxicology*, 18, 9857933. https://doi.org/10.1155/jt/9857933.

Battacone, G., et al. (2012). Excretion pattern of aflatoxin M1 in milk of goats fed a single dose of aflatoxin B1. *Journal of Dairy Science*, 95(5), 2656–2661. https://doi.org/10.3168/jds.2011-5003.

Beresford, N. A. and Howard, B. J. (2007). Transfer of radionuclides to food producing animals. *Journal of Environmental Radioactivity*, 98(1), 1–3. https://doi.org/10.1016/j.jenvrad.2007.07.001.

Bernard, A., et al. (2002). The Belgian PCB/Dioxin incident: analysis of the food chain contamination and health risk evaluation. *Environmental Research*, 88(1), 1–18. https://doi.org/10.1006/enrs.2001.4274.

Bernard, A., et al. (1999). Food contamination by PCBs and dioxins. *Nature*, 401(6750), 231–232. https://doi.org/10.1038/45717.

BfR (2013). *Pyrrolizidine alkaloids in herbal teas and teas.* Available from: https://www.bfr.bund.de/cm/349/pyrrolizidine-alkaloids-in-herbal-teas-and-teas.pdf (Accessed 12 January 2026).

BfR (2024). *Feed is key to compliance with maximum PFAS levels in food of animal origin. Opinion 033/2024.* Available from: https://www.bfr.bund.de/cm/349/feed-is-key-to-compliance-with-maximum-pfas-levels-in-food-of-animal-origin.pdf (Accessed 12 January 2026).

Carter, R. L. and Lipman, N. S. (2018). Feed and bedding. In: Weichbrod, R. H., et al., (Eds.), *Management of animal care and use programs in research, education, and testing*. 2nd ed. Boca Raton, FL: CRC Press/Taylor & Francis; Chapter 27.

Carvalhaes, G. K., et al. (2002). Lime as the source of PCDD/F contamination in citrus pulp pellets from Brazil and status of the monitoring program. *Chemosphere*, 46, 9–10, 1413–1416. https://doi.org/10.1016/S0045-6535(01)00263-6.

De Bont, R., et al. (2004). A survey of 3 PCB and dioxin contamination episodes. From contamination of food items to body burdens. In: Darby, W. J., et al. (Eds.), *Annual reviewof nutrition* (pp. 301–342). Boca Raton, FL: CRC Press.

EFSA (European Food Safety Authority), (2004). Opinion of the Scientific Panel on contaminants in the food chain (CONTAM) related to Aflatoxin B1 as undesirable substance in animal feed. *EFSA Journal*, 2(3), 239. https://doi.org/10.2903/j.efsa.2004.239.

EFSA CONTAM Panel (2009). Scientific Opinion on cadmium in food. *EFSA Journal*.

EFSA Panel on Contaminants in the Food Chain (CONTAM) and EFSA Panel on Food Contact Materials, Enzymes, Flavourings and Processing Aids (CEF) (2010a). Scientific opinion on melamine in food and feed. *EFSA Journal*, 8(4), 1573.

EFSA CONTAM Panel (2010b). Scientific Opinion on Polybrominated diphenyl ethers (PBDEs) in food. *EFSA Journal*, 9(5), 2156. https://doi.org/10.2903/j.efsa.2011.2156.

EFSA CONTAM Panel (2010c). Scientific opinion on lead in food. *EFSA Journal*, 8(4), 1570. https://doi.org/10.2903/j.efsa.2010.1570.

EFSA CONTAM Panel (2011a). Scientific opinion on pyrrolizidine alkaloids in food and feed. *EFSA Journal*, 9(11), 2406. https://doi.org/10.2903/j.efsa.2011.2406.

EFSA CONTAM Panel (2011b). Scientific opinion on the risks for animal and public health related to the presence of T-2 and HT-2 toxin in food and feed. *EFSA Journal*, 9(12), 2481. https://doi.org/10.2903/j.efsa.2011.2481.

EFSA CONTAM Panel (2012). Scientific opinion on the risk for public health related to the presence of mercury and methylmercury in food. *EFSA Journal*, 10(12), 2985. https://doi.org/10.2903/j.efsa.2012.2985.

EFSA CONTAM Panel (2013). Scientific Opinion on risks for animal and public health related to the presence of nivalenol in food and feed. *EFSA Journal*, 11(6), 3262. https://doi.org/10.2903/j.efsa.2013.3262.

EFSA CONTAM Panel (2018). Risk for animal and human health related to the presence of dioxins and dioxin-like PCBs in feed and food. *EFSA Journal*, 16(11), 5333. https://doi.org/10.2903/j.efsa.2018.5333.

EFSA CONTAM Panel (2020). Risk to human health related to the presence of perfluoroalkyl substances in food. *EFSA Journal*, 18(9), 6223. https://doi.org/10.2903/j.efsa.2020.6223.

EFSA (2024). EFSA Panel on Contaminants in the Food Chain (CONTAM): Dieter Schrenk, et al. (2024) Guidance for the assessment of detoxification processes in feed. *EFSA Journal*, 22, e8528. https://doi.org/10.2903/j.efsa.2024.8528.

EMA (CVMP) Summary reports of residue depletion studies for individual veterinary medicinal products. Available from: https://www.ema.europa.eu/en/veterinary-regulatory-overview/research-development-veterinary-medicines/maximum-residue-limits-mrl/maximum-residue-limit-assessment-reports (Accessed 12 January 2026).

European Commission (2020). *Commission Recommendation (EU) 2022/1431 on the monitoring of perfluoroalkyl substances in food.* Available from: https://eur-lex.europa.eu/legal-content/EN/TXT/PDF/?uri=CELEX:32022H1431. (Accessed 12 January 2026).

Fernandes Alwyn, R., et al. (2023). The transfer of environmental contaminants (Brominated and Chlorinated dioxins and biphenyls, PBDEs, HBCDDs, PCNs and PFAS) from recycled materials used for bedding to the eggs and tissues of chickens. *Science of the Total Environment*, 892, 164441.

FAO/WHO Codex Alimentarius (2004). Codex alimentarius code of practice on good animal feeding - CAC/RCP 54–2004. Available from: https://www.fao.org/feed-safety/resources/resources-details/en/c/1054052/.

Fries, G. F. (1996). A review of the significance of animal food products as potential pathways of human exposures to dioxins. *Journal of Animal Science*, 73(6), 1639–1650. https://doi.org/10.2527/1995.7361639x.

Goff, J. P. (2018). Invited review: mineral absorption mechanisms, mineral interactions that affect acid–base and antioxidant status, and diet considerations to improve mineral status. *Journal of Dairy Science*, 101(4), 2763–2813. https://doi.org/10.3168/jds.2017-13112.

Gossner, C. M.-E., et al. (2009). The Melamine incident: implications for international food and feed safety. *Environmental Health Perspectives*, 117(12), 1803–1808. https://doi.org/10.1289/ehp.0900949.

Hargin, K. D. and Shears, G. (2013). Regulatory Control and monitoring of heavy metals and trace elements in foods. In: Rose, M. and Fernandes, A. (Eds.), *Chapter 2 in Persistent organic pollutants and toxic metals in foods*. London: Woodhead Publishing.

Heres, L., et al. (2010). Tracing and analytical results of the dioxin contamination incident in 2008 originating from the Republic of Ireland. *Food Additives and Contaminants, Part A*, 27(12), 1733–1744.

Hoogenboom, L. A. P., et al. (2006). Carry-over of dioxins and PCBs from feed and soil to eggs at low contamination levels influence of mycotoxin binders on the carry-over from feed to eggs. *Food Additives and Contaminants*, 23, 518–527. https://doi.org/10.1080/02652030500512037.

Hoogenboom, L. A. P., et al. (2015). Accumulation of polychlorinated dibenzo-p-dioxins, dibenzofurans, and biphenyls in livers of young sheep. *Chemosphere*, 122, 137–144. https://doi.org/10.1016/j.chemosphere.2014.11.030.

Hoogenboom, R., et al. (2015). European developments following incidents with dioxins and PCBs in the food and feed chain. *Food Control*, 50, 670–683. https://doi.org/10.1080/19440049.2010.522598.

IAEA (2010). *Handbook of parameter values for the prediction of radionuclide transfer in terrestrial and freshwater environments.* Technical Reports Series No. 472. Available from: https://www-pub.iaea.org/MTCD/Publications/PDF/trs472_web.pdf (Accessed 12 January 2026).

JECFA (2011). *Evaluation of certain contaminants in food: seventy-third report of the Joint FAO/WHO Expert Committee on Food Additives (JECFA).* WHO Technical Report Series. Available from: https://www.who.int/publications/i/item/9789241209601 (Accessed 12 January 2026).

JECFA (2018). *Safety evaluation of certain contaminants in food – WHO FOOD ADDITIVES SERIES: 74 FAO JECFA Monographs 19 bis Prepared by the eighty-third meeting of the Joint FAO/WHO Expert Committee on Food Additives Aflatoxins.* WHO Food Additives Series. Available from: https://iris.who.int/bitstream/handle/10665/276868/9789241660747-eng.pdf?ua=1. (Accessed 12 January 2026).

JECFA (various years). *Residues of veterinary drugs in food: FAO/WHO JECFA monographs* – e.g. for tylosin, tetracyclines, macrolides, avermectins. Available from https://www.fao.org/food/food-safety-quality/scientific-advice/jecfa/jecfa-vetdrugs/en/ (Accessed 12 January 2026).

Kocyigit, E, et al. (2023). Plant toxic proteins: their biological activities, mechanism of action and removal strategies. *Toxins (Basel)*, 15(6), 356. https://doi.org/10.3390/toxins15060356.

Leeman, W. R., et al. (2007). Transfer of chemicals from feed to animal products: the use of transfer factors in risk assessment. *Food Additives and Contaminants - Nutrition*, 24(1), 1–13. https://doi.org/10.1080/02652030600815512.

Lorber, M. (2002). A pharmacokinetic model for estimating exposure of Americans to dioxin-like compounds in the past, present, and future. *Science of the Total Environment*, 288(1–2), 81–95. https://doi.org/10.1016/S0048-9697(01)01119-6.

Malisch, R. (2000). Increase of the PCDD/F-contamination of milk, butter and meat samples by use of contaminated citrus pulp. *Chemosphere*, 40(9–11), 1041–1053. https://doi.org/10.1016/S0045-6535(99)00352-5.

Malisch, R. and Kotz, A. (2014). Dioxins and PCBs in feed and food-Review from European perspective. *Science of the Total Environment*, 491–492, 2–10. https://doi.org/10.1016/j.scitotenv.2014.03.022.

Mantovani, A. (2012). Chapter 18 - Chemical risk assessment of animal feed. In: Fink-Gremmels, J. (Ed.), *Animal feed contamination* (pp. 449–464). London: Woodhead Publishing Series in Food Science, Technology and Nutrition. https://doi.org/10.1533/9780857093615.4.449.

Marnane, I. (2012). Comprehensive environmental review following the pork PCB/dioxin contamination incident in Ireland. *Journal of Dynamic Environmental Monitoring*, 14, 2551–2556. https://doi.org/10.1039/C2EM30374D.

Ravindran, V. (2014). Nutrition of meat animals; Poultry. *Encyclopedia of Meat Sciences (Second Edition)*, 2014, 463–470. https://doi.org/10.1016/B978-0-12-384731-7.00024-6.

Rigby, H., et al. (2023). Transfer of polychlorinated, polybrominated and mixed-halogenated dioxins, furans and biphenyls, polychlorinated naphthalenes and alkanes, polycyclic aromatic hydrocarbons and chlorobenzenes to the milk of dairy cattle from controlled ingestion of industrial and municipal bioresources recycled to agricultural land. *Science of the Total Environment*, 886, 163546. https://doi.org/10.1016/j.scitotenv.2023.163546.

Rose, M., et al. (2010). Dietary exposure to metals and other elements in the 2006 UK total diet study and some trends over the last 30 years. *Food Additives and Contaminants*, 27(10), 1380–1404. https://doi.org/10.1080/19440049.2010.496794.

Rose, M. (2012). Chapter 7: hazardous chemicals as animal feed contaminants and methods for their detection. In: Fink-Gremmels, J. (Ed.), *Animal feed contamination: effects on livestock and food safety* (pp. 117–130). The Netherlands: Utrecht University.

Rose, M. (2023). Chapter 8. Pollutants, residues and other contaminants in foods obtained from marine and fresh water. In: Knowles, M., et al. (Eds.), *Present knowledge in food safety: a risk-based approach through the food chain 1st edition - October 3, 2022*. London: Elsevier.

Rose, M., et al. (2012). Transfer and uptake of polychlorinated dibenzo-p-dioxins and furans (PCDD/Fs) and polychlorinated biphenyls (PCBs) into meat and organs from indoor and outdoor reared pigs. *Food Additives and Contaminants*, 29(3), 431–448. https://doi.org/10.1080/19440049.2011.635346.

Tlustos, C., et al. (2012). The dioxin contamination incident in Ireland, 2008: analytical results and congener patterns. *Food Additives and Contaminants, Part A*, 29, 128–138. https://doi.org/10.1080/19440049.2011.615030.

Traag, W. A., et al. (2006). Residues of dioxins and PCBs in eggs, fat and livers of laying hens following consumption of contaminated feed. *Chemosphere*, 65(9), 1518–1525. https://doi.org/10.1016/j.chemosphere.2006.04.001.

Tretola, M., et al. (2025). Review: nutritional, safety, and environmental aspects of former foodstuff products in ruminant feeding. *Animal*, 19(2), 101512. https://doi.org/10.1016/j.animal.2025.101512.

Urugo, M. M. and Tringo, T. T. (2023). Naturally occurring plant food toxicants and the role of food processing methods in their detoxification. *International Journal of Food Science*, 2023, 9947841. https://doi.org/10.1155/2023/9947841.

Van Eijkeren, J. C. H., et al. (2006). A simple steady-state model for carry-over of aflatoxins from feed to cow's milk. *Food Additives and Contaminants*, 23(8), 833–838. https://doi.org/10.1080/02652030600779890.

Van Larebeke, N., et al. (2001). The Belgian PCB and dioxin incident of January-June 1999: exposure data and potential impact on health. *Environmental Health Perspectives*, 109(3), 265–273. https://doi.org/10.1289/ehp.01109265.

Veldman, A, et al. (1992). Carry-over of aflatoxin from cows' food to milk. *Animal Science*, 55(2), 163–168. https://doi.org/10.1017/S0003356100037417.

Verdegem, M., et al. (2023). The contribution of aquaculture systems to global aquaculture production. *Journal of the World Aquaculture Society*, 54, 206–250. https://doi.org/10.1111/jwas.12963.

Yang, R., et al. (2009). Milk adulteration with melamine in China: crisis and response. *Quality Assurance and Safety of Crops & Foods*, 1(2), 111–116. https://doi.org/10.1111/j.1757-837X.2009.00018.x.

Zhou, Y., et al. (2024). One Health governance: theory, practice and ethics. *Science in One Health*, 3, 100089. https://doi.org/10.1016/j.soh.2024.100089.

Part 3

Technologies to improve batch control

Chapter 7

Molecular techniques in identifying microorganisms in feed: microbiome analyses

Steven C. Ricke and Elena G. Olson, University of Wisconsin-Madison, USA

1 Introduction

Pathogen contaminants, such as *Salmonella,* have received the most attention in feed manufacturing and the food animal industry (Maciorowski et al., 2004; Obe et al., 2023). Indeed, the presence of such pathogens, albeit likely at low levels, is a critical concern. However, animal feeds and pet foods contain a highly diverse microbiome, including bacteria, fungi, and bacteriophages, with many of these organisms not necessarily being pathogenic to animals (Maciorowski et al., 2001a; Oguz, 2011; Kazimierska et al., 2021; Olson et al., 2022, 2023). This diversity is to be expected, given the multiple potential sources through which animal feeds can encounter microorganisms during production, feed milling, distribution, and animal feeding (Maciorowski et al., 2007; Ricke, 2018; Olson et al., 2023). Microbial exposure of cereal grains used in feed manufacturing may originate as early as crop production when soil and cereal grains come into contact with microorganisms during the growth of agronomic crops (Ricke, 2018). It is well known that the soil microbial community is complex, making it a potential source of diverse microbial populations that may influence feed hygiene (Delgado Baquerizo et al., 2018;

http://dx.doi.org/10.19103/AS.2026.0172.25

Ricke, 2018). This diversity is potentially dependent on several conditions, such as soil fertilizer management, routine use of pesticides and herbicides, environmental changes, and organic versus non-organic tillage practices, among other factors (Azarbad, 2022). Land application of livestock waste can also cause shifts in the soil microbial ecology and introduce microbial antibiotic resistance (Crippen et al., 2019; Yang et al., 2019; Gurmessa et al., 2021; Seyoum et al., 2023). Airborne transmission represents another potential source of microbial contamination, not only during the growing season but also at harvest and during further processing (Pillai and Ricke 2002). Similarly, water used in agricultural applications may serve as a source of microbial contamination, depending on its origin and environmental exposure (Faour-Klingbeil and Todd, 2018; Ricke, 2018).

An overview of the major sources of microbial exposure in animal feeds, along with the application of 16S rRNA sequencing for characterizing microbial communities, is illustrated in Fig. 1. Once cereal grains and other feed ingredients arrive at the feed mill, their microbial diversity can be further influenced by a range of intrinsic and extrinsic factors, potentially altering the microbial composition of the final feed product. Extrinsic factors include feed milling processes, such as thermal treatments (e.g. pelleting) and the addition of antimicrobial or antifungal compounds for microbial control (Wales et al., 2010; Jones, 2011; Dittoe et al., 2018; Ricke, 2003, 2018; Ricke et al., 2019, 2020). Such exposures could certainly alter feed composition by selecting organisms resistant to these external stressors. Environmental exposure within the feed mill, including contact with equipment, the surrounding environment, and personnel, likely contributes to the microbial

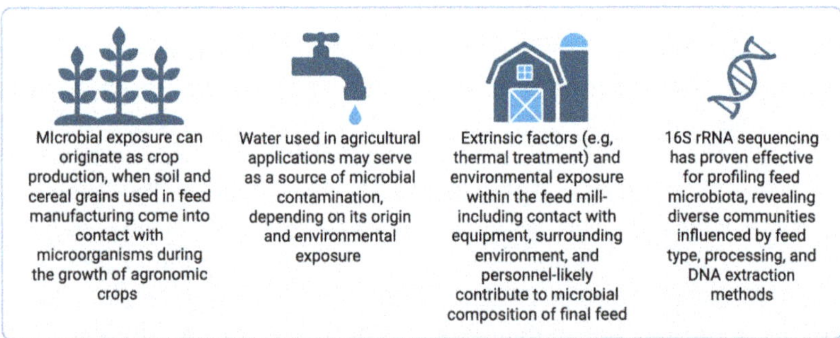

Figure 1 Illustration of microbial exposure routes in animal feeds and the role of 16S rRNA gene sequencing. Microbial contamination can originate from soil, water, and environmental factors during feed processing. 16S rRNA sequencing enables detailed profiling of feed microbial communities and reveals how feed type, processing, and extraction methods influence microbial diversity. Created in BioRender. Olson, E. (2025) https://BioRender.com/npyxtoo.

composition of the final feed (Olson et al., 2023). Consequently, sanitation and biosecurity practices at the feed mill could impact the extent of this microbial contamination (Olson et al., 2023). The level of contamination originating from ingredients being incorporated into the feed mixture during milling adds another potential layer of microbial diversity in the final feed product (Olson et al., 2023). Given the complexity of feed microbial composition and its potential impact on feed quality and animal health, it is increasingly important to develop a comprehensive understanding of microbial communities present throughout the various stages of feed processing and milling. Molecular sequencing approaches targeting variable regions of the 16S rRNA gene – particularly the V3–V4 or V4 regions – have emerged as powerful tools for taxonomic identification of resident microbial populations on foods and are increasingly being applied to assess microbial ecology in other areas of animal agriculture, including feeds (Ricke et al., 2017; Olson et al., 2022; Weinroth et al., 2022; Lyte et al., 2025). This review discusses previous attempts to characterize the feed microbial ecology by cultural methods, followed by an examination of the earlier molecular approaches. This will lead to a discussion on 16S rRNA gene sequencing as a tool to enhance understanding of the microbial ecology of animal feeds, highlighting recent advancements across various feed types.

2 Culture-based assessment of animal feed microorganisms

Historically, assessments of the microbial composition of animal feeds have been conducted using a variety of nonselective and selective media. Not surprisingly, many of the selective media-based studies focused on identifying and enumerating pathogens, such as *Salmonella* (Ricke et al., 1998; Maciorowski et al., 2005, 2006; Feye et al., 2021). Similar to isolation methods for food samples, cultural approaches for *Salmonella* in animal feeds typically involve a nonselective enrichment step followed by a selective enrichment medium that contains elements exclusively associated with the organism's metabolism (Maciorowski et al., 2006; Ricke et al., 2018). Culture methods can be followed up with non-culture-based confirmation, such as serotyping through specific immunological or molecular-based methods (Maciorowski et al., 2005, 2006). Recovering *Salmonella* from complex matrices, such as animal feeds, presents challenges that can compromise the accuracy of culture-based methods; however, these limitations can be addressed through procedural adjustments (Feye et al., 2021). While less extensively studied, similar culture-based approaches can be adapted for detecting other pathogens of interest in animal feeds.

While pathogen isolation and identification methodologies are generally well-defined and can be applied to feeds, the culture-based characterization of nonpathogenic microbial populations in feeds has been less developed from a methodological standpoint. Early research on microbial contamination of feeds utilized standard culture media similar to those applied in non-feed applications, such as microbial assessment of cereal grain quality during storage (Nichols and Leaver, 1966; Cuero et al., 1986; Hanis et al., 1988). Most attempts at assessing microbial quality in feeds applied combinations of nonselective media to determine total bacteria, along with some form of selective media for targeted organisms of interest. For example, Furuta et al. (1980) recovered viable bacterial counts from chicken feed and ingredients by employing heart infusion agar for enumerating all bacteria and deoxycholate-hydrogen sulfide-lactose agar for estimating enteric bacterial populations. Tabib et al. (1981) assessed the microbial quality of broiler, layer, and turkey feeds, as well as feed ingredients, using commercial selective media for coliforms and yeast/molds after diluting samples in Triton X-100 to maximize microbial recovery. Similarly, Cox et al. (1986) used total plate count media, along with Enterobacteriaceae- and *Salmonella*-selective media, to evaluate the effects of steam conditioning and pelleting on commercial poultry feeds.

The use of general commercial nonselective media and selective media has provided some insights into feed microbial ecology. More recently, efforts have been made to refine media and sampling methodologies to accurately represent the types of organisms present in animal and poultry feeds. One issue that plagued the bacterial culturing of animal and poultry feeds was the prevalence of indigenous fungi, which precluded differentiation between bacterial and fungal populations. Grains and animal feeds are well-established reservoirs of a wide variety of fungi, some of which can produce mycotoxins that are deleterious to animal health when consumed (Christensen, 1957; Moreno Romo et al., 1986; Dänicke, 2002; Oguz, 2011; Chavez et al., 2020; Wang et al., 2021). The presence of feed fungi can also lead to overgrowth on both nonselective and selective bacterial plates, interfering with accurate enumeration (Ha et al., 1995a). Consequently, when culturing and isolating organisms from animal feeds, it was critical to utilize media that could selectively inhibit fungal growth on bacterial plates.

Ha et al. (1995a) incorporated the antifungal compound cycloheximide as a selective agent into plate count agar (PCA) to recover animal feed bacteria and inhibit the spread of feed fungi. There was precedent for this as cycloheximide had been previously used in brewery, water, and soil samples as well as for moist grains (Green and Gray, 1950; Nichols and Leaver, 1966; Kozulis and Page, 1968; Newell et al., 1983; Pillai and Pepper, 1991). The authors demonstrated that cycloheximide did not inhibit the growth of pure cultures of *Escherichia coli*, *Listeria monocytogenes*, or *Pseudomonas fluorescens* at any of the tested concentrations. However, high levels (1,000 mg/L) inhibited *Bacillus cereus*, and 600 mg/L suppressed *Staphylococcus aureus* in tryptic soy broth. Indigenous microorganisms were enumerated from feed samples stored short-term (7 days) and long-term (42 days) using PCA plates amended with either 40 ppm chlorotetracycline to inhibit bacteria or varying concentrations of cycloheximide (10 to 300 mg/L) to suppress fungi. Across all samples and storage durations, 300 mg/L of cycloheximide effectively inhibited fungal growth without affecting the recovery of aerobic bacteria. In a follow-up study, Ha et al. (1995b) demonstrated that cycloheximide could also be used as an antifungal agent without impacting *Salmonella* growth. In later studies, Ha et al. (1998a, 1998b) successively used cycloheximide in plating media to enumerate indigenous feed bacteria and *Salmonella* Typhimurium inoculated in feeds containing different protein sources over a 56-day storage period and in feeds containing the antimicrobial buffered propionic acid.

Recovering representative microbial populations from feed matrices also presents challenges for culture-based enumeration. Maciorowski et al. (2002a) noted that the likely stressed nature of some animal feed microorganisms, as well as the differences among various feed matrices, could impact the ability to sufficiently recover viable organisms in diverse physiological states of growth and/or survival. Stressors in feed can originate from feed antimicrobial amendments or physical treatment applications, such as heat, leading to injured and, in turn, difficult-to-recover viable organisms. Maciorowski et al. (2002a) hypothesized that low-nutrient-based agar could be more suitable for recovering organisms subjected to more stressful environmental conditions, such as those found in animal feeds. To test this, they compared tryptic soy agar (TSA) with a low-nutrient plate agar medium (R2A) that had previously been developed for water quality appraisal (Reasoner and Geldreich, 1985; Horgan et al., 1999; Maciorowski et al., 2002a). At the same time, they detected slight differences between the two media. The enumerated populations correlated with each other, indicating that R2A could be used as a substitute medium for TSA. However, they also noted that feed microbiota that are experiencing more severe environmental stresses could accentuate the differences between the two media. In a follow-up study, Maciorowski et al. (2002b) observed a higher recovery of organisms (half a logarithm higher for TSA and one logarithm higher for R2A) from various animal feeds stored at −20°C for 14 months versus their fresh feed counterparts.

Considerable effort has been made to reliably estimate the metabolic products of some feed organisms, such as fungi, given their importance in producing harmful mycotoxins (Chavez et al., 2020). Efforts to quantify mycotoxins directly in foods and feeds have led to the development of a multitude of methodologies, including liquid chromatography, fluorescence imaging, and reflectance imaging (Chavez et al., 2020). For animal and poultry feed bacterial populations, there has been less emphasis on aspects beyond overall microbial quality and the presence of pathogens, which have garnered little-to-no extended research and/or commercial interest. Indeed, the presence of *Salmonella* in feeds has been, and continues to be, a major focal point for animal production systems, such as poultry. Therefore, the development and refinement of methods for detecting and quantifying *Salmonella* in feeds and feed components (Feye et al., 2021; Obe et al., 2023) are crucial. Nonpathogenic feed bacterial populations have received relatively little attention, aside from occasional reports of indicator organisms or overall bacterial counts, as part of assessing bacterial loads and microbial quality in feeds. Generally, standard selective and nonselective media appear to be sufficient for determining the microbial quality of feeds. However, given the complexity and diversity of microbial populations present in feeds and feed ingredients, non-culture-based methods are likely to be more informative for identifying feed microbial communities in their entirety.

3 Initial molecular approaches for assessing feed microbial quality

Molecular-based methods have been applied to animal and poultry feeds for pathogen detection and quantification for several years. Most of this focus, not surprisingly, has been on *Salmonella*. Early work focused on developing and adapting polymerase chain reaction (PCR) assays for the identification of *Salmonella*, with further efforts to quantify individual *Salmonella* genes in feeds using quantitative PCR (qPCR) (Maciorowski et al., 2005; Park et al., 2011; Andino et al., 2014). This technique has proven beneficial not only for detecting and quantifying *Salmonella* but also for assessing the physiological and virulence responses of *Salmonella* when exposed to thermal stress and the ability of this pathogen to survive in feeds. More recently, the use of clustered regularly interspaced short palindromic repeats (CRISPR) typing has been employed to serotype *Salmonella* serovars for animal feed mills (Shariat et al., 2021). Advancements in CRISPR and whole-genome sequencing technologies are enhancing the resolution for distinguishing *Salmonella* serovars and strains while also enabling precise tracking of specific *Salmonella* lineages throughout animal production systems, including feed processing, distribution, and on-farm delivery.

Detection of nonpathogenic bacteria in feeds using PCR approaches has also been suggested. However, unlike the specificity for a particular pathogen, such as *Salmonella*, this requires developing a PCR assay that can detect a relatively broad range of organisms likely to be present in animal feeds. More specifically, this requires primers that can target DNA sequences that represent most of the microorganisms presumably associated with animal feeds and feed ingredients. Based on early work characterizing the molecular-based evolutionary relationships among ribosomal genes of microbial groups, these regions of the genomes proved ideal for identifying and distinguishing different organisms (Woese and Fox, 1977; Pace et al., 1985; Schloss et al., 2009; Escobar-Zapeda et al., 2015). Consequently, the 16S rRNA gene – being a highly conserved region of the microbial genome – provides a reliable target for PCR-based differentiation of organisms, as it is well characterized across a broad range of microbial taxa (Olsen et al., 1986; Amann et al., 1995,1997; Raghava et al., 2000; Theron and Cloete, 2000).

However, consistently recovering DNA from feed matrices for molecular assays such as PCR remains challenging due to their complexity (Maciorowski et al., 2005). Maciorowski et al. (2001b) compared two extraction approaches: an indirect method involving the detachment of bacterial cells from feed particles followed by cell lysis and a direct method aimed at releasing DNA from bacteria attached to the feed matrix. For initial removal of bacteria cells from feed particles, feed samples were diluted in a detergent-based solution (Triton X-100) that had been used previously to recover maximum levels of bacteria and fungi for plate enumeration (Stewart et al., 1977; Tabib et al. 1981). For direct extraction of microbial DNA from feed, two methods were evaluated: a rapid one-day protocol using an EDTA-NaCl solution with sonication in the presence of lysozyme followed by isopropanol precipitation (Porteous et al., 1994; Maciorowski et al., 2001b), and a longer overnight procedure incorporating lysis buffer and isopropanol with an additional precipitation step using 8000 molecular weight polyethylene glycol (PEG) (Widmer et al., 1996). When combined with a detergent step, both extraction methods yielded consistent DNA recovery from animal feeds, as confirmed by visible bands on gel electrophoresis. However, PCR amplification of the small subunit ribosomal gene was inconsistent, with some poultry feed samples failing to produce detectable DNA, suggesting the presence of PCR inhibitors in feed matrices (Maciorowski et al., 2002c). In a follow-up study, Maciorowski et al. (2002d) evaluated seven animal feeds and two feed ingredients – both fresh and stored for 14 months at −20°C – using multiple extraction methods. They concluded that the PEG-based method consistently produced at least one positive result in most samples.

In summary, these early efforts using molecular methods with conserved gene targets suggested that they may have some utility for determining bacterial feed quality. However, some limitations were also noted. Feed matrices posed challenges for consistent DNA recovery, especially when using protocols initially developed for other environmental samples, such as soil. More recently, commercial DNA extraction kits have shown promise in improving both the consistency and yield of DNA recovery from feeds (Piskata et al., 2019). In addition, targeting the 16S rRNA gene via PCR can reveal the presence of bacterial DNA that is presumably representative of the entire microbial community. Still, it offers no taxonomic information on the individual members of that microbial community. Development of denatured gradient gel electrophoresis that relies on gradient gel separation of different sizes of PCR-amplified 16S rRNA gene products that represent unique sequences of various organisms has been used as a means to characterize shifts in microbial populations (Ercolini, 2004; Hanning and Ricke, 2011). This method distinguishes microbial communities by differences in banding patterns, and individual bands can be excised from the gel and sequenced for taxonomic identification. However, because it relies on the visual interpretation of gel bands, the approach introduces a degree of imprecision and provides only a partial representation of the overall microbial community.

4 Microbiome analyses: definitions and general concepts

4.1 Introduction

Advances in sequencing technologies have ushered in a new era of molecular characterization of microorganisms. Building on earlier approaches that targeted conserved ribosomal genes to differentiate organisms, sequencing of 16S rRNA genes has emerged as a powerful tool for generating detailed taxonomic profiles of mixed microbial communities in food production systems (Ricke et al., 2017). Microbiome sequencing technologies have been rapidly adopted across multiple sectors of food animal production, from farm environments to meat processing. These advancements have been thoroughly reviewed elsewhere; therefore, only the general concepts will be covered in this review (Dittoe et al., 2022; Ricke et al., 2017, 2022a, 2022b; Weinroth et al., 2022; Lyte et al., 2025; Olson et al., 2025).

The utility of microbiome sequencing continues to grow as more studies are conducted and reported. In pre-harvest poultry production, sequencing of the gastrointestinal tract (GIT) microbiota has offered more profound insights into the interactions between microbial consortia and the host, revealing potential links to animal performance and overall well-being in both broilers and laying hens (Dittoe et al., 2022; Ricke et al., 2022b). Similarly, microbiome analyses have been developed and applied to poultry processing plants to identify changes in the microbial ecology of chicken carcasses as they progress through the various stages of processing, experiencing temperature fluctuations and exposure to antimicrobials (Ricke et al., 2022a; Brown et al., 2025). However, as microbiome datasets grow in complexity, interpreting and effectively communicating the resulting bioinformatic information presents a significant challenge for practical application by non-specialist audiences (Thompson et al., 2017; Feye et al., 2020; Ricke et al., 2022c).

4.2 Microbiome 16S rDNA sequencing technologies - general concepts

Sequencing microbial communities using 16S rRNA genes involves several key steps. These steps include the effective recovery of DNA representative of the microbial population, the selection of the appropriate variable region of the 16S rRNA gene for sequencing, and subsequent data analysis for taxonomic identification and diversity assessment. Multiple commercial kits for DNA extraction are available; however, the choice depends on the type of sample and sample source. Sequencing via the 16S rRNA gene relies on a component of the 30S small subunit of the ribosome, which is present in all bacteria (Amato, 2017). This gene contains nine variable regions (V1 through V9) that have been identified as being unique to specific bacterial groups and can serve as the basis for taxonomic classification (Robinson et al., 2016). This gene region consists of approximately 1500 base pairs; however, the more commonly used commercial sequencing technologies, such as the Illumina MiSeq, are capable of sequencing only shorter frames of 100–350 base pairs of DNA (Robinson et al., 2016; Amato, 2017).

Because the V1–V9 regions of the 16S rRNA gene vary in their ability to detect specific taxa, the choice of the target region for short-read sequencing depends on the microbial groups of interest and the sample source (Yu and Morrison, 2004; Comeau et al., 2017; Ricke et al., 2017). Sunthornthummas et al. (2025) emphasize the importance of careful primer selection and have proposed that a multi-primer approach may provide more representative coverage and reduce bias in microbiome sequencing. Although short-read sequencing has long been the standard for microbiome studies, the growing availability of long-read platforms is making it increasingly feasible to perform full-length 16S rRNA gene sequencing – from V1 to V9 – for more comprehensive taxonomic identification (Lyte et al., 2025; Olson et al., 2025). These advancements are expected to reduce sequencing costs, which will likely promote broader use of microbiome analysis in materials such as feed samples and support the development of standardized protocols for sample collection and processing. As sequencing becomes more widespread, it will also contribute to the expansion of microbiome databases across a broader range of animal feeds and feed ingredients.

4.3 Bioinformatics

After sequencing, raw data are processed for taxonomic identification by clustering sequences into operational taxonomic units (OTUs), which are typically defined by 95%–97% similarity. These OTUs can then be classified into taxonomic categories such as phylum, class, order, family, genus, and species using established databases (DeSantis et al., 2006; Schloss, 2010; Kozich et al., 2013; Jovel et al., 2016; Robinson et al., 2016; Amato, 2017). More recently, amplicon sequence variants (ASVs) are being substituted for OTUs to allow for the distinction between actual biological variation and sequencing errors (Callahan et al., 2017; Weinroth et al., 2022). The computational processing of raw sequence data into OTUs or ASVs is known as bioinformatics, enabling the conversion of sequence data into digital formats for taxonomic identification and the statistical comparison of microbial communities across treatments (Luscombe et al., 2001; Ricke et al., 2017). Once taxonomic classification is complete, additional statistical analyses can be performed to evaluate microbial diversity across samples or treatment conditions (Gensollen and Blumberg, 2017). Alpha diversity metrics describe the level of microbial taxonomic diversity within a sample or single microbial community (Ricke et al., 2017). Conversely, beta diversity analyses compare microbial communities between samples by identifying taxa that are shared or unique across different treatments or conditions (Robinson et al., 2016).

Customized software programs or analysis pipelines – such as mothur and Quantitative Insights into Microbial Ecology (QIIME2) – are commonly used to assemble, organize, and analyze sequence data for diversity comparisons and to generate visual representations of microbial community structures (Schloss et al., 2009; Schloss, 2010; Caporaso et al., 2010; Gonzalez and Knight, 2012; Huse et al., 2014; Nilakanta et al., 2014; Bolyen et al., 2019). In addition to visual representations, these pipelines support statistical analyses of diversity metrics. Alpha diversity effects are typically analyzed using non-parametric analysis of variance (ANOVA) with pairwise comparisons, as described by Kruskal and Wallis (1952). Beta diversity is commonly assessed using permutational multivariate analysis of variance (PERMANOVA) to evaluate overall differences among groups (Anderson, 2017). Taxonomic analysis can employ machine learning algorithms to classify ASVs against reference databases, as implemented in QIIME2, enabling identification at the family and genus levels and facilitating differential abundance analysis (Weinroth et al., 2022). It is essential to note that differential abundance is relative – taxa are measured in proportion to one another within the sample, so their values collectively sum to 100% (Mandal et al., 2015). Analysis of Composition of Microbiomes (ANCOM), addresses this compositional nature by using log-ratio transformations and controls for false discovery rates in differential abundance testing (Mandal et al., 2015).

As sequencing technologies advance, they are expected to provide more comprehensive data on both the taxonomy and diversity of microbial communities for bioinformatic analysis. These improvements will likely coincide with enhancements and expanded features in the computational pipelines used to process raw sequence data. Together, these developments are expected to facilitate a more nuanced understanding of microbial communities and their ecological interactions. When combined with advances in non-microbiome approaches such as metabolomics, these tools can offer deeper insights into microbial interactions and the functional roles of individual community members (Chatman et al., 2024). However, effectively applying and interpreting these complex datasets will require expanded educational programs and training to support practical understanding and use (Thompson et al., 2017; Feye et al., 2020; Olson et al., 2022; Ricke et al., 2022c).

5 Microbiome applications for animal and poultry feeds

5.1 Introduction

The application of microbiome sequencing technologies has significantly expanded in livestock research over the past few years (Weinroth et al., 2022; Lyte et al., 2025). However, the use of microbiome analyses in animal feeds has remained relatively limited to date (Olson et al., 2022; Ge et al., 2025). This limitation likely stems from a low interest in nonpathogenic feed microbiota and uncertainty regarding the practical relevance of such data for producers. As more microbiome data emerge from live animal production studies, this perspective may begin to shift. The role of gastrointestinal microbiota in gut ecosystem maturation and overall animal health is becoming better understood, and thus it is increasingly important to identify other factors that contribute to this microbial development. One of those factors could be the feed and its indigenous microbiota. There is precedent for this, as *Salmonella*-contaminated feed can introduce this pathogen into the GIT of susceptible animals, indicating that feed may also serve as a vehicle for introducing other microorganisms, including nonpathogens – particularly in young animals with immature gut microbiota. Consequently, developing a microbiome database of feed microbiomes may have some utility for understanding their contribution to the maturation of the gastrointestinal microbial ecology (Olson et al., 2022). This understanding may influence animal nutrition and performance in ways that extend beyond the feed's chemical composition.

5.2 Molecular identification of organisms in feeds and feed ingredients

Early molecular identification of feed microorganisms often combined traditional culturing methods with sequencing of isolated colonies. For example, Olson et al. (2020) investigated the microbiota of corn-based commercial poultry feed by isolating morphologically distinct colonies on aerobic agar. Eleven distinct colonies were selected for further characterization. DNA was extracted from each culture, and paired-end libraries were prepared and sequenced using an Illumina HiSeq platform (Illumina, San Diego, CA). Whole genome sequencing identified the isolates as belonging to several genera, including *Klebsiella*, *Kosakonia*, *Pantoea*, *Stenotrophomonas*, and *Enterococcus*.

Building on their whole-genome sequencing work, Olson et al. (2021) evaluated whether 16S rRNA gene sequencing could serve as a more rapid method to approximate the microbial composition of commercial poultry feeds. DNA was extracted from morphologically distinct colonies, and a library targeting the V4 region of the 16S rRNA gene was constructed. Sequencing was conducted using an Illumina MiSeq platform (San Diego, CA), and taxonomic profiles and diversity metrics were generated using the QIIME2-2020.2 pipeline (Boylen et al., 2019; Olson et al., 2021). Based on V4 region short-read sequencing, 25 distinct genera were detected from 11 morphologically distinct colonies, including 13 Gram-positive and 12 Gram-negative genera, suggesting taxonomic heterogeneity within a colony. Taxonomic heterogeneity has been noted previously for both nonselective and selective plate media when colonies are pooled (Kim et al., 2017; Bodie et al., 2025). Taxonomic differences were also observed between the two studies. Olson et al. (2021) suggested that observed differences could be attributed to factors such as variation in DNA extraction methods or a high degree of sequence homology. Additionally, the short-read sequencing targeted only the V4 region of the 16S rRNA gene, in contrast to the full-length sequencing used in the earlier study (Olson et al., 2020, 2021).

While short-read 16S rRNA gene sequencing typically resolves taxonomy at the genus level, Olson et al. (2020) demonstrated that long-read sequencing of the full-length 16S rRNA gene (approximately 1500 bp) and the translation initiation factor-2 gene (approximately 2750 bp) enabled species-level identification of feed isolates. Despite its lower taxonomic resolution, the organisms identified through short-read microbiome sequencing were consistent with those previously detected using culture-based methods. Depending on the research focus, once microbial prevalence patterns are detected in feed samples, whole-genome sequencing can be employed as a follow-up to identify the target organisms of interest with greater precision. This approach is beneficial for studies aiming to characterize how feed microbiota respond to specific treatments, potentially leading to the identification of specific indicator organisms for feed hygiene monitoring (Olson et al., 2021).

5.3 Microbiome analysis of feeds and feed ingredients - DNA extraction

A critical first step before conducting microbiome sequencing on samples such as feeds and feed ingredients is the effective recovery of microbial community DNA. DNA extraction from foods and feeds presents consistent challenges, whether the goal is to detect specific pathogens using PCR or to characterize microbial communities more broadly (Maciorowski et al., 2005; Piskata et al., 2019; Weinroth et al., 2022; Ge et al., 2025). Extracting DNA for microbiome analysis poses additional challenges, as it requires both a sufficient concentration and high-quality DNA suitable for amplicon-based sequencing. Weinroth et al. (2022) emphasized two key criteria when selecting a DNA extraction method: its effectiveness across diverse cell types and compatibility with various sample matrices. Common approaches to cell lysis include mechanical, physical, chemical, and enzymatic processes. However, achieving this requires a careful balance – ensuring efficient lysis and recovery of both Gram-positive and Gram-negative bacteria while minimizing DNA degradation (Lyte et al., 2025). For example, while mechanical treatment of samples, such as bead beating accompanied by chemical treatment, may be required to recover DNA from certain groups of bacteria, there is a risk of overexposure that can lead to DNA shearing (Lyte et al., 2025). Additionally, the presence of inhibitors and low cell biomass must be considered when developing a DNA extraction protocol (Weinroth et al., 2022).

To date, research on evaluating and optimizing DNA extraction methods for animal feed samples has been limited. Olson et al. (2024a) compared five different DNA extraction methods using a variety of commercial poultry and pelleted feed samples. Three of the methods were based on the Qiagen stool kit (Qiagen, Valencia, CA), each incorporating specific modifications: one used the standard kit protocol, another added Lysate Matrix B (MP Biomedicals, Solon, OH) with a bead-beating step, and a third replaced the spin filter columns with Celite. Two additional methods included a one-day direct extraction with bead beating and lysozyme, adapted from Porteous et al. (1994), and the PEG-based method adapted from Widmer et al. (1996), which used sonication instead of bead beating. After assessing DNA concentration and purity, all samples were stored at −20°C until sequencing. Sequencing was based on the V4 region of the 16S rRNA gene and analyzed on an Illumina MiSeq (San Diego, CA). Taxonomic alignment profiles, as well as alpha and beta diversity, were assessed using the QIIME2-2021.4 pipeline. Overall, Olson et al. (2024a) concluded that both microbial diversity and community composition were influenced by the DNA extraction method used. While modifications to the Qiagen kits improved DNA purification, combining lysozyme and bead beating with the one-day and PEG-based extraction protocols enhanced bacterial diversity and appeared to improve the recovery of the microbial community. These findings may be essential for capturing both over- and underrepresented organisms in feed samples. However, caution is warranted before adopting generalized protocols, as Olson et al. (2024a) emphasized the need to standardize extraction methods before implementing them routinely. The standardization will require evaluating a broader range of feed types and ingredients, as well as incorporating a defined internal control microbiota for meaningful comparisons.

5.4 Microbiome analysis of feeds and feed ingredient microbial populations

As microbiome analysis methods continue to evolve, they offer increasing potential for evaluating how environmental conditions affect feeds and feed ingredients by enabling comprehensive assessment of microbial community responses. For example, Solanki et al. (2019) used microbiome analyses to evaluate microbial community responses in stored wheat grains following phosphine fumigation. They targeted the 16S rRNA gene for bacteria and the internal transcribed spacer (ITS) region for fungi. Fumigation did not significantly affect fungal diversity or abundance across three sampling points over six months, with dominant genera such as *Alternaria* and *Stemphylium* consistently comprising more than 90% of the fungal community. However, the bacterial microbiome alpha and beta diversity differed markedly at the 6-month sample time point compared to the first week and the 3-month sample times, with the genera *Bacillus* and *Erwinia* becoming prominent at 6 months and several genera only being detected at the 6-month sample time point. When qPCR was used to quantify the significant pathogenic fungi, *Alternaria*, *Aspergillus*, and *Fusarium,* increases in *Fusarium* were detected at 6 months. All samples were positive for *Alternaria,* while *Aspergillus* only sporadically occurred. Samples also contained various mycotoxins associated with these genera. Taken collectively, the authors suggested that phosphine fumigation altered the microbial community in stored wheat grain, which could, in turn, impact functional diversity.

If storage and treatment conditions influence the microbiome of feeds and feed ingredients, it is also likely that ingredient sources and variations in feed nutrient composition contribute to the formation of distinct microbial communities. Olson et al. (2024b) compared nine different commercial poultry feeds with 16S rRNA microbiome sequencing. Feed sources included various combinations of grower, starter/grower, probiotic, or medicated feeds, as well as pelleted and crumbled diets. Over 40 distinct phyla and over 600 distinct genera were detected in all nine feeds. However, the poultry feeds combined with probiotics yielded the least even and rich microbial compositional profiles, as well as the least phylogenetic diversity, compared to the other six feed types. When core microbiome taxonomic profiles were assessed, several potential pathogens were detected, including *Campylobacter*, *Helicobacter*, and *Listeria*, along with numerous organisms that have been identified as commensal members of the poultry GIT, such as *Lactobacillus*, *Bacteroides*, *Lachnospiraceae*, and *Ruminococcus*, among others. Based on these findings, the authors concluded that microbiome analysis could serve as a functional indicator of feed microbial quality. They also noted that the inclusion of probiotics may influence the feed microbiota, which is essential when assessing overall feed integrity. Furthermore, the observed similarities between feed and chick gastrointestinal microbiomes suggest that the microbial composition of feed may play a role in shaping gut colonization during early development.

More recently, in-depth microbiome analyses have been conducted to determine the functionality of resident microbial populations in animals. Ge et al. (2025) employed microbiome and shotgun metagenomic sequencing to investigate the microbial composition and resistomes in animal feeds. They compared cattle feed, dry dog food, and poultry feed for sequencing, incorporating a commercial mock microbial community to optimize their protocols for DNA extraction. Additionally, they added peptide nucleic acid blockers to remove chloroplast and mitochondrial reads from the sequencing. However, when they compared the use of blockers versus post-sequencing *in silico* filtering, similar results were achieved for chloroplasts and mitochondria. While both 16S rRNA microbiome sequencing and shotgun metagenomics yielded comparable insights into overall microbial diversity, metagenomics provided a greater resolution at the species level. Dry dog food was found to harbor a more diverse microbial community compared to cattle and poultry feeds. Using metagenomic sequencing, ten antimicrobial resistance genes were identified, including those conferring resistance to beta-lactamases, erythromycin, lincomycin, pristinamycin, tylosin, fosfomycin, phenicol, and quinolone. The authors concluded that both 16S rRNA and metagenomic sequencing provided valuable insights into feed core microbiomes and antimicrobial resistance, supporting their use in identifying potential pathogen interventions and strategies to mitigate antimicrobial resistance. A summary of recent studies utilizing 16S rRNA sequencing and related molecular approaches to evaluate feed microbiota is provided in Table 1. Given the lower economic cost of 16S rRNA gene microbiome sequencing, it will likely remain an effective initial screening tool for establishing microbiome baseline profiles. Depending on the question, more expensive methods, such as metagenomics, can be used as a follow-up to initial studies.

Table 1 Summary of selected studies applying 16S rRNA gene sequencing and related molecular approaches to assess microbial populations in animal feeds and feed ingredients. The table outlines the methodology and key findings of each study

Study	Approach	Key Findings
Olson et al. (2020)	Whole genome sequencing of pure cultures from poultry feed	Identified multiple genera including *Klebsiella*, *Enterococcus*, *Kosakonia*, *Pantoea*, *Stenotrophomonas*
Olson et al. (2021)	16S rRNA V4 sequencing from colonies on feed	Detected 25 genera from 11 colony morphotypes; highlighted taxonomic heterogeneity
Olson et al. (2024a)	Comparison of five DNA extraction methods for poultry feeds	Extraction method affected microbial diversity; bead beating and lysozyme improved results
Olson et al. (2024b)	16S rRNA V4 sequencing of nine different poultry feeds	Probiotic feeds had the lowest diversity; detected both pathogens and commensals
Solanki et al. (2019)	16S rRNA and ITS sequencing of stored wheat after fumigation	Fumigation impacted bacterial diversity and abundance over time, while fungal community was not impacted
Ge et al. (2025)	16S rRNA and shotgun metagenomics of cattle, poultry, and dog feed	Dog food had the highest diversity; detected AMR genes and validated both sequencing methods

6 Conclusion and future trends in research

Before microbiome analyses can be routinely applied to feeds and feed ingredients, several key challenges must be addressed. Further research is needed to develop universal and standardized DNA extraction protocols. In parallel, identifying core microbiomes commonly present in feeds would aid in designing mock communities to serve as internal sequencing controls. Achieving this will require extensive microbiome profiling across diverse feed types and ingredients. Another critical hurdle is establishing effective sampling protocols. Due to the bulk nature of feeds, obtaining representative samples – particularly for detecting low-abundance organisms, such as *Salmonella* – can be challenging. This challenge also extends to collecting samples that accurately represent the overall microbial composition of the feed. Efforts in other environments, such as GIT studies, have aimed to address pseudoreplication by distinguishing true biological replication from non-independent sampling (Reish et al., 2024). From a statistical perspective, power analyses have been used to determine the minimum number of samples needed to detect actual treatment effects (Xia et al., 2018). However, factors such as variability in sequencing depth, microbial interactions, and dominance by specific taxa must be taken into account. Preliminary computer simulations may be necessary to support accurate power calculations (Arnold et al., 2011; Chen, 2020). Integrating these statistical approaches with practical sampling strategies – such as those suggested for bagged feed error assessment by Combs et al. (2024) – could be implemented to develop standardized protocols for sampling feeds in microbiome studies.

In conclusion, the application of microbiome analyses to animal and poultry feeds holds valuable potential for evaluating the microbial quality of feeds. It provides a comprehensive approach to characterizing the entire microbial community present in feed, extending beyond the capabilities of traditional methods. As sequencing technologies advance – particularly with the incorporation of long-read platforms – the ability to achieve deeper taxonomic resolution of feed microbial populations is expected to increase significantly. Applying metagenomics and metabolomics enables the functional characterization of feed microbial consortia, deepening our understanding of their ecology and the potential influence on the gastrointestinal microbiota following feed consumption (Olson et al., 2022). This knowledge is especially relevant as alternative feed sources, such as black soldier flies and their gut microbiomes, are being explored for their effects on feed safety (Khamis et al., 2020). Additionally, feed microbiome data can reveal how treatments and storage conditions alter microbial communities, helping to identify new indicator organisms beyond traditional marker strains, such as *E. coli* and *Enterococcus* species (Ge et al., 2020). While these insights enhance our understanding of feed microbial ecology, integrating other omics approaches will be essential for generating more comprehensive interpretations.

7 Where to look for further information

- Lyte, J.M., et al. 2025. Do we need a standardized 16S rRNA gene amplicon sequencing analysis protocol for poultry microbiota research? *Poult. Sci.* 104: 105242. https://doi.org/10.1016/j.psj.2025.105242.
- Maciorowski, K.G., et al. 2007. Effects of poultry and livestock feed with bacteria and fungi. *Animal Feed Sci. Technol.* 133: 109–136.
- Olson, E.G., et al. 2023. Chapter 27. Microbial pathogen contamination of animal feed. In: L. Anelich, et al. (Eds.) *Present Knowledge in Food Safety: A Risk-Based Approach Through the Food Chain*. International Life Sciences Institute, Academic Press, Elsevier. pp. 378–393. https://doi.org/10/1016/B978-0-0-12-819470-6.00023-8.
- Olson, E.G., et al. 2022. Application of microbial analyses to feeds and potential implications for poultry nutrition. *Poult. Sci.* 101: 101789. https://doi.org/10.1016/j.psj.2022.101789.
- Ricke, S.C., et al. 2017. Chap. 19. Unravelling food production microbiomes: concepts and future directions. In: S.C. Ricke, et al. (Eds.) *Food and Feed Safety Systems and Analysis*. Elsevier Inc., San Diego, CA. pp. 347–374. https://doi.org/10.1016/B978-0-12-811835-1.00019-1.

- Weinroth, M.D., et al. 2022. Considerations and best practices in animal science 16S ribosomal RNA gene sequencing microbiome studies. *J. Anim. Sci.* 100(2): skab346. https://doi.org/10.1093/jas/skab346.

8 References

Amann, R., et al. 1997. Modern methods in subsurface microbiology: in situ identification of microorganisms with nucleic acid probes. *FEMS Microbiol. Rev.* 20: 191–200.

Amann, R.I., et al. 1995. Phylogenetic and in situ detection of individual microbial cells without cultivation. *Microbiol. Rev.* 59: 143–169.

Amato, K.R. 2017. An introduction to microbiome analysis for human biology applications. *Am. J. Human Biol.* 29 (e22931), 1–17. https://doi.org/10.1002/ajhb.22931.

Anderson, M.J. 2017. Permutational multivariate analysis of variance (PERMANOVA). In: N. Balakrishnan, (Eds.) *Wiley StatsRef: Statistics Reference Online*. John Wiley & Sons, Ltd., Hoboken, NJ. pp. 1–15. https://doi.org/10.1002/9781118445112.stat07841.

Andino, A., et al. 2014. Survival of *Salmonella enterica* in poultry feed is strain dependent. *Poult. Sci.* 93: 441–447. https://doi.org/10.3382/ps.2013-03401.

Arnold, B.F., et al. 2011. Simulation methods to estimate design power: an overview for applied research. *BMC Med. Res. Methodol.* 11: 94 http://www.biomedcentral.com/1471-2288/11/94.

Azarbad, H. 2022. Conventional vs. organic agriculture–Which one promotes better yields and microbial resilience in rapidly changing climate? *Front. Microbiol.* 13:903500. https://doi.org/10.3389/fmicb.2022.903500.

Bodie, A.R., et al. 2025. Microbial diversity of selective and non-selective media for *Campylobacter jejuni, coli* and *lari* in poultry post-chill whole bird carcass rinsates. *Discover Bact.* 2:8. https://doi.org/10.1007/s44351-025-00017-7.

Bolyen, E., et al. 2019. Reproducible, interactive, scalable and extensible microbiome data science using QIIME2. *Nat. Biotechnol.* 37: 852–857. https://doi.org/10.1038/s41587-019-0209-9.

Brown, J.A., et al. 2025. Chapter 18. Illumina MiSeq 16S rRNA gene library preparation for poultry processing microbiome analyses. In: A. Bridier (Ed.) *Foodborne Bacterial Pathogens: Methods and Protocols* Vol. 2852, 2nd edition. Methods in Molecular Biology. Springer Science, New York, NY. pp. 273–288. https://doi.org/10.1007/978-1-016-4100-2_18.

Callahan, B., et al. 2017. Exact sequence variants should replace operational taxonomic units in marker-gene data analysis. *ISME J.* 11: 2639–2643 https://doi.org/10.1038/ismej.2017.119.

Caporaso, J.G., et al. 2010. QIIME allows analysis of high-throughput community sequencing data. *Nat. Methods* 7: 335–336.

Chatman, C.C., et al. 2024. Meat animal biologics discovery opportunities from the gut microbiome: application of metabolomics. *Meat Muscle Biol.* 8(1): 18261, 1–11. https://doi.org/10.22175/mmb.18261.

Chavez, R.A., et al. 2020. A review of the methodology of analyzing aflatoxin and fumonisin in single corn kernels the potential impacts of these methods on food security. *Foods* 9: 297. https://doi.org/10.3390/foods9030297.

Chen, L. 2020. powmic: an R package for power assessment in microbiome case-control studies. *Bioinformatics* 36: 3563–3565. https://doi.org/10.1093/bioinformatics/btaa197.

Christensen, C.M. 1957. Deterioration of stored grains by fungi. *Bot. Rev.* 23: 108–134.

Combs, J.C., et al. 2024. Assessing error upon sampling bagged feed. *J. Regul. Sci.* 12: 1–31.

Comeau, A.M., et al. 2017. Microbiome helper: a custom and streamlined workflow for microbiome research. *mSystems* 2. https://doi.org/10.1128/mSystems 00127–16. e00127–16.

Cox, N.A., et al. 1986. Effect of the steam conditioning and pelleting process on the microbiology and quality of commercial-type poultry feeds. *Poult. Sci.* 65: 704–709.

Crippen, T.L., et al. 2019. How management practices within a poultry house during successive flock rotations change the structure of the soil microbiome. *Front. Microbiol.* 10:2100. https://doi.org/10.3389/fmicb.2019.02100.

Cuero, R.G., et al. 1986. The influence of gamma irradiation and sodium hypochlorite sterilization on maize seed corn microflora and germination. *Food Microbiol.* 3: 107–113.

Dänicke, S. 2002. Prevention and control of mycotoxins in the poultry productio chain: a European view. *World's Poult. Sci. J.* 58: 451–474.

Delgado-Baquerizo, M., et al. 2018. A global atlas of the dominant bacteria found in soil. *Science* 359: 320–325.

DeSantis, T.Z., et al. 2006. Greengenes, a chimera-checked 16S rRNA gene database and workbench compatible with ARB. *Appl. Environ. Microbiol.* 72: 5069–5072.

Dittoe, D.K., et al. 2022. Impact of the gastrointestinal microbiome and fermentation metabolites on broiler performance. *Poult. Sci.* 101: 101786. https://doi.org/10.1016/j.psj.2022.101786.

Dittoe, D.K., et al. 2018. Organic acids and potential for modifying the avian gastrointestinal tract and reducing pathogens and disease. *Front. Vet. Sci.* 5: 216. https://doi.org/10.3389/fvets.2018.00216.

Ercolini, D. 2004. PCR-DGGE fingerprinting novel strategies for detection of microbes in food. *Microbiol. Meths.* 56: 297–314.

Escobar-Zapeda, A., et al. 2015. The road to metagenomics: from microbiology to DNA sequencing technologies and bioinformatics. *Front. Genet.* 6: 1–15. https://doi.org/10.3389/fgene.2015.00348.

Faour-Klingbeil, D. and Todd, E.C.D. 2018. The impact of climate change on raw and untreated wastewater use for agriculture, especially in arid regions: a review. *Foodborne Pathog. Dis.* 15: 61–72. https://doi.org/10.1089/fpd.2017.2389.

Feye, K.M., et al. 2021. Chapter 2. Isolation of *Salmonella* spp. from animal feed. In: H. Schatten (Ed.) *Salmonella Methods and Protocols - Methods in Molecular Biology*, 3rd edition. Humana Press, Springer Science+Business Media LLC, New York, NY. pp. 7–16. https://doi.org/10.1007/978-1-0716-0791-6_2.

Feye, K.M., et al. 2020. Survey analysis of computer science, food science, and cybersecurity skills and coursework of undergraduate and graduate students interested in food safety. *J. Food Sci. Educ.* 19: 240–249. https://doi.org/10.1111/1541-4329.12200.

Furuta, K., et al. 1980. Bacterial contamination in feed ingredients, formulated chicken feed and reduction of viable bacteria by pelleting. *Lab. Anims.* 14: 221–224.

Ge, B., et al. 2020. Prevalence and antimicrobial susceptibility of indicator organisms *Escherichia coli* and *Enterococcus* spp. isolated from U.S. animal food, 2005, 2011. *Microorganisms* 8: 1048. https://doi.org/10.3390/microorganisms8071048.

Ge, B., et al. 2025. Exploring animal food microbiomes and resistomes via 16S rRNA gene amplicon sequencing and shotgun metagenomics. *Appl. Environ. Microbiol.* 91: 1–14. https://doi.org/10.1128/aem.02230-24.

Gensollen, T. and Blumberg, R.S. 2017. Correlation between early-life regulation of the immune system by microbiota and allergy development. *J. Allergy and Clin. Immunol.* 139: 1084–1091.

Gonzalez, A. and Knight, R. 2012. Advancing analytical algorithms and pipelines for billions of microbial sequences. *Curr. Opin. Biotechnol.* 23(1): 64–71.

Green, S.R. and P.P. Gray. 1950. A differential procedure applicable to bacteriological investigation in brewing. *Wallerstein Labs. Communs.* 13: 357–366.

Gurmessa, B., et al. 2021. Variations in bacterial community structure and antimicrobial resistance gene abundance in cattle manure and poultry litter. *Environ. Res.* 197: 11101. https://doi.org/10.1016/j.envres.2021.111011.

Ha, S.D., et al. 1998a. Survivability of indigenous feed microflora and a *Salmonella typhimurium* marker strain in poultry feed treated with buffered propionic acid. *Anim. Feed Sci. Technol.* 75: 145–155.

Ha, S.D., et al. 1998b. Indigenous feed microflora and *Salmonella typhimurium* marker strain survival in poultry feed with varying levels of protein. *Anim. Feed Sci. Technol.* 76: 23–33.

Ha, S.D., et al. 1995a. Cycloheximide as a media amendment for enumerating bacterial populations in animal feeds. *J. Rapid Methods Automat. Microbiol.* 4: 95–105.

Ha, S.D., et al. 1995b. Growth response of *Salmonella* spp. to cycloheximide amendment in media. *J. Rapid Methods Automat. Microbiol.* 4: 77–85.

Hanis, T., et al. 1988. Effect of gamma irradiation on survival of natural microflora and some nutrients in cereal meals. *Cereal Chem.* 65: 381–383.

Hanning, I. and Ricke, S.C. 2011. Prescreening methods of microbial populations for the assessment of sequencing potential. In: Y.M. Kwon and S.C. Ricke (Eds.) *Methods in Molecular Microbiology 733 - High-Throughput Next Generation Sequencing: Methods and Applications*. Springer Protocols, Humana Press, New York, NY. pp. 159–170. https://doi.org/10.1007/978-1-61779-089-8_11.

Horgan, S.E., et al. 1999. Use of a low nutrient culture medium for the identification of bacteria causing severe ocular infection. *J. Med. Microbiol.* 48: 701–703.

Huse, S.M., et al. 2014. VAMPS: a website for visualization and analysis of microbial population structures. *BMC Bioinformat.* 15(1): 41. https://doi.org/10.1186/1471-2105-15-41.

Jones, F.T. 2011. A review of practical *Salmonella* control measures in animal feed. *J. Appl. Poult. Sci. Res.* 20: 102–113. https://doi.org/10.3382/japr.2010-00281.

Jovel, J., et al. 2016. Characterization of the gut microbiome using 16S or shotgun metagenomics. *Front. Microbiol.* 7: 459, 17. https://doi.org/10.3389/fmicb.00459.2016.

Kazimierska, K., et al. 2021. Evaluation of nutritional value and microbiological safety in commercial dog food. *Vet. Res. Comm.* 45: 111–128. https://doi.org/1007/s11259-021-09791-6.

Khamis, F.M., et al. 2020. Insights in the global genetics and gut microbiome of black soldier fly, *Hermetia illucens*: Implications for animal feed safety control. *Front. Microbiol.* 11: 1538. https://doi.org/10.3389/fmicb.2020.01538.

Kim, S.A., et al. 2017. Assessment of chicken carcass microbiome responses during processing in the presence of commercial antimicrobials using a next generation sequencing approach. *Sci. Rep.* 7: 43354. https://doi.org/10.1038/srep43354.

Kozich, J.J., et al. 2013. Development of a dual-index sequencing strategy and curation pipeline for analyzing amplicon sequence data on the MiSeq Illumina sequencing platform. *Appl. Environ. Microbiol.* 79: 5112–5120.

Kozulis, J.A. and Page, H.E. 1968. A new universal beer agar medium for the enumeration of wort and beer microorganisms. *Proc. Amer. Soc. Brew. Chem.* 26: 52–58.

Kruskal, W.H. and Wallis, W.A. 1952. Use of ranks in one criterion variance analysis. *J. Am. Stat. Assoc.* 47: 583–621. https://doi.org/10.1080/01621459.1952.10483441.

Luscombe, N.M., et al. 2001. What is bioinformatics? A proposed definition and of the field. *Methods Inf. Med.* 40: 346–358.

Lyte, J.M., et al. 2025. Do we need a standardized 16S rRNA gene amplicon sequencing analysis protocol for poultry microbiota research? *Poult. Sci.* 104: 105242. https://doi.org/10.1016/j.psj.2025.105242.

Maciorowski, K.G., et al. 2007. Effects of poultry and livestock feed with bacteria and fungi. *Anim. Feed Sci. Technol.* 133: 109–136. https://doi.org/10.1016/j.anifeedsci.2006.08.006.

Maciorowski, K.G., et al. 2006. Cultural and immunological detection methods for *Salmonella* spp. in animal feeds: a review. *Vet. Res. Comm.* 30: 127–137.

Maciorowski, K.G., et al. 2005. Polymerase chain reaction detection of foodborne *Salmonella* spp. in animal feeds. *Crit. Rev. Microbiol.* 31: 45–53. https://doi.org/10.1080/10408410590912970.

Maciorowski, K.G., et al. 2004. Incidence and control of food-borne *Salmonella* spp. in poultry feeds: a review. *World's Poult. Sci. J.* 60: 446–457. https://doi.org/10.1079/WPS200428.

Maciorowski, K.G., et al. 2002a. Low nutrient R2A culture medium for bacterial enumeration from poultry feeds. *J. Rapid Meth. Automat. Microbiol.* 10: 59–68.

Maciorowski, K.G., et al. 2002b. Comparison of media and a liquid detergent extraction step on bacterial recovery from animal feeds. *J. Environ. Sci. Health* B37: 255–264.

Maciorowski, K.G., et al. 2002c. Rapid assessment of poultry feed microbial quality using polymerase chain reaction detection of microbial ribosomal gene sequences. *J. Rapid Meth. Automat. Microbiol.* 10: 9–18.

Maciorowski, K.G., et al. 2002d. Polymerase chain reaction detection of bacterial ribosomal genes from fresh and stored animal feeds. *J. Sci. Food Agric.* 82: 1398–1404. https://doi.org/10.1002/jsfa.1199.

Maciorowski, K.G., et al. 2001a. Presence of bacteriophages in animal feed as indicators of fecal contamination. *J. Environ. Sci. Health* B36: 699–708.

Maciorowski, K.G., et al. 2001b. Adaptation of nucleic acid extraction methods for animal feeds. *J. Rapid Meth. Automat. Microbiol.* 9: 217–227.

Mandal, S., et al. 2015. Analysis of composition of microbiomes: a novel method for studying microbial composition. *Microb. Ecol. Health Dis.* 26: 27663. https://doi.org/10.3402/mehd.v26.27663.

Moreno Romo, M.A. and Fernandez, G.S. 1986. Mycoflora of commercial poultry mixed feeds. *Poult. Sci.* 65: 284–287.

Newell, S.Y., et al. 1983. Bacterial response to presence of eukaryote inhibitors in water from a coastal marine environment. *Mar. Environ. Res.* 10: 147–157.

Nichols, A.A. and Leaver, C.W. 1966. Methods of examining damp grain at harvest and after sealed and open storage: Changes in the microflora of damp grain during sealed storage. *J. Appl. Bacteriol.* 29: 566–581.

Nilakanta, H., et al. 2014. A review of software for analyzing molecular sequences. *BMC Res. Notes* 7: 1–9. https://doi.org/10.1186/1756-0500-7-830.

Obe, T., et al. 2023. Controlling *Salmonella*: strategies for feed, the farm, and the processing plant. *Poult. Sci.* 102: 103086. https://doi.org/10.1016/j.psj.2023.103086.

Oguz, H. 2011. A review from experimental trials on detoxification of aflatoxin in poultry feed. *Eurasian J. Vet. Sci.* 27: 1–12.

Olsen, G.J., et al. 1986. Microbial ecology and evolution: a ribosomal RNA approach. *Annual Rev. Microbiol.* 40: 337–365.

Olson, E.G., et al. 2025. Meat animal biologics discovery opportunities from the gut microbiome: 16S sequencing in meat animal biologics discovery. *Meat and Muscle Biology*. 9(1): 20022, 1–21. https://doi.org/10.22175/mmb.20022"doi.org/10.22175/mmb.20022.

Olson, E.G., et al. 2024a. Microbiome analyses of poultry feeds: Part I. Comparison of five different DNA extraction methods. *J. Environ. Sci. Health B.* 59: 378–389. https://doi.org/10.1080/03601234.2024.2353002.

Olson, E.G., et al. 2024b. Microbiome analyses of poultry feeds: Part II. Comparison of different poultry feeds. *J. Environ. Sci. Health B.* 59: 448–482. https://doi.org/10.1080/03601234.2024.2361596.

Olson, E.G., et al. 2023. Chapter 27. Microbial pathogen contamination of animal feed. In: L. Anelich, et al. (Eds.) *Present Knowledge in Food Safety: A Risk-Based Approach Through the Food Chain*. International Life Sciences Institute, Academic Press, Elsevier. pp. 378–393. https://doi.org/10/1016/B978-0-0-12-819470-6.00023-8.

Olson, E.G., et al. 2022. Application of microbial analyses to feeds and potential implications for poultry nutrition. *Poult. Sci.* 101:101789. https://doi.org/10.1016/j.psj.2022.101789.

Olson, E.G., et al. 2021. Identification of bacterial isolates from commercial poultry feed via 16S rDNA. *J. Environ. Sci. Health B.* 56: 272–281. https://doi.org/10.1080/03601234.2020.1868236.

Olson, E., et al. 2020. Draft genome sequences of 11 bacterial strains isolated from commercial corn-based poultry feed. *Microbiol. Resour. Announc.* 9: e00170-20. https://doi.org/10.1128/ MRA.00170-20.

Pace, N.R., et al. 1985. Analyzing natural microbial populations by rRNA sequences. *ASM News* 51, 4–12.

Park, S.H., et al. 2011. Detection of *Salmonella* spp. survival and virulence in poultry feed by targeting the *hilA* gene. *J. Appl. Microbiol.* 111: 426–432.

Pillai, S.D. and Ricke, S.C. 2002. Aerosols from municipal and animal wastes: Background and contemporary issues. *Can. J. Microbiol.* 48: 681–696.

Pillai, S.D. and Pepper, I.L. 1991. Transposon Tn5 as an identifiable marker in rhizobia: Survival and genetic stability of Tn5 mutant bean rhizobia under temperature stressed conditions in desert soils. *Microb. Ecol.* 21: 21–33.

Piskata, Z., et al. 2019. The quality of DNA isolated from processed food and feed via different extraction procedures. *Molecules* 24: 1188. https://doi.org/10.3390/molecules24061188.

Porteous, L.A., et al. 1994. An effective method to extract DNA from environmental samples for polymerase chain reaction amplification and DNA fingerprint analysis. *Curr. Microbiol.* 29: 301–307.

Raghava, G.P.S., et al. 2000. Fingerprinting method for phylogenetic classification and identification of microorganisms based on variation in 16s rRNA gene sequences. *Bio-Techniques* 29: 108–110, 112, 114, 116.

Reasoner, D.J. and Geldreich, E.E. 1985. A new medium for the enumeration and subculture of bacteria from potable water. *Appl. Environ. Microbiol.* 49: 1–7.

Reish, H.M., et al. 2024. A host of issues: Pseudoreplication in host-microbiota studies. *Appl. Environ. Microbiol.* 90: 1–5. https://doi.org/10.1128/aem.01033-24.

Ricke, S.C. 2003. Perspectives on the use of organic acids and short chain fatty acids as antimicrobials. *Poult. Sci.* 82: 632–639.

Ricke, S.C. 2018. Chap. 8. Feed Hygiene. In: J. Dewulf and F. Van Immerseel (Eds.) *Biosecurity in Animal Production and Veterinary Medicine – From Principles to Practice*. ACCO (Academische Coöperative Vennootschap cvba), Leuven, Belgium. pp. 177–209.

Ricke, S.C., et al. 1998. Applicability of rapid methods for detection of *Salmonella* spp. in poultry feeds: a review. *J. Rapid Methods Automat. Microbiol.* 6: 239–258.

Ricke, S.C., et al. 2017. Chap. 19. Unravelling food production microbiomes: concepts and future directions. In: S.C. Ricke, et al. (Eds.) *Food and Feed Safety Systems and Analysis*. Elsevier Inc., San Diego, CA. pp. 347–374. https://doi.org/10.1016/B978-0-12-811835-1.00019-1.

Ricke, S.C., et al. 2018. Molecular-based identification and detection of *Salmonella* in food production systems: Current perspectives. *J. Appl. Microbiol.* 125: 313–327. https://doi.org/10.1111/jam.13888.

Ricke, S.C., et al. 2019. Formaldehydes in feeds and interaction with the poultry gastrointestinal tract microbial community. *Front. Veterinary Sci.* 6: 188. https://doi.org/10.3389/fvets.2019.00188.

Ricke, S.C., et al. 2020. Formic acid as an antimicrobial for poultry production: a review. *Front. Vet. Sci.* 7: 563. https://doi.org/10.3389/fvets.2020.00563.

Ricke, S.C., et al. 2022a. Practical opportunities for microbiome analyses and bioinformatics in poultry processing. *Poult. Sci.* 101: 101787. https://doi.org/10.1016/j.psj.2022.

Ricke, S.C., et al. 2022b. Microbiome applications for laying hen performance and egg production. *Poult. Sci.* 101: 101784. https://doi.org/10.1016/j.psj.2022.101784.

Ricke, S.C., et al. 2022c. Communicating the utility of the microbiome and bioinformatics to small flock poultry growers. *Poult. Sci.* 101: 101788. https://doi.org/10.1016/j.psj.2022.101788.

Robinson, C.K., et al. 2016. Intricacies of assessing the human microbiome in epidemiological studies. *Ann. Epidemiol.* 26: 311–321.

Schloss, P.D. 2010. The effects of alignment quality, distance calculation method, sequence filtering and region on the analysis of 16S rRNA gene-based studies. *PLoS Comput. Biol.* 6: e1000844. https://doi.org/10.1371/journal.pcbi.1000844.

Schloss, P.D., et al. 2009. Introducing mothur: open-source, platform-independent, community-supported software for describing and comparing microbial communities. *Appl. Environ. Microbiol.* 75: 7537–7541.

Seyoum, M.M., et al. 2023. Long-term impacts of conservation pasture management in manuresheds on system-level microbiome and antibiotic resistant genes. *Front. Microbiol.* 14: 1227006. https://doi.org/10.3389/fmicb.2023.1227006.

Shariat, N.W., et al. 2021. Incidence of *Salmonella* serovars isolated from commercial animal feed mills in the United States and serovar identification using CRISPR analysis. *J. Appl. Microbiol.* 130: 2141–2146. https://doi.org/10.1111/jam.14933.

Solanki, M.K., et al. 2019. Shifts in the composition of the microbiota of stored wheat grains in response to fumigation. *Front. Microbiol.* 10: 1098. https://doi.org/10.3389/fmicb.2019.01098.

Stewart, R.G., et al. 1977. The effect of various antifungal agents on aflatoxin production and growth characteristics of *Aspergillus parasiticus* and *Aspergillus flavus* in liquid medium. *Poult. Sci.* 56: 1630–1635.

Sunthornthummas, S., et al. 2025. Unveiling the impact of 16S rRNA gene intergenomic variation on primer design and gut microbiome profiling. *Front. Microbiol.* 16: 1573920. https://doi.org/10.3389/fmicb.2025.1573920.

Tabib, Z., et al. 1981. Microbiological quality of poultry feed and ingredients. *Poult. Sci.* 60: 1392–1397.

Theron, J. and Cloete, T.E. 2000. Molecular techniques for determining microbial diversity and community structure in natural environments. *Crit. Rev. Microbiol.* 26: 37–57.

Thompson, D.R., et al. 2017. Chap. 20. Student cross-training opportunities for combining food, transportation, and critical infrastructure cybersecurity into an academic food systems education program. In: S.C. Ricke, et al. (Eds.) *Food and Feed Safety Systems and Analysis*. Elsevier Inc., San Diego, CA pp. 375–391. https://doi.org/10.1016/B978-0-12-811835-1.00020-8.

Wales, A.D., et al. 2010. Chemical treatment of animal feed and water for the control of *Salmonella*. *Foodborne Paths. Dis.* 7: 1–15. https://doi.org/10.1089/fdp.2009.0373.

Wang, Y., et al. 2021. Contamination with fumonisin B and deoxynivalenol is a threat to egg safety and contributes to gizzard ulcerations of newborn chickens. *Front. Microbiol.* 12: 676671. https://doi.org/10.3389/fmicb.2021.676671.

Weinroth, M.D., et al. 2022. Considerations and best practices in animal science 16S rRNA gene sequencing microbiome studies. *J. Anim. Sci.* 100: 1–18. https://doi.org/10.1093/jas/skab346.

Widmer, F., et al. 1996. Sensitive detection of transgenic plant marker gene persistence in soil microcosms. *Mol. Ecol.* 5: 603–613.

Woese, C.R. and Fox, G.E. 1977. Phylogenetic structure of the prokaryotic domain: the primary kingdoms. *Proc. Natl. Acad. Sci. U. S. A.* 74: 5088–5090.

Xia, Y., et al. 2018. Power and Sample Size Calculations for Microbiome Data. Pages 129-166 in *Statistical Analysis of Microbiome Data* with R. Y. Xia, J. Sun, and D.-G. Chen eds. Springer Singapore, Singapore.

Yang, Y., et al. 2019. Soil bacterial biodiversity is driven by long-term pasture management, poultry litter, and cattle manure inputs. *PeerJ* 7: e7839. https://doi.org/10.7717/peerj.7839.

Yu, Z. and Morrison, M. 2004. Comparisons of different hypervariable regions of *rrs* genes for use in fingerprinting of microbial communities by PCR-denaturing gradient gel electrophoresis. *Appl. and Environ. Microbiol.* 70: 4800–4806.

Chapter 8

Techniques for detecting viral pathogens in animal feed products and manufacturing surfaces

Jordan T. Gebhardt, Kansas State University, USA

1 Introduction

The spread of pathogens is a complex problem facing medical experts, veterinarians, epidemiologists, and public health officials. Within livestock production systems, many routes of pathogen spread exist, including direct and indirect transmission pathways, with several being quite intuitive, such as the movement of infected animals resulting in pathogen spread to naïve populations exposed to infected animals. Other, more complex transmission pathways exist and have been addressed through advancements in biosecurity practices. Such practices have been widely implemented on farms globally with a high degree of success. Biosecurity practices largely are not 'silver bullet'

http://dx.doi.org/10.19103/AS.2026.0172.11

solutions. Generally speaking, a single biosecurity practice often can reduce the risk of pathogen spread, but in many situations does not completely eliminate the hazard. This is due to the complex nature in which pathogens spread. Addressing one hazard does not prevent other hazards from resulting in the risk of exposure to susceptible animals and resulting infection. One such example is through the isolation and diagnostic testing of reproductive animals before entry into a swine production facility using samples such as blood samples or oral fluids. This practice is aimed to address the risk of direct animal–animal transmission associated with the introduction of new animals into the facility. Quarantine and a robust diagnostic testing procedure greatly reduce the risk of this animal–animal transmission pathway. However, incorporating robust procedures does not completely eliminate the risk of pathogen introduction – if farm employees are not following good biosecurity practices such as changing clothes, disinfection of supplies, equipment, or other fomites, the farm is still at significant risk of being exposed to a variety of pathogens. This is the core of biosecurity. Implementing multiple strategic interventions, each being a valuable means to reduce the risk of pathogen transmission, but each of them alone is not likely to offer full protection against a wide variety of threats.

Successful biosecurity programs incorporate multiple complementary strategies and should be routinely evaluated to understand current limitations and where opportunities for improvement exist. With recent advances in diagnostic tools, the ability to perform this evaluation for continuous improvement of processes is now easier, more rapid, and cost-effective than ever. However, there remain significant challenges to overcome in regard to developing, conducting, and interpreting information for the development and refinement of biosecurity practices. The goal of this chapter will be to discuss a brief history of viral pathogens in animal feed ingredients, complete feed, and feed delivery vehicles and highlight strategies that can be considered to improve feed biosecurity practices through environmental sampling.

2 History of viral pathogens in animal feed

Safety of animal feed has been a topic of focus for many years. Animal food processing facilities have known about microbial pathogens such as *Salmonella* species and have been addressing this concern for decades. With a clear understanding of bacterial pathogens being biological hazards, biosecurity practices have been widely implemented, particularly in the companion animal and poultry feed processing industries. Specifically for viruses, there was little attention directed towards feed as being a sizable hazard until the introduction of porcine epidemic diarrhea virus (PEDV) into the United States in 2013, with rapid spread throughout the US industry throughout 2014 and beyond. For more details, please see elsewhere in this book. While no definitive origin has ever been identified nor a clear understanding of what factors contributed to the rapid dissemination once introduced into the United States, swine feed has been a topic of great discussion and research efforts. There undoubtedly is anecdotal and even epidemiological evidence that, under certain conditions, complete swine feed and particularly certain high-risk porcine-origin ingredients have been associated with some of the early cases of PEDV transmission. Research and publications were numerous during this time, and a variety of publications focused on viral mitigation practices, which is discussed in greater detail elsewhere. Through these field observations and controlled research, an abundance of information was generated. This was a very novel concept at the time – livestock feed being implicated in the transmission of viral pathogens.

Following this, a plethora of research by industry and academia focused on feed biosecurity with an emphasis on viral pathogens, extending beyond PEDV and to also include other viruses such as porcine reproductive and respiratory syndrome virus (PRRSV) as well as viruses considered foreign animal diseases or surrogate viruses believed to have similar characteristics. Research focused largely on mitigation strategies to reduce the ability of viruses to remain viable within feed and feed ingredients. Such strategies include point-in-time mitigation strategies such as thermal processing or irradiation. Further research explored the use of feed additives to reduce viral transmission in feed ingredients or complete feed. These research efforts have transformed the swine feed industry, and more broadly the feed industry as a whole.

Understanding of viral transmission via the feed supply was further questioned following the rapid expansion in geographic range of African swine fever virus (ASFV). While ASFV has been present in several geographies for nearly 100 years, the speed of dissemination reached an unprecedented pace in 2018 when ASFV was first detected in China. Following the initial introduction and detection, many countries in Southeast Asia quickly became infected. This region has a tremendously large swine population, with an estimated 50% of the pigs globally being located in China (Kim et al., 2023). Not only was this an exceptionally large population of naïve animals, but concern quickly arose given the global dependence on this region for the production of a wide variety of products such as animal feed ingredients. A recent survey published by Faccin et al. (2024) evaluated the usage of feed ingredients in the US swine industry from countries and/or regions affected by ASFV or classical swine fever virus (CSFV). This survey included 7 swine production systems located in the United States, representing a sow inventory of approximately 1.5 million, approximately 25% of the US sow inventory at the time the survey was conducted. The survey included only emphasis on non-bulk ingredients packaged in 1 metric ton or smaller packaging. Import of bulk ingredients was not evaluated in the currently described survey. Across diets manufactured for all phases of production, approximately one third of ingredients in the complete diet originated from countries currently affected by ASFV (survey as of summer 2024). This seems like an exceptionally large percentage; however, with further evaluation, these ingredients are low inclusion rate ingredients such as individual vitamins within a vitamin premix, specialty feed additives, etc., that in many times are necessary for dietary inclusion but are at low inclusion rates. Collectively, the one third of all ingredients in swine diets represented in this survey being non-bulk ingredients from ASFV/CSFV affected countries represents less than 1% of the volume of the complete feed. Nonetheless, the next phase of feed biosecurity with a focus on non-bulk ingredients has focused on improving the understanding of manufacturing, transport, and storage conditions once arriving in a region.

Moving beyond non-bulk ingredients, there has been significant work that has been performed indicating that viral pathogens such as PEDV and ASFV survive exceptionally well in soybean-based feed ingredients such as soybean meal (Dee et al., 2018; Dee et al., 2022). While soy products are not routinely imported into the United States from countries affected by viral diseases such as ASF or CSF, work has been completed to document that this does occur (Patterson et al., 2020; Blomme et al., 2021). The supply chain of feed ingredient trade and distribution globally is extremely complex. Traceability of bulk ingredients has a greater level of challenge compared to non-bulk ingredients, so although considered a risk of virus transmission with robust scientific evidence to support this hypothesis, further biosecurity practices have not been implemented for bulk ingredients to the same manner that have been done for non-bulk ingredients as described by Faccin et al. (2024). Global approaches to feed biosecurity have adapted over time, with recent emphasis being placed on ingredient importation practices and further exploring handling and quarantine procedures once arriving in a country or region free from a particular pathogen such as ASF.

3 Applications of diagnostic sampling and analysis in animal feed production

Advances in diagnostic tools in recent years have propelled the use of modern technologies to improve biosecurity practices by addressing particular focus areas. As firsthand lessons have been learned with feed ingredient and complete feed biosecurity, modern diagnostic tools have been shown to be extremely valuable to build baseline knowledge of biological hazards and implement corrective measures.

Before discussing specific diagnostic tools available for use and potential utility, it is warranted to review areas within animal feed production where these technologies can be incorporated. Table 1 describes the three primary sample types used within feed biosecurity, along with a description of how they can be

Table 1 Description of diagnostic applications commonly used for feed biosecurity.

Test	Use case	Advantages	Limitations
Environmental sampling	1. Detection of gaps in physical biosecurity practices. 2. Flexible, with variety of use scenarios based on desired information	1. PCR = rapid & cost effective 2. Ease of collection	1. Challenging to interpret PCR results since does not indicate if viable virus present or not
Feed ingredients	1. Understand ingredient contamination prior to arrival at feed mill.	1. PCR = rapid & cost effective 2. Ease of collection 3. Isolates contamination to ingredient processing, separate from contamination within feed mill.	
Complete feed	1. Understand ingredient contamination + contamination during feed manufacturing process.	1. PCR = rapid & cost effective 2. Ease of collection 3. Incorporates all possible routes of contamination up to final feed to be presented to animals.	

beneficial in improving understanding of biosecurity, as well as key advantages and the major limitation to consider.

Of the sampling types, there are applications and use cases each for environmental sampling, feed ingredients, and complete feed depending on the anticipated goals and desired decisions to be made. As with any diagnostic test, the result should generate information that alone or in combination with additional information leads to a decision. Specifics of how each of these sample types can contribute to practical decision making are described in greater detail below.

The most common diagnostic test used today in feed biosecurity is real-time PCR diagnostic testing. Further specifics of this technique are also described in greater detail later in this chapter. The key limitation needed to be considered when using it is that results only indicate whether there is detectable genetic material – a PCR test does not indicate whether viable, infectious virus is present or not. This is absolutely critical to understand when using this technique. The advantages are clear and numerous, however, making real-time PCR the backbone of feed biosecurity diagnostic testing programs.

3.1 Feed ingredient sampling and analysis

Animal feed is comprised of a series of ingredients, each providing desired nutrients that ultimately lead to the desired nutritional profile designed to meet the specific needs for maintenance, growth, reproductive, and other productive purposes. The origin of these ingredients is widespread, often originating from multiple countries and even hemispheres globally. As such, with feed ingredients traveling vast geographic distances, it is possible that if a contamination event were to occur, novel pathogens could be spread if the pathogen remained viable during transit and storage and were present in high enough quantities to cause infection when provided to animals. This has been a significant focus area for ASF to understand the potential risk for ingredients originating from regions affected (Dee et al., 2018). In the case of certain pathogens, control of diagnostic testing is tightly controlled. This is the current status for testing of ASF in the United States, and any testing of incoming feed ingredients is not permissible. With pathogens of high regulatory importance, close control of processes is critical to avoid false positive diagnostic results, which would have devastating impacts on global trade.

With other pathogens, more routine and frequent diagnostic testing is available as a tool. In these situations, feed ingredients can be sampled and tested to determine if viral pathogens are present. Using this technique, information can be generated to determine if a contamination step occurred at some time prior to the time the sample was collected. Depending on the ingredient, this could be at a variety of points in the process, but importantly it is known that it was a contamination step that occurred before the ingredient was sampled. This can help isolate where the contamination could have occurred, and with further focus on the previous steps along with additional diagnostic interventions can help isolate opportunities to improve the manufacturing, storage, and transportation biosecurity. The methods used to sample feed ingredients can vary significantly, particularly when considering that ingredients may be stored in a variety of forms such as conventional small bags (roughly 25 kg), 1 metric ton totes, or bulk. Table 2 describes a general framework to consider sampling of feed ingredients of a variety of forms, and the conceptual framework would apply to sampling complete feed also which will be discussed further below.

Decisions generated through diagnostic testing of feed ingredients help isolate processes that occur prior to ingredient utilization in mixing complete diets. Most often, molecular diagnostic techniques such as PCR are utilized, and limitations of this are critical to consider and are described in greater detail later in this chapter.

3.2 Finished feed sampling and analysis

Similar to ingredient sampling, finished feed sampling can be a useful tool to evaluate for the presence of pathogens such as viruses. Decisions generated

Table 2 Description of sampling procedures used to sample feed ingredients and complete feed.

Packaging form	Equipment	Sampling steps	Sample generated/storage
Small packaging (bags up to 25 kg)	1. Single tube trier (preferred) 2. Representative sample can be taken by opening individual bags if tube trier is not available. 3. Clean storage bag with appropriate labelling.	1. Sample 10 representative locations within batch of ingredients/complete feed. 2. Place all samples in single container, mix to create representative composite sample.	1. Composite sample, sample amount can vary as long as sample is representative of ingredient/complete feed. 2. Generally, a 200 to 400 gram sample is appropriate. 3. Storage conditions dependent upon pathogen of interest. For viral pathogens, generally refrigeration preferred (4°C). 4. Expedite shipment to diagnostic laboratory for further processing.
Tote (typically up to 1 metric ton)	1. Sleeved feed probe 2. Clean storage bag with appropriate labelling.	1. Sample 10 representative locations within batch of ingredients/complete feed. 2. Place all samples in single container, mix to create representative composite sample.	1. Composite sample; quantity sufficient to represent the batch. 2. Storage and shipment conditions as described above.
Bulk	1. Sleeved feed probe OR clean sample collection container if sampling via cut stream. 2. Clean storage bag with appropriate labelling.	1. Two primary approaches can be taken to collect samples from bulk ingredients or complete feed. 1) Sampling using similar methodology to described above when sampling from tote using sleeved feed probe, or 2) sampling via cut stream as bulk material is being unloaded. 2. Using either approach, sample 10 representative locations within batch of ingredients/complete feed. 3. Place all samples in single container, mix to create representative composite sample.	1. Composite sample; quantity sufficient to represent the batch. 2. Storage and shipment conditions as described above.

through diagnostic testing of complete feed help provide an understanding of the risk present for animals that soon could consume the feed. Multiple processes could result in any identified pathogen presence in complete feed, including contamination occurring for raw materials or contamination during the feed manufacturing process. Thus, when complete feed samples have detectable viral contamination, it is necessary to further understand the point of origin and whether the contamination occurred during the feed manufacturing process or at some point prior for a specific feed ingredient. Similar to feed ingredients, real-time PCR is the most common diagnostic technique used to test finished feed, with similar limitations to be discussed further.

3.3 Environmental surface sampling and analysis

Environmental surface sampling is a technique that can be extremely flexible and applicable to many real-world situations. The basic concept of environmental sampling is to understand whether viral pathogens are present on a surface, with that surface being anything ranging from the tires of a truck, feed contact surfaces within a feed manufacturing facility, or flooring within an animal housing facility. Virtually any type of surface can easily be sampled. Common sampling locations specific to feed biosecurity include surfaces within the feed mill and surfaces associated with ingredient or complete feed transportation, i.e. trucks. When used within a feed mill, environmental sampling coupled with real-time PCR can give a strong indication of whether feed was manufactured containing viral pathogens. In a research setting, it has been documented that when feed is inoculated with viruses such as PEDV or ASF, viral contamination becomes widely distributed within the feed mill and can be found on a wide variety of feed contact and non-feed contact surfaces (Schumacher et al., 2017; Elijah et al., 2021). Furthermore, under experimental conditions when surfaces from an experimental feed manufacturing facility are stored following the manufacture of ASF-contaminated feed, the detection of ASF DNA using real-time PCR is extremely stable for many months following the initial contamination event (Houston et al., 2023a). This is logical given the very high stability of ASF DNA, and again, when using a PCR-based assay does not confirm or give any indication about pathogen viability, but instead indicates that at some point previously surface contamination occurred. Similar results have been observed for feed dust collected from animal feeders used in experimental in vivo swine bioassay experiments (Khanal et al., 2021).

An advantage of using environmental sampling for the purpose of detecting viral contamination within a feed mill is that feed dust represents a more homogenous representation of the potential contamination risk profile for the feed previously manufactured compared to a single or even multiple point samples of ingredient or finished feed. Within a feed mill, dust is generated, and through previous controlled research with PEDV (Schumacher et al., 2017; Gebhardt et al., 2018) and ASF (Elijah et al., 2021) it is well understood that dust can serve as a very good sample type given an apparent viral affinity for dust particles. The exact mechanism of viral attachment to dust particles and the practical application of this information remains un-elucidated, but dust collected from within a feed mill is a great location to collect environmental samples to determine if feed manufactured previously was contaminated with the viral pathogen of concern.

Another common and useful application for environmental sampling to supplement feed biosecurity programs is surface sampling of transportation vehicles such as ingredient delivery vehicles bringing raw materials to a feed mill and trucks transporting finished feed to farms. Using environmental sampling of external and internal (i.e. within truck cab) surfaces, surface decontamination efforts can be prioritized to maximize efficacy of disinfection and to ensure no surfaces are missed. An example of the utility of this will be described in greater detail later within this chapter.

There are many variations of specific practices and methodologies to collect environmental samples. These practices many times are adapted to the specific use case so creating a standardized framework applicable to all situations would be an oversimplification. Nonetheless, using practices outlined would be a viable starting point for the implementation of an environmental sample collection program. A selection of possible sample collection procedures is described further in Table 3.

In addition to the two methods described earlier, other sampling devices have been effectively used for this purpose including fiber paint roller covers, dry sweeper pads (manufactured by Swiffer, others available). There are advantages and disadvantages with each of these sampling devices. Key highlights to note are the accessibility of cotton gauze surgical sponges which can be found very easily and would be immediately available in a disease outbreak scenario. Stick Sponges are widely used in the food processing industry largely for sampling of bacterial pathogens and procedures can be adapted for sampling of feed manufacturing facilities. Fiber paint roller covers have been successfully used with a long pole attachment to reach surfaces with limited accessibility, including loadout facilities at feedmills where a surface to be sampled may be several meters above ground and outside of reach using other sampling devices. Using any of these sampling devices, successful results may be achieved. Key principles to incorporate include (1) select a reasonable sample collection device for the desired surface types and location, (2) prepare the sample collection device in a sterile manner, (3) sample approximately 20 × 20 cm surface which would be believed to be at greatest risk of viral contamination, (4) place the sample in a secure tube or bag, and (5) promptly

Table 3 Description of environmental sampling procedures.

Sampling device	Advantage	Sampling device preparation	Sampling method
Cotton gauze surgical sponge	1. Readily available; can be found at any pharmacy.	1. Using sterile technique, place one 10 cm × 10 cm cotton gauze surgical sponge in sterile conical tube (50 mL conical tube works well; other tube sizes could be used) 2. Apply 10 mL of sterile phosphate buffered saline to cotton gauze surgical sponge. 3. Sampling device now ready to be used; prepared sampling device can be stored at room temperature or at refrigerated conditions (4°C) for up to 1-2 weeks until use.	1. Don clean pair of gloves, remove pre-moistened cotton gauze surgical sponge from tube, sample a surface approximately 20 cm × 20 cm by using 10 horizontal passes and 10 vertical passes to ensure effective uptake of any material present on surface. 2. Place cotton gauze back into storage container, seal cap and ensure appropriately labeled. 3. Storage conditions dependent upon pathogen of interest. For viral pathogens, generally refrigeration preferred (4°C). 4. Expedite shipment to diagnostic laboratory for further processing.
Stick Sponge (variety of manufacturers)	1. Widely used in food processing industry for detection of bacterial contamination.	1. Samples can be purchased dry or premoistened with variety of sampling buffers. 2. Recommended to use pre-moistened stick sponges to improve effectiveness of sampling. For virus detection, phosphate buffered solution or 0.9% NaCl have been shown to be effective.	1. Sampling, storage, and shipment conditions as described above.

process and submit for diagnostic evaluation using techniques such as PCR. These principles should serve as the foundation of an environmental sample collection plan, with further specifics refined to optimize the outcome based on situation-specific considerations.

4 Challenges in detecting viral pathogens

Many diagnostic tests are developed, optimized, and routinely use clinical samples such as animal tissue, blood, fecal material, etc. as the diagnostic specimen. In many of these situations, the quantity of viral antigen present in the sample is exceptionally high. For example, a cycle threshold value (Ct) for a fecal sample from a baby pig acutely infected with PEDV may be 15 or lower. From a diagnostic perspective, with such a high quantity of viral antigen, an effective diagnostic test can have terrific diagnostic sensitivity to conclude the sample contains PEDV RNA. However, with other sample types such as feed ingredients, a variety of challenges exist making diagnostic analysis and interpretation challenging.

4.1 Low concentration of viral particles

One of the primary challenges with ingredient, complete feed, and environmental surface analysis is that in many situations there may be a low quantity of virus present. The process by which a feed ingredient, complete feed, or surface is contaminated can vary dramatically. In many scenarios, a small amount of pathogen is believed to contaminate a surface or material. Thus, where and how the sample is collected along with the significant potential for dilution results in many situations a low amount of viral contamination present within the diagnostic sample. Then, once the sample is collected, the process by which, in many cases, samples are processed further contributes to this dilution. This is particularly challenging given that a sample of dry feed ingredient, complete feed, and, in many cases, environmental samples need to undergo a series of laboratory processing steps to be appropriate diagnostic samples to undergo analysis using techniques such as PCR. Through these steps, often a diluent such as phosphate buffered saline or other buffer solution is used to create a liquid suspension appropriate for further analysis. During this process, however, the sample, with possibly already a low concentration of virus particles, is further diluted, making detection further complicated. Specific recommendations will be provided below to optimize this process based on recent research. Nonetheless, a fundamental challenge with these sample types are, in many cases, the level of viral contamination may be quite low, pushing the limits of utility for many diagnostic tests.

4.2 Non-homogenous nature of viral contamination

Another challenge with sampling from ingredient, complete feed, and/or feed manufacturing surfaces is the reality that contamination often is not distributed evenly. The process of sample collection must include a representative area to maximize the likelihood of the virus being incorporated into the sample, thus if present in sufficient quantities able to be detected using the preferred detection methodology. However, that is often not the reality of viral distribution in feedstuffs or on environmental surfaces. As with collecting any sample, proper sample collection techniques are critical. For feed ingredients or complete feed, creating a homogenous sample from a bulk holding is challenging. However, when performed in a strategic manner, bulk sampling can effectively detect viral contamination (Jones et al., 2019). Recognizing the non-homogenous nature of feed ingredient contamination is important to consider and to create a robust and representative sampling scheme. Because of this, environmental sampling of dust-accumulating locations can help overcome the issues associated with non-homogenous distribution of biological contamination within feed ingredients and complete feed.

4.3 Diagnostic interference from other compounds in diagnostic specimen

When a proper sample has been collected, subsequent analysis using methodologies using the techniques to be described in greater detail below can encounter some challenges. One such challenge is that compounds present in the sample may interfere with the assay. Examples include PCR inhibitors that interfere with the polymerase chain reaction and result in reduced diagnostic sensitivity. Research has explored this using environmental samples using PCR-based techniques to detect PEDV (Houston et al., 2023a) as well as ASF (Kwon et al., 2024a) and have clearly shown that the presence of organic matter can interfere with the diagnostic results. Methods to combat this challenge will be discussed in greater detail later in this chapter.

For other diagnostic methods such as cell culture, compounds naturally present within the feed ingredient, complete feed, or on an environmental surface may directly result in cytotoxicity to the cell monolayer, making interpretation of cytopathic effect very challenging. The nature of raw material, complete feed, and environmental surface samples inherently has a more complex set of naturally occurring compounds that have the potential to interfere with the diagnostic test. Limited published research has been conducted in this area, although recent work has aimed to improve these techniques to optimize the value of the information generated as a result of these laboratory-based assays using these sampling matrices.

5 Categories of detection methods

5.1 Direct methods

Direct methods of virus detection include molecular-based techniques and immunological techniques. Of the two, molecular-based techniques are by far the most common methodology used for the detection of viral pathogens in the sample types described. Molecular-based techniques include the use of PCR, loop-mediated isothermal amplification, and sequencing techniques such as Next Generation sequencing.

Of the molecular-based techniques, PCR is undoubtedly the most common diagnostic assay used for the purpose of feed biosecurity. The main advantages of PCR techniques are the speed at which testing can be performed. Modern PCR techniques can be performed in a wide variety of conditions. Specialized diagnostic laboratories are not needed to generate meaningful results. Although specialized laboratories with robust quality control procedures and internal validation steps in place are considered the gold standard, using less equipped facilities can still successfully generate information meaningful to improvements in biosecurity practices. Furthermore, pen-side diagnostic tests continue to be developed to facilitate rapid antigen detection and diagnosis, and are an opportunity for continued work to validate using feed ingredient, complete feed, and environmental sampling matrices.

A downside of PCR-based techniques is the lack of information regarding the ability for any material in that sample to be able to cause infection if exposed to a susceptible host. The PCR test detects a specific fragment of viral genetic material, but does not indicate whether an intact virus particle capable of replicating within a host is present. There have been recent advancements in assays termed 'viability' PCR which use the combination of standard PCR techniques along with a dye as a means to better understand whether the virus structure is intact. Research has implemented this technique for a wide variety of research purposes (Puente et al., 2020; Balistreri et al., 2024). Although promising, more information must be generated before this viability PCR is feasible for widespread use beyond research settings.

Immunological methods are not used nearly as commonly for feed biosecurity applications. Techniques such as ELISA, IFA, and lateral flow ELISA techniques all have utility for other diagnostic purposes in animal health. Unfortunately, these techniques are generally not suitable for the specific needs of pathogen detection in feed ingredients or on environmental surfaces.

5.2 Indirect methods

For indirect methods, the technique is not designed to detect a particular portion of the virus – such as PCR techniques amplifying viral genetic material to be detected via fluorescence. Indirect methods such as cell culture or virus isolation techniques and biological assays (often referred to as a bioassay model) instead are designed to detect intact, viable virus which through the process of replication in these systems create a change to the host. The advantage of these methods is that they can detect viable virus, a significant advantage compared to molecular-based techniques such as PCR. A significant downside of indirect methods is they require a much greater degree of time to reach diagnostic results compared to more rapid techniques such as PCR. Within indirect techniques, there also are limitations with diagnostic sensitivity. It is well recognized that the limit of detection is higher for virus isolation compared to a bioassay, meaning that for a sample to be identified to contain infectious virus that concentration generally needs to be greater for detection via virus isolation compared to detection via development of clinical signs in an animal bioassay.

There are many significant details that could be explored in much greater depth in this area. Largely, these details are outside the scope of this reference. For the practical purposes of tools most realistic based on currently available techniques for improving feed biosecurity, indirect methods are more challenging to incorporate and have substantial limitations, namely cost and time required. Depending on the specific question to be answered, they may be the preferred choice. If detection of viable virus vs. inactivated viral fragments is needed, these indirect techniques are still one of the best options available. Furthermore, as is the case with any diagnostic test, there are specific situations where the level of clinical appropriateness may vary. As an example, if aiming to understand if viral contamination would be present in the cab of a feed delivery vehicle, a PCR-based assay would be a reasonable tool available to answer this question. The diagnostic result generated would indicate whether there was a contamination step that occurred to lead to ASF DNA being present. It does not mean infectivity, but it does indicate that contamination with viral fragments did occur, and in some situations, that alone may warrant changes in biosecurity practices. A separate situation, such as applying a disinfectant to the exterior surface of a feed delivery truck, if environmental sampling were to be used after that disinfection process and a PCR-based technique is used, the resulting answer would have very little meaning. When exploring hazard identification and understanding where contamination is occurring, PCR-based techniques are highly valuable. When aiming to evaluate and validate disinfection efficacy, PCR-based techniques are of much more limited value, and additional techniques such as viability PCR or cell-culture-based methods are more appropriate diagnostic tools suitable for that purpose. Nonetheless, understanding the question aiming to be answered is a critical first step in the investigation process, and then an appropriate sample type and diagnostic technique should be identified.

6 Refinement of practices for sample collection and analysis

Recent research has explored methodologies to improve the value of diagnostic results using PCR techniques in animal feed and on environmental surfaces. Early research in this area evaluated different techniques for the detection of PEDV on environmental surfaces (Houston et al., 2023b) and found that when organic material is present, particularly soil, diagnostic sensitivity is reduced if no additional sample processing steps are taken. However, if a centrifugation step is incorporated prior to PCR extraction, diagnostic sensitivity is comparable across a range of different organic matter contamination situations. Furthermore, Houston et al. further found that using a syringe filter as a means to reduce sample contamination due to organic matter reduces detection of PEDV RNA. From that research, it was concluded that all environmental samples should undergo a centrifugation step to remove potential PCR inhibitors, and this practice is often used in diagnostic laboratories, but this research further emphasized that this process should be a routine aspect of sample processing standard operating procedures.

Building upon this research, further research focused on optimizing diagnostic sample collection and processing techniques for environmental surfaces contaminated with ASFV. Research explored multiple sample collection devices and processing conditions for stainless steel surfaces contaminated with ASFV in the presence of no organic matter contamination (Kwon et al., 2024a) or with the presence of organic material contamination (Kwon et al., 2024b). Through this research, it was identified that under field conditions, multiple sample collection devices can all generate satisfactory results, although a sponge stick containing a buffer such as DNA/RNA shield (Zymo Research, Irvine, CA) can give slightly greater levels of ASFV DNA detection via PCR compared to using a cotton gauze surgical sponge moistened with standard phosphate buffered saline (PBS). Nonetheless, in field conditions with limited access to diagnostic supplies, the simplified cotton gauze + PBS can be successfully implemented. Consistent with the previous research with PEDV, this research with ASFV also found that in the presence of organic matter, particularly soil contamination, incorporating a centrifugation step is necessary to maintain satisfactory diagnostic sensitivity. The speed of centrifugation (700 vs. 10 000 g) and time of centrifugation (5 vs. 10 vs. 15 min) demonstrated minimal difference in ASFV DNA detection via PCR, but importantly, any of those conditions can successfully improve diagnostic sensitivity compared to no centrifugation, particularly in the presence of soil within the environmental sample. Through this research, critical information has been generated to provide scientific merit to the diagnostic sampling processes.

Further research has aimed to understand how environmental sample handling conditions affect detection of ASFV DNA via PCR (Kwon et al., 2025). This research demonstrated that across many levels of organic matter contamination with using cotton gauze + PBS or a sponge stick with DNA/RNA shield, the number of freeze-thaw cycles largely had limited impact on ASFV DNA detection, and even after three freeze/thaw cycles, ASFV DNA was still detectable in every sample combination. Furthermore, storing environmental samples for up to 7 days at 4°C or room temperature still allowed for ASFV DNA detection, although the quantity of viral DNA was reduced for many combinations of organic matter and sampling device, particularly when stored at room temperature compared to storing at 4°C. Practical takeaways from this research is that flexibility does exist where environmental samples can be stored, shipped, and undergo multiple freeze/thaw cycles with minimal impact on detection of ASFV DNA, and diagnostic interpretation should be valid under a fairly wide range of sample handling procedures.

7 Best practices for sample collection, analysis and interpretation

An overview of the processes for feed ingredient, complete feed, and environmental surface sampling for detection of viral pathogens is shown in Fig. 1. A greater summary of best practices for feed ingredient, complete feed, and environmental sample collection, processing, diagnostic testing, and interpretation is found at the www.ksuswine.org website. Resources have been

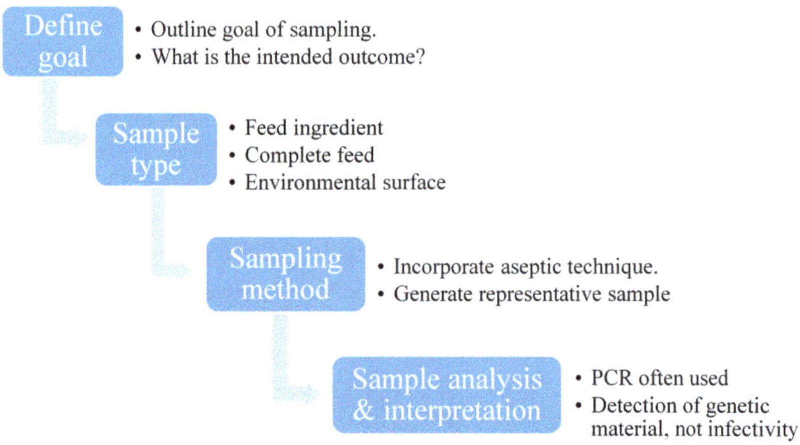

Figure 1 Conceptual framework of ingredient, complete feed, and environmental sampling to detect viral contamination.

developed including factsheets, step-by-step standard operating procedures, and videos demonstrating these techniques.

8 Examples of utility to improve biosecurity programs

To demonstrate the value of diagnostic testing for the purposes of improving biosecurity practices, we will explore a few field examples of these techniques being utilized. In an investigation by Gebhardt et al. (2022), diagnostic testing was used to identify gaps in feed biosecurity practices which led to actionable improvements translating to reduced evidence of ASF contamination. Following the introduction of ASF in Vietnam in early 2019, a collaboration was initiated to use diagnostic tools which had been developed and refined through experiences with PEDV in the United States and apply that knowledge to a swine production and feed manufacturing company located in a region of active ASF circulation. Using feed ingredient, complete feed, and environmental surface sampling methods in conjunction with real-time PCR diagnostics, several key findings were identified. The first key finding of the investigation was focused on finished feed delivery trucks. Due to the known risk of fomite transmission of viruses such as ASF, vehicle sanitation protocols were robust, including exterior sanitation of every feed delivery vehicle entering the feed mill, including high-pressure water and application of an effective disinfectant at the appropriate concentration with an effective contact time. Environmental samples collected from the exterior of the vehicles following this sanitation largely suggested a high degree of sanitation efficacy, evidenced by lack of detection of even viral genetic material (i.e. PCR negative results). However, it was very quickly identified that the cabs of feed delivery vehicles was a location with significant presence of ASF contamination. With contamination identified within the truck cab, contamination was subsequently identified in a high foot traffic location where truck drivers congregate within the feed mill. Through identification of this gap in biosecurity, this allowed the key decision makers to immediately implement enhanced biosecurity practices where in addition to the exterior surfaces of all vehicles undergoing sanitation efforts when arriving at the feed mill, the truck cab was also disinfected. Following this identification of clear biosecurity gap and implementation of corrective actions, no further PCR-positive samples were identified from within the truck cabs.

Another key finding from the described ASF sampling efforts in Vietnam was that collectively, there often are a relatively small number of PCR-positive samples identified. Of the 2,861 samples collected (2,719 environmental surface samples, 142 feed ingredient, water, and complete feed samples), a total of 39 samples were PCR positive, indicating the presence of detectable ASF DNA. This represents approximately 1.4% of all samples, which many would consider as being very low and thus not meaningful enough to warrant the effort and expense. However, although a lower proportion of samples are often PCR-positive, they still can generate a tremendous amount of meaningful information to drive improvements in biosecurity practices.

More recently, a similar effort was conducted to evaluate ASF contamination on transport vehicles in the Philippines (Bartolome et al., 2025). Similar to the findings of feed delivery vehicles described above, the current report indicates that interior cab surfaces of transportation vehicles are a key area where ASF DNA can be detected, thus warranting further consideration regarding practices to prevent viral contamination from occurring, as well as potential mitigation practices.

In both of these case reports, a significant limitation is that the results generated using environmental sampling in conjunction with real-time PCR can indicate whether viral DNA is present in a sample – but cannot differentiate fragments of viral DNA from viable, infectious virus particles. Nonetheless, the learnings can still be used to drive improvements in biosecurity practices, but it illustrates how continued advancement in diagnostic techniques is needed to fully understand the true risk associated with feed ingredient, complete feed, and surface contamination.

9 Considerations for application

As highlighted, techniques can successfully be deployed to improve biosecurity practices including evaluation of feed ingredients, complete feed, and a wide variety of environmental surfaces. Specifics of when and how these techniques should be used are an area with little solid information. Structured risk assessments have been conducted and reported focused mainly on ingredient origin and risk of pathogen transmission via feed as well as transportation networks such as finished feed delivery to farms. However, there has not been a great level of focus directed towards understanding how strategic sampling can be incorporated to further understand these hazards and structured interventions. This presents a research opportunity that will be addressed in greater detail below.

The techniques described all have a cost, and the value of the information generated must outweigh the cost of generating said information. As with many biosecurity practices, quantifying the cost-benefit ratio can be very challenging and is fraught with assumptions. Nonetheless, prior to implementing a widespread sampling effort, appropriate diligence must be considered to understand how the evaluation of biosecurity practices, potential further practices to be implemented based on findings of sampling efforts, and expected outcomes balance.

When considering diseases of high regulatory importance such as foreign animal diseases, there often are highly structured regulations that dictate what testing can be performed, appropriate sample types, validated testing procedures, among others. This often creates a significant limitation for any sort of feed ingredient, complete feed, or environmental surface sampling efforts. As an example, currently in the United States any type of surveillance for ASF in feed ingredients coming from countries or regions affected by ASF is not permitted. Even if feed mills wanted to conduct this testing, it would not be possible based on regulatory oversight due to the significant implications of a scenario such as a false positive diagnostic test. However, in other regions such as Southeast Asia, the access to diagnostic testing for viruses such as ASF is much more available with very little, if any, regulatory oversight. Thus, while the techniques described have demonstrated value to improve biosecurity practices, consideration needs to be directed towards access to these diagnostic techniques as well as an understanding of the implications of testing results, while considering many of the described tests often are not validated for sample types applicable for feed biosecurity and have unknown diagnostic sensitivity and specificity characteristics. Frameworks exist for the validation of diagnostic tests such as that of the WOAH and are widely used for diagnostic specimens such as blood, serum, and tissues of potentially infected animals; however, such processes have not been explored in substantial detail for feed ingredients, complete feed, or environmental samples.

10 Future trends in research

Research has been ongoing in this area. Several recent publications have provided critical clarity regarding sample collection, processing, and analytical techniques. The greatest limitation currently facing practitioners is the limited information available to rapidly and accurately assess whether infectious virus is present within a sample. Opportunity exists to further develop diagnostic techniques such as viability PCR, but commercially deployable assays currently are not available and such techniques remain only available in research settings. The greatest research need continues to be the development and validation of diagnostic techniques – with particular focus needing to be directed towards assessment of viral infectivity potential.

With tools available to be used, further research should focus on understanding where efforts should be focused within the feed supply chain to drive the greatest opportunity for maintaining animal health. Specifically, understanding of sampling strategies, sample locations, and sample sizes necessary to make meaningful inference largely has been derived from limitations of feasibility and available resources. There have been very little, if any, modeling efforts to understand how to prioritize diagnostic sampling efforts within the feed supply chain to maximize positive outcomes. Historically, this has been very challenging to consider, but recent advances in technology to track individual feed ingredients, trucks, and complete feed through a feed manufacturing system to individual feed bin on a specific farm greatly increase the amount of information available to quantify the contact network of the feed supply chain. Further research should focus on developing these contact networks including raw materials, finished feed, and transportation of feed ingredients and complete feed with the purpose to objectively evaluate where the greatest risks occur, how diagnostic sampling efforts could provide key information to feedback and improve the system, and how intervention efforts could be prioritized and implemented.

Other research areas include expanding the use of point-of-care diagnostic techniques to increase the speed of diagnostic results generation while maintaining a high level of analytical sensitivity and specificity. Further enhancements in data analysis and machine learning efforts would be valuable to integrate into diagnostic networks to enhance the usefulness of diagnostic data. While much has been learned in the past 15 years regarding viral pathogens in feed, many research questions remain and serve as important focus areas to continue improving feed biosecurity processes.

11 Conclusion

Feed ingredient, complete feed, and the surfaces associated with manufacture and transportation of feed products represent the frontier of biosecurity strategies for viral pathogens. Historically built with bacterial pathogens in mind, feed biosecurity programs proved woefully ineffective particularly in the swine industry during the introduction and rapid dissemination of PEDV in the US in 2013 and beyond. Key lessons have been learned for the betterment of the feed industry as a whole. Particularly useful is the use of modern diagnostic techniques such as sampling of feed ingredients, complete feed, and/or feed manufacturing surfaces as a means to identify gaps in existing biosecurity practices so improvements can be made.

12 Where to look for further information

In addition to the content described, there are several organizations that offer resources applicable to this topic:

- United States Food and Drug Administration (www.fda.gov).
- American Feed Industry Association (www.afia.org).
- World Organization for Animal Health (www.woah.org).
- Kansas State University Feed Biosecurity Resources (www.ksuswine.org).

13 References

Balestreri, C., et al. (2024), 'Unexpected thermal stability of two enveloped megaviruses, Emiliania huxleyi virus and African swine fever virus, as measured by viability PCR', *Virol J*, 21(1):1. https://doi.org/10.1186/s12985-023-02272-z.

Bartolome, M. J., et al. (2025), 'Environmental DNA (eDNA) contamination patterns of African swine fever virus (ASFV) in swine transport vehicles in the Philippines', *Veterinarni Medicina*, 70(5):156–165. https://doi.org/10.17221/84/2024-VETMED.

Blomme, A. K., et al. (2021), 'Assessment of soy-based imports into the United States and associated foreign animal disease status', *Transbound Emerg Dis*, 69:137–148. https://doi.org/10.1111/tbed.14284.

Dee, S. A., et al. (2018), 'Survival of viral pathogens in animal feed ingredients under transboundary shipping models', *PLoS One*, 13(3):e0194509. https://doi.org/10.1371/journal.pone.0194509.

Dee, S. A., et al. (2022), 'Evidence of viral survival in representative volumes of feed and feed ingredients during long-distance commercial transport across the continental United States', *Transbound Emerg Dis*, 69:149–156. https://doi.org/10.1111/tbed.14057.

Elijah, C. G., et al. (2021), 'Evaluating the distribution of African swine fever virus within a feed mill environment following manufacture of inoculated feed', *PLoS One*, 16(8):e0256138. https://doi.org/10.1371/journal.pone.0256138.

Faccin, J. E. G., et al. (2024), 'Survey of biosecurity practices applied to feed ingredients imported from ASF/CSF positive countries in the U.S. swine industry', *Kansas Agricultural Experiment Station Research Reports* 10(6). https://doi.org/10.4148/2378-5977.8652.

Gebhardt, J. T. (2022), Sampling and detection of African swine fever virus within a feed manufacturing and swine production system. *Transbound. Emerg. Dis.* 69(1):103-114. doi:10.1111/tbed.14335.

Gebhardt, J. T. (2018), Evaluation of the effects of flushing feed manufacturing equipment with chemically treated rice hulls on porcine epidemic diarrhea virus cross-contamination during feed manufacturing. *Journal of Animal Science* 96(10): 4149–4158. https://doi.org/10.1093/jas/sky295.

Houston, G. E., et al. (2023a), 'Evaluating the impact of organic matter and sample processing techniques on RNA detection using environmental samples', *J Swine Health Prod*, 31(2),87–91. https://doi.org/10/54846/jshap/1311.

Houston, G. E., et al. (2023b), Detection of African swine fever virus in feed and feed mill environment following extended storage', *Transbound Emerg Dis*, 3455128. https://doi.org/10.1155/2023/3455128.

Jones, C., et al. (2019), Validation of sampling methods in bulk feed ingredients for detection of swine viruses', *Transbound Emerg Dis*, 67(1):1–5. https://doi.org/10.1111/tbed.13326.

Khanal, P., et al. (2021), 'Detection of African swine fever virus in feed dust collected from experimentally inoculated complete feed using quantitative PCR and virus titration assays', *Transbound Emerg Dis*, 69:97–102. https://doi.org/10.1111/tbed.14176.

Kim, S. W., et al. (2023), 'Current status of global pig production: an overview and research trends', *Anim Biosci*, 37(4):719–729. https://doi.org/10.5713/ab.23.0367.

Kwon, T., et al. (2024a), 'Development and optimization of sampling techniques for environmental samples from African swine fever virus-contaminated surfaces with no organic contaminants', *Front Vet Sci*, 11:1425928. https://doi.org/10.3389/fvets.2024.1425928.

Kwon, T., et al. (2024b), 'Improved African swine fever detection for environmental samples in the presence of organic contaminants', *Transbound Emerg Dis*, 8841168. https://doi.org/10.1155/tbed/8841168.

Kwon, T., et al. (2025), 'The effect of freeze-thaw and storage on African swine fever virus detection in environmental samples', *Front Vet Sci*, 12:1570575. https://doi.org/10.3389/fvets.2025.1570575.

Patterson, G., et al. (2020), 'Quantification of soya-based feed ingredient entry from ASFV-positive countries to the United States by ocean freight shipping and associated seaports', *Transbound Emerg Dis*, 68(4):2603–2609. https://doi.org/10.1111/tbedi.13881.

Puente, H., et al. (2020), 'Rapid selective detection of potentially infectious porcine epidemic diarrhea coronavirus exposed to heat treatments using viability RT-qPCR', *Front Microbiol*, 11:1911. https://doi.org/10.3389/fmicb.2020.01911.

Schumacher, L. L., et al. (2017), 'Characterizing the rapid spread of porcine epidemic diarrhea virus (PEDV) through an animal food manufacturing facility', *PLoS One*, 12(11):e0187309. https://doi.org/10.1371/journal.pone.0187309.

Chapter 9

Techniques for detecting mycotoxins in animal feed products

Yiqiang Chen, China Agricultural University, China

1 Introduction

Mycotoxins are structurally diverse secondary metabolites produced by toxigenic fungi (e.g. *Aspergillus, Fusarium, Penicillium*) under optimal temperature and humidity, and are commonly present in cereals and animal feed, contaminating approximately 60–80% of crops worldwide (Latham et al., 2023). These toxins can lead to both acute and chronic toxicity in animals (Ostrý, 1999, Wei et al., 2023). With the rapid development of the feed industry, detecting mycotoxins has become especially crucial, as it is closely related to animal health, feed, and food safety (Li et al., 2025, Imade et al., 2021). There are various types of mycotoxins, including mycotoxins like aflatoxins (AFs), deoxynivalenol (DON), zearalenone (ZEN), fumonisins (FUM), and ochratoxins (OTA). Due to their potential cytotoxicity, reproductive toxicity, neurotoxicity, carcinogenicity, and immunosuppressive effects, Mycotoxins not only affect animal growth and performance, but some toxins (such as AFs) can also enter the human food chain through animal-derived food products (e.g. meat, eggs, milk), thereby impacting human health (Alameri et al., 2023, Zhang and Banerjee, 2020). For example, AFs are considered strong carcinogens, and long-term exposure may lead to severe health issues such as liver cancer (Mahmoud et al., 2025). The

http://dx.doi.org/10.19103/AS.2026.0172.15

combined effect of AFB1 and the hepatitis B virus can greatly raise the risk of liver cancer (Xing et al., 2025). Furthermore, the widespread contamination of mycotoxins has also seriously affected the agricultural economy, leading to reduced crop yields and rising food prices (Anumudu et al., 2024). Therefore, establishing an accurate mycotoxin detection system is not just a necessary way to keep animals healthy but also a key strategy for promoting sustainable development in the industry.

Timely detection of mycotoxins is crucial for protecting animal health in feed and food safety management. For instance, a study found that using liquid chromatography-tandem mass spectrometry (LC-MS/MS) showed that various regulated mycotoxins are commonly detected in feed around the world. While most samples had low levels of toxins, a few went over the regulatory limits, with maximum levels reaching OTA 187.9 μg/kg, ZEN 7681 μg/kg, and DON exceeding 5000 μg/kg in feed (Muñoz-Solano et al., 2024). Meanwhile, research on monitoring biomarkers in animals is making progress. This helps us better understand how animals metabolize mycotoxins and the effects on their health. Recently, as we've learned more about the dangers of mycotoxins, we've made great strides in detection technologies. Modern detection techniques include chromatography, immunochemical methods, and emerging biosensors and spectroscopic technologies. Using these technologies not only boosts the sensitivity and accuracy of mycotoxin detection but also speeds up the process. For example, microfluidic paper-based analytical devices (μPAD) based on competitive enzyme-linked immunosorbent assay (ELISA) technology can spot multiple mycotoxins in corn flour in just 20 min. This method looks promising because it's fast, sensitive, and cheap (Feng et al., 2023). Additionally, molecular imprinting technology is gradually being applied for rapid detection of mycotoxins, using synthetic molecularly imprinted polymers as specific recognition elements along with platforms like electrochemical sensors for efficient mycotoxin detection in food (Wang et al., 2024).

2 Sample collection and pretreatment

2.1 Sample collection

When collecting samples from field-grown grains (e.g. corn, wheat) intended for feed production, random multi-point sampling is adopted. Typically, 5–10 subsamples are collected from different areas of the field (considering factors like soil type and irrigation conditions) and mixed to form a composite sample, reflecting the initial contamination status before storage. For grain or feed stored in warehouses or silos, stratified sampling is implemented. Samples are collected from different storage locations (e.g. near walls, center) and heights (upper, middle, lower layers) to avoid biases caused by localized contamination

(Janik et al., 2021). Traditional methods, such as the 'three layers, five points' method in the GB/T 14699.1 standard, are still widely used in smaller and mid-sized laboratories due to their low cost and simple operation. During feed production (e.g. mixing, pelleting stages), interval sampling is recommended: 1 sample is collected every 30 min from the production line outlet, with 6–8 samples in total mixed into a composite sample to represent the batch of feed.

In laboratory settings, samples undergo further homogenization to ensure uniformity. For granular feed, modern spectral pre-scanning technology is used to identify potential contamination hotspots (not limited to visible mold), guiding targeted subsampling. Additionally, robotic three-dimensional sampling systems combined with near-infrared pre-screening enhance efficiency and spatial coverage, reducing unnecessary testing. A combination of random and systematic sampling methods is often applied to improve sample representativeness for subsequent analysis (Feng et al., 2023). The sampling methods for feed and feed raw materials are shown in Table 1.

2.2 Sample preservation

The preservation of samples also affects the stability of mycotoxins and the subsequent test results. The degradation or changes in mycotoxins during the sample preservation process can cause inaccurate test results. Generally, samples should be refrigerated or frozen as soon as possible after collection to reduce their degradation rate. During preservation, you should avoid direct sunlight and high temperatures, and prevent cross-contamination, keeping the samples in their original state (Wang et al., 2024). Using modern ultra-low temperature preservation (−80°C vacuum packaging) and nano-titanium dioxide (TiO_2 NPs) as a photostabilizer significantly extends the half-life of the aflatoxin B1 and reduces the degradation rate (Nasiri Poroj et al., 2023, Rhouati et al., 2017). In addition, packaging technology using inert gases like N_2 effectively inhibits the oxidation of lipids in high-fat feeds, ensuring the stability of vomitoxin. Some studies suggest appropriate sub-packaging before sample preservation to cut down the risks from frequently opening (Pandey et al., 2025).

2.3 Sample grinding and mixing

The grinding and mixing of samples are crucial for making sure the samples are uniform. Because mycotoxins often aren't evenly spread out in feed, grinding can effectively improve the uniformity of the samples, which boosts detection accuracy. Usually, a mechanical grinder is used to grind the samples, controlling the grinding time and speed to prevent the samples from overheating or getting contaminated with other stuff. Additionally, after grinding, it's a good

Table 1 Workflow of mycotoxin sample collection in animal feed supply chain

Sampling scene	Sampling method	Key operation	Sampling quantity	Purpose
Field (grain)	Random multi-point sampling	Divide field into 5–10 uniform regions; collect 500 g subsample from each region (avoiding edge/root areas)	2.5–5 kg composite sample	Reflect initial contamination before storage
Warehouse/silo (grain/feed)	Stratified sampling	Divide storage space into 3 height layers (upper: 1/3 height, middle: 2/3 height, lower: 10 cm above bottom); collect 300 g subsample from 3 locations (wall, middle, center) per layer	2.7–3.6 kg composite sample	Avoid localized contamination bias
Feed production line	Interval sampling	Collect sample from outlet every 30 min; ensure sample is representative of mixed/pelleted feed	1.2–1.6 kg composite sample (6–8 subsamples × 200 g)	Represent batch feed quality

idea to use a specific mixing method to blend samples from different batches to get rid of any differences that might come from samples from different sources (Feng et al., 2023).

2.4 Extraction and purification

When detecting mycotoxins, extraction and purification are key to making sure the detection is sensitive and specific. Common extraction methods include liquid–liquid extraction and solid-phase extraction, which can effectively separate mycotoxins from other interfering substances (Pandey et al., 2025). For example, the liquid–liquid extraction method can improve the recovery rate of target mycotoxins through the use of selective solvents. Also, techniques like immunoaffinity columns can be used in the purification process to further remove impurities and enhance the accuracy of the final detection (Alameri et al., 2023). To optimize the extraction and purification processes, current research is focused on boosting how efficiently we can extract mycotoxins and cutting down detection costs (Xue et al., 2025).

The diversification of collaboration in technology and industrial applications has also driven the development of detection technologies. From the traditional SPE-HPLC combination to fully automated platforms for robotic sampling, magnetic solid-phase extraction, and LC/MS/MS, as well as paper spray ionization mass spectrometry (PSI-MS) for rapid field screening, the integration of different technologies meets the detection needs across various scenarios. At the same time, the application of green, cost-effective technologies like citrus pectin bioadsorbents and microwave-assisted extraction (MAE) not only reduces costs but also improves extraction efficiency, offering more cost-effective solutions for feed companies. The continuous innovation in sample collection and pretreatment technologies provides more accurate and efficient methods for detecting mycotoxins in feed, which helps ensure feed quality and animal health. The specific separation and purification methods are shown in Table 2.

3 Detection methods for mycotoxins in feed and feedstuffs

3.1 Chromatographic separation technology

Chromatographic separation technology is an important method for detecting mycotoxins in animal feed products. Common chromatographic methods include thin layer chromatography (TLC), liquid chromatography (LC), and gas chromatography (GC). Each of these techniques has its unique advantages and applications, effectively meeting the analysis needs of different types of samples.

Table 2 Comparison of extraction and purification methods for mycotoxin detection in animal feed and feedstuffs

Extraction and purification methods	Core principle	Main applicable mycotoxins	Advantages
Liquid-Liquid Extraction (LLE)	Separates target substances through differences in their distribution coefficients in different solvents via multiple extractions	Lipid-soluble toxins such as AFs, ZEN, T-2 toxin, and HT-2 toxin	Simple operation, low equipment requirements, and low cost
Solid-Phase Extraction (SPE)	Separates impurities based on the specific adsorption-desorption between target substances and solid-phase adsorbents	Toxins with moderate to strong polarity such as OTA, DON, and FUM	High purification efficiency, reduces matrix interference, and uses less solvent
Immunoaffinity Column Cleanup (IAC)	Selectively adsorbs target toxins using the principle of specific antigen-antibody binding	Specific single or family toxins such as AFs, ZEN, DON, and OTA	Extremely high specificity, significantly improving detection accuracy
Magnetic Solid-Phase Extraction (MSPE)	Achieves rapid separation using magnetic nanomaterials as adsorption carriers under an external magnetic field	Most common toxins such as AFs, ZEN, DON, and FUM	Fast separation speed, reusable, and easy to automate
Microwave Assisted Extraction (MAE)	Accelerates the extraction efficiency of solvents for target substances using microwave heating	Thermally stable toxins such as FUM, AFs, and ZEN	Short extraction time (usually <30 min) and high efficiency

Note: Table data are summarized from references cited in this chapter.

Chromatography is a commonly used method for detecting mycotoxins. The detection limit (LOD) and quantification limit (LOQ) vary depending on the method, the type of toxin, and the sample matrix. Please refer to Table 3 for specific details.

3.1.1 Thin layer chromatography (TLC)

TLC is a traditional separation technique used to separate and detect non-volatile compounds like mycotoxins in food and feed (Zhang and Banerjee, 2020). It works on the different affinities of sample components for the stationary and mobile phases, allowing for chromatographic separation. Although TLC is not more efficient or sensitive than other modern chromatographic methods, it is simple to operate, low in cost, and fast, which makes it great for preliminary screening and qualitative analysis (Lapris et al., 2024). In the detection of mycotoxins, TLC can serve as a convenient method, especially in

Table 3 Comparison of chromatographic separation techniques for mycotoxin detection in animal feed and feedstuffs

Chromatographic method	Limit of detection (LOD)	Limit of quantification (LOQ)	Advantages
Thin Layer Chromatography (TLC)	1–10 μg/kg	5–20 μg/kg	Simple operation, low cost
High Performance Liquid Chromatography (HPLC)	0.1–1 μg/kg	0.5–5 μg/kg	High sensitivity, good repeatability
Gas Chromatography-Mass Spectrometry (GC-MS)	0.05–0.5 μg/kg	0.2–2 μg/kg	Suitable for volatile mycotoxins
Liquid Chromatography-Tandem Mass Spectrometry (LC-MS/MS)	0.01–0.1 μg/kg	0.05–0.5 μg/kg	High specificity, enables multi-toxin detection simultaneously

Note: Table data are summarized from references cited in this chapter.

resource-limited laboratories. Research shows that TLC requires specific sample preparation and cleanup steps based on the physical and chemical properties of the mycotoxins being detected (Agriopoulou et al., 2020). However, TLC's ability for quantitative analysis is relatively weak, and it may be interfered with in complex samples (Singh and Mehta, 2020). Currently, it is mainly used for preliminary screening in grassroots laboratories in developing countries; therefore, when high-precision quantification is required, it's often combined with other chromatographic techniques to improve the accuracy and reliability of detection (Hassanien et al., 2021).

3.1.2 Liquid chromatography (LC)

LC is one of the most commonly used techniques in modern mycotoxin analysis, often employed for the separation of heat-stable, non-volatile compounds and substances with varying polarities. Related compounds can also be separated using LC, and it does not necessarily require the derivatization steps needed for gas chromatography (Santos et al., 2022). HPLC is favored for its high sensitivity, good selectivity, accurate identification, and short analysis time (Shanakhat et al., 2018). Liquid chromatography is typically combined with mass spectrometry (MS) technology to form liquid chromatography-mass spectrometry (LC-MS), which boosts sensitivity and specificity. Its efficient separation capability and superior quantification ability make it the gold standard for detecting mycotoxins in food and feed safety (Malachová et al., 2018). Most fungi that produce mycotoxins can simultaneously produce several different mycotoxins, and agricultural products may be contaminated by different fungal toxins at the same time (Smith et al., 2016). Most fungi that produce mycotoxins can simultaneously produce several different mycotoxins, and agricultural products may be contaminated by different fungal toxins at the same time (Nasaruddin et al., 2022, Liu et al., 2022). There are various detectors for mycotoxin detection, such as triple quadrupole, ion trap, time-of-flight, and orbitrap mass spectrometers. Two types of analyzers can also be combined to form hybrid analysis systems. Existing screening data indicate that the triple quadrupole system (QqQ) using multiple reaction monitoring (MRM) mode is the most commonly used mass spectrometry detector (De Girolamo et al., 2022). Studies have shown that LC-MS/MS is one of the most reliable and sensitive methods for analyzing multiple fungal toxins at once (Bessaire et al., 2019, Abdallah et al., 2017). Furthermore, compared to traditional methods using conventional detectors, LC-MS/MS offers high selectivity and sensitivity, more reliable analyte identification, and a wider range of matrices (Pascale et al., 2019). Liquid chromatography offers advantages in analysis speed and separation efficiency, playing an increasingly important role in food and feed safety monitoring. HPLC technology's main limitations are its portability and practical issues concerning matrix effects, sample types, preparation, and calibration (Singh and Mehta, 2020).

3.1.3 Gas chromatography (GC)

Since the 1970s, GC coupled with mass spectrometry (MS) has been used to analyze various contaminants in food matrices. GC is primarily used for the separation and analysis of volatile compounds, which makes it suitable for certain types of mycotoxins. Compared to liquid chromatography, GC typically requires samples to undergo derivatization (Alshannaq and Yu, 2017) to enhance their volatility, which is suitable for analyzing mycotoxins like AFs (Zhang and Banerjee, 2020). GC is known for its high separation efficiency and quick analysis speed, but it has strict requirements for sample pretreatment, which means careful selection is needed in real-world applications. The low volatility and high polarity of mycotoxins are why gas chromatography isn't often used for analyzing them in food or feed. Recently, most analytical laboratories have moved from using gas chromatography-mass spectrometry (GC-MS) to gas chromatography-tandem mass spectrometry (GC-MS/MS) to boost sensitivity, improve accuracy, and cut down on interference from matrix effects (Dib et al., 2023). The combination of gas chromatography and mass spectrometry (GC-MS) can further enhance detection sensitivity and specificity, making it really important for food and feed safety monitoring (Maggira et al., 2022). The GC-MS/MS method has been used to detect mycotoxins in processed grain products (Rodríguez-Carrasco et al., 2014) and in wheat and wheat flour (Rodríguez-Carrasco et al., 2012). All in all, gas chromatography is an efficient and reliable choice for detecting mycotoxins, particularly suitable for studying the behavior and characteristics of volatile mycotoxins.

3.2 Rapid detection technology

With the increasing demand for monitoring mycotoxins in animal feed and stuffs, the development of rapid detection technology is becoming increasingly important. These technologies not only improve detection efficiency but also reduce detection costs without sacrificing accuracy. For instance, high-performance liquid chromatography (HPLC) is accurate but can be time-consuming and expensive, that's why rapid detection methods like enzyme-linked immunosorbent assay (ELISA), lateral flow immunoassay (LFIA), and biosensors are now the focus of research. The advantages of these methods are their sensitivity, simplicity, and speed, making them widely applicable in field detection and laboratory analysis.

3.2.1 Enzyme-linked immunosorbent assay (ELISA)

ELISA is a popular method for detecting mycotoxins because of its high sensitivity and specificity. The basic principle of ELISA is to use the specific binding between antigens and antibodies, amplifying the signal through enzyme-labeled antibodies to achieve quantitative detection of mycotoxins. For example, there's an ELISA method designed to detect AFB1 in feed samples, with a detection limit of 3.4 μg/kg and reliable repeatability and recovery rates (Wang et al., 2021). Additionally, ELISA has also demonstrated excellent detection capabilities for other mycotoxins such as DON and ZEN (Anumudu et al., 2024). Commercial ELISA kits can reach recoveries of 85–98% in corn samples, with both detection limits (LOD) and quantification limits (LOQ) below the maximum residue limits (MRL) set by the EU; for instance, one kit has an LOD of 0.1 μg/kg and an LOQ of 0.3 μg/kg (Maggira et al., 2022). Solcan et al. also used the ELISA method to detect residues of AFB in chicken liver samples (Solcan et al., 2013). Dual toxin detection can also be performed to simultaneously detect DON and ZEN, with detection limits of 0.1 ppm for DON and 0.05 ppm for ZEN, and recoveries of 85–98% for DON and 92–110% for ZEN (Zhang et al., 2011). However, one limitation of ELISA is that antibodies can cross-react with other similar compounds (for example, the cross-reactivity rate between zearalenone and α-zearalenol is about 15%), and samples with high pigment levels need extra purification steps to get rid of matrix interference (Zhang et al., 2023). We still need to optimize cross-reactivity and interference issues in complex samples to enhance the applicability and accuracy of ELISA in actual samples.

3.2.2 Lateral flow immunoassay (LFIA)

Lateral flow immunoassay (LFIA), also known as immunochromatography test strips, is a rapid and convenient detection technology suitable for quick on-site testing of mycotoxins (Liu et al., 2020b). LFIA works by having the sample flow to the test strip, where it reacts with labeled antibodies to create a visible signal (Song et al., 2014). LFIA's accuracy mainly relies on the signal label. Traditionally, gold nanoparticles (GNPs) are the most widely used labels for generating visual signals (Li et al., 2019), while other materials such as magnetic (MNPs), carbon (CNPs), and quantum dots (QDs) have also been used as labels (Wang et al., 2009, Zhang et al., 2017b, Li et al., 2020, Anfossi et al., 2018). This method is easy to use, gives quick results, and doesn't require complex equipment. For example, the LFIA method developed for AFB1 can finish testing in just 15 min and has good sensitivity and accuracy (Wang et al., 2021). LFIA has significant application value in on-site testing in farmland and animal husbandry, providing farmers with quick updates on mycotoxin contamination, so they can take the right steps to keep animals healthy and ensure food safety. However, LFIA also faces challenges such as sample matrix interference and insufficient quantitative capability, which future research needs to tackle.(Krska and Molinelli, 2009).

3.2.3 Biosensors

Biosensors are an emerging detection technology that have gained a lot of attention for being fast, sensitive, and highly selective. Biosensors can detect specific biomolecules by integrating biorecognition elements (such as antibodies, cells, or enzymes) with physical or chemical sensors, thereby converting biological signals into electrical signals (Perumal and Hashim, 2014). Different types of sensors can be used for mycotoxin detection, like electrochemical (potentiometric, amperometric, and impedance), optical (surface plasmon resonance (SPR) and fluorescence), and piezoelectric (quartz crystal microbalance (QCM)) (Oliveira et al., 2019). Common materials include nucleic acids, peptides, enzymes, antibodies, and cells, but other bio-inspired elements such as recombinant antibodies, aptamers, and molecularly imprinted polymers (MIPs) are also an option (Hedieh et al., 2017, Vasapollo et al., 2011). Furthermore, to boost biosensor sensitivity, metal nanoparticles, carbon nanotubes (CNTs), nanofibers, and quantum dots are widely utilized because of their biocompatibility, physicochemical properties, and high surface area-to-volume ratio (Doria et al., 2012, Altintas et al., 2017). For instance, sensors targeting mycotoxins such as DON and ZEN have been developed, demonstrating their potential application in grain and feed quality control (Majer-Baranyi et al., 2021). These sensors not only speed up detection but also cut costs, making them great for large-scale field use. Additionally, the ability of biosensors to detect multiple things at once is a big plus for monitoring the co-contamination of different mycotoxins. Future research should aim to boost the sensitivity and stability of biosensors to better meet real-world needs.

3.3 New technologies for analysis and detection of mycotoxins

In recent years, there has been significant progress in mycotoxin detection technologies. With the increasing focus on food safety, various new technologies are being increasingly applied to the detection of mycotoxins. These new technologies not only enhance the sensitivity and accuracy of detection but also speed up the testing process, making them easier to use in real-world applications. For example, technologies like electronic noses, fluorescence polarization, and aggregation-induced emission have been widely researched and applied for the identification and analysis of mycotoxins to meet the needs of the food and feed industries.

3.3.1 Electronic nose

The electronic nose (e-nose) is made up of several non-specific chemical sensors that can capture different volatile organic compounds (VOCs) and identify the characteristics of gas molecules. Through a pattern recognition system, it can detect odors and give a rough idea of the types of metabolites produced (Leggieri et al., 2021). The electronic nose technology for detecting fungal infections is based on identifying specific volatile organic compounds associated with the growth of fungi on grains. The growth and biochemical patterns of fungi that produce mycotoxins can change the chemical makeup of VOCs, and you can see a correlation between VOCs and the concentration of mycotoxins in food (Ottoboni et al., 2018). The electronic nose technology has been successfully used to detect OTA in dried cured meats (Lippolis et al., 2016), AFs and FUM in corn (Ottoboni et al., 2018), and DON in wheat bran and durum wheat (Lippolis et al., 2018). But the electronic nose still struggles to detect multiple mycotoxins in complex environments, as most mycotoxins are non-volatile organic compounds, which poses a challenge for electronic nose detection (Alshannaq and Yu, 2017), We need more research to improve sensor materials and data processing algorithms to better serve different industries (Cheli et al., 2023).

3.3.2 Fluorescent polarization

Fluorescence polarization (FP) technology is a method for detecting mycotoxins based on the competition between the fluorophore-labeled analyte and the tracer for specific antibody binding sites (Janik et al., 2021). The amount of tracer bound is inversely proportional to the amount of free analyte in the sample, which means the polarization value is inversely proportional to the analyte concentration (Valenzano et al., 2014). This technology is advantageous because it's highly sensitive and easy to use. The fluorescence polarization method, especially in rapid screening and quantitative analysis, can quickly and effectively detect mycotoxins. When combined with immune responses, fluorescence polarization technology demonstrates strong application potential, particularly in the detection of specific mycotoxins such as Afs (Pandey et al., 2025).

In FP technology, there's no need for time-consuming pre-analysis steps like multiple washings or separating free substances from antibody-bound analytes (Huang et al., 2020a). Research shows that FP immunoassays are used to detect various mycotoxins in food, such as ZEN in corn (Zhang et al., 2017a), DON in wheat products (Lippolis et al., 2006), AFB in corn (Zhang et al., 2018), and OTA in rice (Huang et al., 2020a). However, FP methods have some limitations in sensitivity and accuracy compared to HPLC methods. This might be because antibodies can cross-react with other fungal metabolites and components in food (Alshannaq and Yu, 2017).

3.3.3 Aggregation-induced emission

The aggregation-Induced emission (AIE) technology is an emerging fluorescence detection technique that has been applied to the detection of mycotoxins in recent years. This technology enhances detection sensitivity by using probes to excite strong fluorescence signals in the aggregated state (Zhu et al., 2019, Li et al., 2017). AIE dyes that exhibit high fluorescence emission characteristics in the aggregated state include 9,10-diphenylvinyl anthracene (DSA), silicon-containing heterocyclic diene, tetraphenylethylene (TPE), and their derivatives (Wang and Liu, 2018). Studies have found that AIE materials show great selectivity and sensitivity when it comes to detecting mycotoxins, and researchers have successfully developed AIE dye-based sensors to detect OTA in wine and coffee (Zhu et al., 2019), as well as AFB1 in peanut oil and doubanjiang (a type of fermented broad bean paste) (Xuhan et al., 2018). This technology effectively detects low levels of mycotoxins and is stable and reproducible, offering a new solution for monitoring food safety (Wang et al., 2024).

3.3.4 Molecularly Imprinted Polymers (MIPs)

Molecularly Imprinted Polymers (MIPs) are a new type of material designed to mimic natural recognition molecules such as antibodies and biological receptors, possessing specificity similar to antibody-antigen interactions (Vasapollo et al., 2011). They boost the sensitivity of mycotoxin detection by selectively recognizing certain molecules. During the molecular imprinting process, cross-linked polymers are formed by co-polymerizing functional monomers and cross-linking agents. This occurs in the presence of the analyte, like mycotoxins, which acts as a template (Krska et al., 2005). The main advantages of MIPs lie in their high selectivity and affinity for the target molecules used in the imprinting process, their ability to withstand high temperatures and pressures, and their inertness to bases, acids, metal ions, and organic solvents. Additionally, they have low synthesis costs and long storage times, keeping their recognition ability for several years at room temperature (Vasapollo et al., 2011, Michael and Anja, 2016). MIPs effectively capture and bind to target mycotoxins, improving detection accuracy. In recent years, the use of MIPs for detecting mycotoxins has been on the rise. Studies have shown that MIPs have been used to analyze AFB in wheat (Guo et al., 2019), OTA in beer and wine (Pacheco et al., 2015), and ZEN in grains (Lucci et al., 2010, Huang et al., 2020b), showing significant development potential in mycotoxin detection (Alshannaq and Yu, 2017). By combining molecularly imprinted polymers with other sensor technologies, researchers have been able to achieve rapid and sensitive detection of multiple mycotoxins, offering fresh ideas for real-world applications (Wang et al., 2024).

3.3.5 Application of nanotechnology in detection

The application of nanotechnology in the detection of mycotoxins holds great promise. The application of nanotechnology in feed mycotoxin detection mainly shows up in the development of novel nano-biosensors. Sensors based on nanomaterials greatly boost detection performance thanks to their unique physical and chemical properties (Rai et al., 2015). Biosensors have built-in advantages such as portability, simplicity, and real-time monitoring, making them suitable for on-site analysis (Liu et al., 2020a). Nanomaterial-based sensors can specifically identify key toxins such as AFB1 and ZEN by modifying surfaces and amplifying signals. For example, gold nanoparticles (Au NPs) can produce color changes through the localized surface plasmon resonance (LSPR) effect after binding with AFB1. When combined with smartphone colorimetric analysis, detection can be completed within 8 min with a sensitivity of 0.05 ng/mL, which is a 50-fold improvement over traditional ELISA methods (Rai et al., 2015). By using the properties of nanomaterials, researchers have developed various detection methods based on nano-sensors, which not only improve detection sensitivity but also allow for quick detection of mycotoxins. Nanotechnology can make detection methods more efficient by boosting signals and increasing surface area. Recent studies have shown that nanomaterials exhibit good potential in the recognition and quantitative analysis of mycotoxins, opening up new possibilities for the development of future detection technologies (Xue et al., 2025). The emergence of portable detection devices has made detecting mycotoxins more flexible and efficient. These devices typically use emerging sensing technologies, such as biosensors and nano-sensors, that can quickly and accurately detect mycotoxins on-site. The development of portable

Table 4 Comparison of performance characteristics of mycotoxin detection methods in animal feed and feedstuffs

Detection method	Sensitivity (LOD)	Specificity	Application scenarios	Time consumption
TLC	1–10 μg/kg	Moderate	Preliminary screening, grassroots laboratories	1–2 h
HPLC	0.1–1 μg/kg	High	Routine detection, laboratory settings	2–4 h
ELISA	0.01–0.1 μg/kg	High (prone to cross-reactivity)	Batch screening, rapid detection	0.5–2 h
LFIA	0.1–1 μg/kg	Moderate	On-site rapid testing	10–30 min
Biosensors	0.001–0.01 μg/kg	High	Real-time monitoring, on-site/laboratory	5–30 min
Electronic nose	10–50 μg/kg	Low	Preliminary contamination screening	10–20 min

Note: Table data are summarized from references cited in this chapter.

detection devices gives a big boost to the feed and food industries, helping producers and regulatory agencies to timely identify and control the risks of mycotoxins, thereby ensuring food safety and animal health (Vardali et al., 2023). To provide a comprehensive overview of the performance characteristics of different mycotoxin detection methods, Table 4 summarizes their sensitivity, specificity, applicability, cost, and time consumption, facilitating selection based on specific detection needs.

4 Current status of mycotoxin contamination in animal feed

4.1 Contamination scenarios in different regions

In China, testing of 1569 feed samples from various provinces between 2016 and 2017 revealed that DON had the highest detection rate (exceeding 83%), followed by ZEA (88%) and AFB1 (74.5%). The average concentration range of DON was 450.0–4381.5 µg/kg, while that of ZEA was 2.3–729.2 µg/kg, and AFB1 was 1.3–10.0 µg/kg. These data indicate that DON has become the primary mycotoxin contaminant in Chinese feed (Ma et al., 2018). Similarly, a study conducted in Taiwan, China, from 2015 to 2017 showed that the positive rate of DON reached 91.4%, with the positive rates of ZEA and AFs being 70.2% and 58.0%, respectively. Moreover, 91.3% of the samples were contaminated with multiple mycotoxins (Yang et al., 2019).

In Southeast Asia, a survey conducted in central Thailand from 2015 to 2020 presented distinct contamination characteristics: ZEA had the highest contamination rate, accounting for approximately 50%; the positive rates of AFs and FUM remained between 20% and 30%; and the contamination rates of T-2 toxin and DON were relatively low, both below 10% (Kananub et al., 2021). A study in South Korea from 2009 to 2016 indicated that the contamination level of ZEA in compound feeds and feed raw materials was low, with very few samples exceeding the standard (Chang et al., 2017). Another South Korean study focusing on the contamination of fumonisin B1 and B2 found that the contamination rate of FB1 in cattle feed was as high as 100%, while the contamination rates of FB1 and FB2 in poultry feed were 78% and 40%, respectively (Seo et al., 2013).

In Africa, research in Nigeria has exposed the severe situation of mycotoxin contamination in local poultry feed. Among them, AFB1 and FB1 were particularly severely contaminated, with detection rates reaching 83% and 97%, respectively. Additionally, the coexistence of multiple mycotoxins was extremely common (Akinmusire et al., 2019). A seasonal study on poultry feed in Nigeria further pointed out that the concentrations of AFs and FUM in feed during the dry season were significantly higher than those in the wet season.

Further research identified peanut meal and corn as the main sources of these two mycotoxins (Esan et al., 2024).

4.2 Prevalence of co-contamination

Co-contamination by multiple mycotoxins has become a widespread issue in the global feed industry, significantly complicating risk assessment. In relevant studies in China, over 75% of feed samples were co-contaminated with two or three mycotoxins (Ma et al., 2018); in the testing of corn samples in the southeastern United States, only 18.29% of the samples contained a single mycotoxin, while more than 80% contained two or more mycotoxins (Pokoo-Aikins et al., 2024); in poultry feed in Nigeria, the coexistence rate of AFs and FUM was as high as 80% (Akinmusire et al., 2019), and seasonal surveys there even found that some feed samples contained at least four types of mycotoxins (Esan et al., 2024).

Co-contamination of mycotoxins not only significantly increases the health risks to animals but also may cause toxicological additivity or synergism, thereby adversely affecting the immune function of animals and reducing their production performance (Čolović et al., 2019). Therefore, establishing a comprehensive monitoring system for multiple mycotoxins and implementing scientific risk management measures have become crucial links in ensuring feed safety and animal health.

4.3 Contamination differences among feed types and raw materials

There are significant differences in the levels of mycotoxin contamination among different feed types and raw materials. A study in Thailand pointed out that the contamination level of AFs in compound feeds was significantly higher than that in feed raw materials, among which soybean meal, corn, and rice bran were the main contaminated raw materials (Kananub et al., 2021). Research in Nigeria also reached a similar conclusion, clearly identifying peanut meal and corn as the main contributors to AFs and FUM (Esan et al., 2024, Akinmusire et al., 2019).

Furthermore, studies in the United States have found that the presence of mycotoxins in corn samples is also closely related to changes in nutrient components. For example, corn contaminated with AFs shows a significant reduction in fat content, while corn contaminated with DON experiences a decrease in protein content (Pokoo-Aikins et al., 2024).

In summary, the scientific selection of feed raw materials and the appropriateness of storage conditions are key factors in controlling mycotoxin contamination. In practical production, measures such as the rational proportioning of feed raw materials and the optimization of storage management processes can effectively reduce the risk of mycotoxin contamination.

5 Technology progress of masked mycotoxins

5.1 *Metabolic mechanism and formation of masked mycotoxins*

In recent years, studies have revealed that plants and other organisms can convert mycotoxins into conjugated or modified forms through metabolic processes. These toxins with special structures are referred to as masked mycotoxins, which are difficult to be effectively identified by conventional detection methods due to their structural properties. As a result, the actual exposure level of mycotoxins in the environment is severely underestimated, posing significant challenges to food safety assessment (Berthiller et al., 2013, Cirlini et al., 2012).

Plants, as living organisms with active metabolic capacity, can modify and transform mycotoxins with the help of specific enzyme systems, among which glycosyltransferases (e.g. UDP-glucosyltransferases) play a core role. Such enzymes can catalyze common mycotoxins like DON and ZEN to undergo conjugation reactions with small-molecule compounds such as glucose, forming glycoside conjugates with higher stability. Typical examples include deoxynivalenol-3-glucoside (DON-3-Glc) and zearalenone-14-O-glucoside (ZEN-14-Glc), which are easily missed in conventional extraction and detection processes (Berthiller et al., 2013, Kovalsky Paris et al., 2014).

In addition to glycoside conjugates, masked mycotoxins also include non-extractable conjugates – complexes formed by strong chemical bonds with plant cell wall components (e.g. cellulose, lignin). Due to their strong binding force, traditional sample pretreatment methods cannot effectively separate and extract them, further increasing the difficulty of detection (Cirlini et al., 2012).

Taking the metabolic transformation of ZEN as an example, relevant studies have found that the UDP-glucosyltransferase HvUGT14077 in barley can specifically catalyze ZEN to generate two glucosides with different structures: ZEN-14-O-β-glucoside and the novel ZEN-16-O-β-glucoside. Notably, the content of the novel ZEN-16-O-β-glucoside in barley root tissues is significantly higher than that of ZEN-14-O-β-glucoside (Kovalsky Paris et al., 2014). More critically, after this novel masked toxin enters the human body, it can be rapidly hydrolyzed by intestinal flora, releasing the toxic parent toxin ZEN again. This finding indicates that it may pose a potential threat to human health (Kovalsky Paris et al., 2014, Gratz, 2017).

Similarly, DON-3-Glc, the masked form of DON, has been widely reported in various crops, but there are still controversies regarding its toxicity and metabolic stability, and the relevant mechanisms require further in-depth research (Gratz, 2017, Pierron et al., 2016).

The formation of masked mycotoxins is not an 'exclusive process' of plant metabolism. Fungal metabolism, in vivo transformation in animals, and food processing links can all lead to structural modification of mycotoxins, resulting in more diverse masked or modified toxins (Kovač et al., 2018, Berthiller et al., 2013). In summary, the formation mechanism of masked mycotoxins involves metabolic pathways of multiple organisms (including plants, fungi, and animals) as well as artificial links such as food processing, with glycosylation reactions and specific conjugation reactions as the core transformation processes. This complex formation (Santos et al., 2022) key issue to be addressed urgently in the current field of food safety.

5.2 Progress in detection technology of masked mycotoxins

Masked mycotoxins have strong structural diversity and complex conjugation forms, which place high demands on detection technologies. Traditional chromatographic analysis and immunoassay methods are insufficient in specificity and sensitivity, making it difficult to achieve effective identification and accurate quantification of these compounds (Berthiller et al., 2013, Cirlini et al., 2012). In recent years, with the rapid development of analytical technology, LC-MS/MS has gradually become the mainstream method for masked mycotoxin detection due to its unique advantages.

LC-MS/MS combines efficient separation capabilities with high-sensitivity detection advantages. It can not only realize the simultaneous separation of parent toxins and masked mycotoxins but also clarify the molecular structure of masked mycotoxins through the structural identification function of mass spectrometry, effectively addressing the limitations of traditional methods (Cirlini et al., 2012, Kovač et al., 2018).

In practical detection, the optimization of sample pretreatment technology is the key to improving detection efficiency. Studies have shown that combining IAC or QuEChERS method (a rapid, simple, economical, effective, rugged, and safe sample pretreatment technology) with LC-MS/MS can significantly improve the extraction rate and purification effect of mycotoxins and their masked derivatives in food samples such as rice (Santos et al., 2022). Through this technology combination, researchers have successfully detected masked mycotoxins with AFs and ochratoxin A as the parent toxins in actual grain samples, and the detection results have met the requirements of food safety detection in terms of sensitivity and accuracy (Santos et al., 2022).

In addition, the application of high-end mass spectrometry technologies has further improved the accuracy of masked mycotoxin detection. For example, ion trap mass spectrometry, time-of-flight mass spectrometry (TOF-MS), and high-resolution orbitrap mass spectrometry (Orbitrap-MS) can achieve accurate qualitative and quantitative analysis of low-concentration masked mycotoxins through high-resolution mass analysis, and are particularly suitable for the structural identification of newly discovered masked mycotoxins (Kovač et al., 2018).

6 Future trends in research

With the continuous development of detection technologies for mycotoxins in animal feed and stuffs, future research directions will focus on several key areas to improve the accuracy, efficiency, and environmental impact of detection. Firstly, future research should focus on the integration of multiplex detection technologies to achieve simultaneous detection of multiple mycotoxins. This integration can effectively boost detection throughput and efficiency, cutting down the time spent on sample processing and analysis in traditional methods. Studies have shown that multiplex detection methods based on immunoassay technologies look promising, as these methods can simultaneously detect various mycotoxins and improve the quality of results (Liew and Sabran, 2022). Furthermore, with advancements in analytical technologies, such as liquid chromatography coupled with high-resolution mass spectrometry (LC-HRMS), you can now efficiently screen and analyze multiple mycotoxins in one go, enhancing the accuracy and efficiency of detection (Lapris et al., 2024).

6.1 Integration of multiplex detection technologies

Integrating multiplex detection technologies will be a key focus for future mycotoxin detection. Traditional detection methods often focus on single mycotoxin detection; as food safety requirements increase, the sensitivity and accuracy of detection are becoming more crucial. The application of immunoassay technologies in multiplex detection can significantly improve detection efficiency and reduce costs by simultaneously detecting multiple toxins. Recent studies have shown that sensors based on nanomaterials and multiplex immunodetection platforms are really helping to make this happen (Adunphatcharaphon et al., 2022). For example, through nanostructured array technology, it is possible to simultaneously detect various mycotoxins, and the rapid development of this technology makes it feasible to achieve multiplex screening in a single analysis process, thereby helping us better understand mycotoxins and how they affect animal and human health (Lapris et al., 2024).

6.2 Eco-friendly detection methods

With the increasing emphasis on environmental protection and sustainable development, future methods for detecting mycotoxins also need to shift towards eco-friendly approaches. Traditional detection methods often rely on harmful chemical reagents, which can burden the environment during use. Therefore, it's especially important to create low-cost, quick, and eco-friendly detection technologies. Recently, we've seen a rise in detection technologies that use biosensors and nanomaterials, which not only boost detection sensitivity and specificity but also cut down on chemical reagent use. For example, the use of magnetic molecularly imprinted polymers can effectively and selectively extract mycotoxins from feed and conduct subsequent quantitative analysis, showing great potential for application (Pérez-Álvarez et al., 2024). Additionally, using natural ingredients like plant extracts to remove and detect toxins can ensure detection efficiency while minimizing environmental impact (Gao et al., 2023).

6.3 Data sharing and standardization

In the research and application of mycotoxin detection technologies, data sharing and standardization are key to driving tech advancements and applications. Currently, the standards and regulations for mycotoxin monitoring and management vary across countries, which limits how comparable and reproducible the data is. Therefore, setting up a unified data sharing platform and standard detection methods is really important. By integrating research data from different regions and institutions, we can provide a more reliable basis for assessing and managing mycotoxin risks. At the same time, encouraging countries to cooperate by sharing research and tech can speed up the spread and use of new technologies. For example, the MycoKey project is focused on pushing forward global mycotoxin research by establishing international cooperation and data sharing mechanisms, boosting countries' abilities in feed and food safety (Logrieco et al., 2021). Future research should ramp up efforts in this area to tackle the growing problem of mycotoxin contamination.

7 Conclusion

In the safety management of animal feed and feedstuffs, mycotoxin detection technology plays a crucial role. With the continuous development of the feed industry, the types and concentrations of mycotoxins are also changing, which places higher demands on detection technology. Although traditional detection methods are still widely used, the introduction of modern detection technologies driven by scientific advancements offers new possibilities for improving the accuracy and efficiency of detection. This article reviews the current state of mycotoxin detection technologies in animal feed and discusses future research directions.

Firstly, traditional detection methods such as HPLC and ELISA remain effective in many cases. However, these methods are often time-consuming, complex, and require high standards of laboratory equipment. In recent years, with advancements in biosensing technology, nanotechnology, and molecular biology, new detection technologies, like rapid detection devices and portable analytical instruments, have been gradually developed. These modern technologies not only improve the sensitivity and specificity of detection but also enable rapid on-site testing, reducing sample processing time and greatly enhancing detection efficiency.

Secondly, given the diversity of mycotoxins and the constantly changing environmental factors, simply relying on a single detection method may not meet the needs for comprehensive monitoring. Therefore, future research should focus on the integration and innovation of technologies. For example, combining multiple detection technologies to form a composite detection system can enhance the reliability of detection results while improving sensitivity. Additionally, the application of machine learning and artificial intelligence provides new perspectives for data analysis, enabling faster identification and prediction of potential mycotoxin risks.

However, while promoting technological advancements, we also need to pay attention to the differences and contradictions between various research results. Different detection methods may exhibit significant differences in sensitivity, specificity, and applicability, which means researchers need to be careful when publishing results and discuss their findings thoroughly. This not only helps build a stronger scientific consensus but also gives valuable references for future research.

In summary, the detection technology of mycotoxins in animal feed is developing towards efficiency, sensitivity, and convenience. Future research should focus on the integration and innovation of technologies to address the complex challenges of feed safety. By combining various detection methods and introducing modern data analysis techniques, we hope to ensure the safety of animal feed while promoting the sustainable development of the entire industry. With feed safety issues becoming more severe, collaboration and communication among the scientific community, enterprises, and regulatory agencies are particularly important. Only through joint efforts can we effectively address the risks posed by mycotoxins and ensure the safety of the food chain and public health.

8 Where to look for further information

Authoritative international bodies provide comprehensive and regularly updated information on mycotoxin contamination, analytical methods, and regulatory limits. The US Food and Drug Administration (FDA) publishes regulatory guidance and analysis methods for mycotoxins in animal feed and human food. Information is available at https://www.fda.gov/food/natural-toxins-food/mycotoxins; https://www.fda.gov/media/73568/download; https://www.fda.gov/food/chemical-contaminants-pesticides/natural-toxins-food. The AOAC International (Association of Official Analytical Chemists) maintains validated official methods of analysis for mycotoxin detection, including chromatographic and immunoassay-based techniques. Their methods database and standard method performance requirements (SMPRs) for mycotoxins can be accessed at https://www.aoac.org/. For research on masked mycotoxins and plant–fungus interactions, the MycoKey project has generated extensive datasets and publications on the occurrence, formation, and toxicological significance of modified mycotoxins in cereals, with a particular focus on DON and ZEN conjugates. Project information is available at http://www.mycokey.eu/.

9 References

Abdallah, M. F., et al. 2017. Occurrence of multiple mycotoxins and other fungal metabolites in animal feed and maize samples from Egypt using LC-MS/MS. *J Sci Food Agric*, 97(13), 4419–4428.

Adunphatcharaphon, S., et al. 2022. The evolution of multiplex detection of mycotoxins using immunoassay platform technologies. *J Hazard Mater*, 432, 128706.

Agriopoulou, S., et al. 2020. Advances in analysis and detection of major mycotoxins in foods. *Foods*, 9(4), 518.

Akinmusire, O. O., et al. 2019. Mycotoxins in poultry feed and feed ingredients in Nigeria. *Mycotoxin Res*, 35, 149–155.

Alameri, M. M., et al. 2023. Aflatoxin contamination: an overview on health issues, detection and management strategies. *Toxins (Basel)*, 15(4), 246.

Alshannaq, A. & Yu, J. H. 2017. Occurrence, toxicity, and analysis of major mycotoxins in food. *Int J Environ Res Public Health*, 14(6), 632.

Altintas, Z., et al. 2017. *Applications of quantum dots in biosensors and diagnostics*. John Wiley & Sons, Inc.

Anfossi, L., et al. 2018. A lateral flow immunoassay for straightforward determination of fumonisin mycotoxins based on the quenching of the fluorescence of CdSe/ZnS quantum dots by gold and silver nanoparticles. *Mikrochim Acta*, 185(2), 94.

Anumudu, C. K., et al. 2024. A review of the mycotoxin family of fumonisins, their biosynthesis, metabolism, methods of detection and effects on humans and animals. *Int J Mol Sci*, 26(1), 184.

Berthiller, F., et al. 2013. Masked mycotoxins: a review. *Mol Nutr Food Res*, 57(1), 165–186.

Bessaire, T., et al. 2019. Multiple mycotoxins determination in food by LC-MS/MS: an international collaborative study. *Toxins (Basel)*, 11(11), 658.

Chang, H., et al. 2017. The occurrence of zearalenone in South Korean Feedstuffs between 2009 and 2016. *Toxins (Basel)*, 9(7), 223.

Cheli, F., et al. 2023. E-nose technology for mycotoxin detection in feed: ready for a real context in field application or still an emerging technology? *Toxins (Basel)*, 15(2), 146.

Cirlini, M., et al. 2012. Hyphenated chromatographic techniques for structural characterization and determination of masked mycotoxins. *J Chromatogr A*, 1255, 145–152.

Čolović, R., et al. 2019. Decontamination of mycotoxin-contaminated feedstuffs and compound feed. *Toxins (Basel)*, 11(11), 617.

De Girolamo, A., et al. 2022. Overview of recent liquid chromatography mass spectrometry-based methods for natural toxins detection in food products. *Toxins (Basel)*, 14(5), 328.

Dib, A. A., et al. 2023. A comparative review on methods of detection and quantification of mycotoxins in solid food and feed: a focus on cereals and nuts. *Mycotoxin Res*, 39(4), 319–345.

Doria, G. A., et al. 2012. Noble metal nanoparticles for biosensing applications. *Sensors*, 12(2), 1657–1687.

Esan, O. O., et al. 2024. Seasonal variation of mycotoxin levels in poultry feeds and feed ingredients in Oyo State, Nigeria. *Mycotoxin Res*, 40(2), 319–325.

Feng, S., et al. 2023. Rapid detection of three mycotoxins in animal feed materials using competitive ELISA-based origami microfluidic paper analytical device (μPAD). *Anal Bioanal Chem*, 415(10), 1943–1951.

Gao, Z., et al. 2023. The natural occurrence, toxicity mechanisms and management strategies of Fumonisin B1: a review. *Environ Pollut*, 320, 121065.

Gratz, S. W. 2017. Do plant-bound masked mycotoxins contribute to toxicity? *Toxins (Basel)*, 9(3), 85.

Guo, P., et al. 2019. Rapid detection of aflatoxin B(1) by dummy template molecularly imprinted polymer capped CdTe quantum dots. *Anal Bioanal Chem*, 411(12), 2607–2617.

Hassanien, A. A., et al. 2021. Antifungal and antitoxin effects of propolis and its nanoemulsion formulation against Aspergillus flavus isolated from human sputum and milk powder samples. *Vet World*, 14(9), 2306–2312.

Hedieh, M., et al. 2017. Noble metal nanoparticles in biosensors: recent studies and applications. *Nanotechnology Reviews*, 6(3), 301-329.

Huang, X., et al. 2020a. Development of an Ultrasensitive and Rapid Fluorescence Polarization Immunoassay for Ochratoxin A in Rice. *Toxins*, 12(11), 682.

Huang, Z., et al. 2020b. Synthesis and application of magnetic-surfaced pseudo molecularly imprinted polymers for zearalenone pretreatment in cereal samples. *Food Chem*, 308, 125696.

Imade, F., et al. 2021. Updates on food and feed mycotoxin contamination and safety in Africa with special reference to Nigeria. *Mycology*, 12(4), 245–260.

Janik, E., et al. 2021. The existing methods and novel approaches in mycotoxins' detection. *Molecules*, 26(13), 3981.

Kananub, S., et al. 2021. Mycotoxin profiles of animal feeds in the central part of Thailand: 2015–2020. *Vet World*, 14(3), 739–743.

Kovač, M., et al. 2018. Yesterday masked, today modified; what do mycotoxins bring next? *Arh Hig Rada Toksikol*, 69(3), 196–214.

Kovalsky Paris, M. P., et al. 2014. Zearalenone-16-O-glucoside: a new masked mycotoxin. *J Agric Food Chem*, 62(5), 1181–1189.

Krska, R. & Molinelli, A. 2009. Rapid test strips for analysis of mycotoxins in food and feed. *Anal Bioanal Chem*, 393(1), 67–71.

Krska, R., et al. 2005. Advances in the analysis of mycotoxins and its quality assurance. *Food Additi Contam*, 22(4), 345–353.

Lapris, M., et al. 2024. The potential of multi-screening methods and omics technologies to detect both regulated and emerging mycotoxins in different matrices. *Foods*, 13(11), 1746.

Latham, R. L., et al. 2023. Diverse mycotoxin threats to safe food and feed cereals. *Essays Biochem*, 67(5), 797–809.

Leggieri, M. C., et al. 2021. An electronic nose supported by an artificial neural network for the rapid detection of aflatoxin B1 and fumonisins in maize. *Food Control*, 123, 107722.

Li, J., et al. 2017. Aggregation-induced fluorescence probe for monitoring membrane potential changes in mitochondria. *ACS Applied Materials & Interfaces*, 10(15), 12150-12154.

Li, K., et al. 2025. Recent progress of mycotoxin in various food products-human exposure and health risk assessment. *Foods*, 14(5), 865.

Li, R., et al. 2019. Fluorometric lateral flow immunoassay for simultaneous determination of three mycotoxins (aflatoxin B(1), zearalenone and deoxynivalenol) using quantum dot microbeads. *Mikrochim Acta*, 186(12), 748.

Li, Y., et al. 2020. Visible and eco-friendly immunoassays for the detection of cyclopiazonic acid in maize and rice. *Journal of Food Science*, 85(1), 105–113.

Liew, W. P. & Sabran, M. R. 2022. Recent advances in immunoassay-based mycotoxin analysis and toxicogenomic technologies. *J Food Drug Anal*, 30(4), 549–561.

Lippolis, V., et al. 2006. Optimization of a fluorescence polarization immunoassay for rapid quantification of deoxynivalenol in durum wheat-based products. *J Food Prot*, 69(11), 2712–2719.

Lippolis, V., et al. 2016. Rapid prediction of ochratoxin A-producing strains of Penicillium on dry-cured meat by MOS-based electronic nose. *Int J Food Microbiol*, 218, 71–77.

Lippolis, V., et al. 2018. Rapid prediction of deoxynivalenol contamination in wheat bran by MOS-based electronic nose and characterization of the relevant pattern of volatile compounds. *J Sci Food Agric*, 98(13), 4955-4962.

Liu, Y., et al. 2022. Immunoaffinity cleanup and isotope dilution-based liquid chromatography tandem mass spectrometry for the determination of six major mycotoxins in feed and feedstuff. *Toxins (Basel)*, 14(9), 631.

Liu, D., et al. 2020a. Recent progress on electrochemical biosensing of aflatoxins: a review. *TrAC Trends Anal Chem*, 133, 115966.

Liu, Z., et al. 2020b. A smartphone-based dual detection mode device integrated with two lateral flow immunoassays for multiplex mycotoxins in cereals. *Biosens Bioelectron*, 158, 112178.

Logrieco, A., et al. 2021. Perspectives on global mycotoxin issues and management from the MycoKey Maize working group. *Plant Dis*, 105(3), 525–537.

Lucci, P., et al. 2010. Molecularly imprinted polymer solid-phase extraction for detection of zearalenone in cereal sample extracts. *Analytica Chimica Acta*, 672(1-2), 15–19.

Ma, R., et al. 2018. Individual and combined occurrence of mycotoxins in feed ingredients and complete feeds in China. *Toxins (Basel)*, 10(3), 113.

Maggira, M., et al. 2022. Comparative evaluation of three commercial elisa kits used for the detection of aflatoxins B1, B2, G1, and G2 in feedstuffs and comparison with an HPLC method. *Vet Sci*, 9(3), 104.

Mahmoud, Y. A., et al. 2025. Comprehensive review for aflatoxin detoxification with special attention to cold plasma treatment. *Mycotoxin Res*, 41(2), 277-300.

Majer-Baranyi, K., et al. 2021. Biosensors for deoxynivalenol and zearalenone determination in feed quality control. *Toxins (Basel)*, 13(7), 499.

Malachová, A., et al. 2018. Advanced LC-MS-based methods to study the co-occurrence and metabolization of multiple mycotoxins in cereals and cereal-based food. *Anal Bioanal Chem*, 410(3), 801–825.

Michael, A. & Anja, M. 2016. Mycotoxin analysis using imprinted materials technology: recent developments. *J AOAC Int*, 99(4), 861-864.

Muñoz-Solano, B., et al. 2024. Monitoring mycotoxin exposure in food-producing animals (Cattle, Pig, Poultry, and Sheep). *Toxins (Basel)*, 16(5), 218.

Nasaruddin, N., et al. 2022. Assessment of multi-mycotoxin contamination throughout the supply chain of maize-based poultry feed from selected regions of Malaysia by LC-MS/MS. *Food Addit Contam Part A Chem Anal Control Expo Risk Assess*, 39(4), 777–787.

Nasiri Poroj, S., et al. 2023. The synergistic effect of titanium dioxide nanoparticles and yeast isolated from fermented foods in reduction of aflatoxin B1. *Food Sci Nutr*, 11(11), 7109–7119.

Oliveira, I. S., et al. 2019. Biosensors for early detection of fungi spoilage and toxigenic and mycotoxins in food. *Current Opinion in Food Science*, 29, 64-79.

Ostrý, V. 1999. Micromycetes, mycotoxins and human health. *Cas Lek Cesk*, 138(17), 515–521.

Ottoboni, M., et al. 2018. Combining E-nose and lateral flow immunoassays (LFIAs) for rapid occurrence/co-occurrence aflatoxin and fumonisin detection in Maize. *Toxins*, 10(10), 416.

Pacheco, J. G., et al. 2015. Molecularly imprinted electrochemical sensor for ochratoxin A detection in food samples. *Sens Actuators B Chem*, 215, 107–112.

Pandey, A., et al. 2025. A review of bioanalytical methods, bioaccessibility, and toxicokinetics of emerging mycotoxins enniatins and beauvericin in the context of health risk assessment. *Toxicon*, 256, 108288.

Pascale, M., et al. 2019. Performance evaluation of LC-MS methods for multimycotoxin determination. *J AOAC Int*, 102(2), 1708–1720.

Pérez-Álvarez, M. D. C., et al. 2024. Magnetic molecularly imprinted polymers for selective extraction of aflatoxins from feeds. *Toxins (Basel)*, 16(2), 120.

Perumal, V. & Hashim, U. 2014. Advances in biosensors: principle, architecture and applications. *J Appl Bio*, 12(1), 1–15.

Pierron, A., et al. 2016. Intestinal toxicity of the masked mycotoxin deoxynivalenol-3-β-D-glucoside. *Arch Toxicol*, 90(8), 2037–2046.

Pokoo-Aikins, A., et al. 2024. Mycotoxin contamination and the nutritional content of corn targeted for animal feed. *Poult Sci*, 103(12), 104303.

Rai, M., et al. 2015. Emerging nanotechnology for detection of mycotoxins in food and feed. *Int J Food Sci Nutr*, 66(4), 363–370.

Rhouati, A., et al. 2017. Photoinduced discharge of electrons stored in a TiO2-MWCNT composite to an analyte: application to the fluorometric determination of hydrogen peroxide, glucose and aflatoxin B1. *Mikrochim Acta*, 185(1), 26.

Rodríguez-Carrasco, Y., et al. 2012. Multi-mycotoxin analysis in wheat semolina using an acetonitrile-based extraction procedure and gas chromatography-tandem mass spectrometry. *J Chromatogr A*, 1270, 28–40.

Rodríguez-Carrasco, Y., et al. 2014. A survey of trichothecenes, zearalenone and patulin in milled grain-based products using GC-MS/MS. *Food Chem*, 146, 212–219.

Santos, A. R., et al. 2022. Mycotoxins contamination in rice: analytical methods, occurrence and detoxification strategies. *Toxins (Basel)*, 14(9), 647.

Seo, D. G., et al. 2013. Occurrence of fusarium mycotoxin fumonisin B1 and B2 in animal feeds in Korea. *Mycotoxin Res*, 29(3), 159–167.

Shanakhat, H., et al. 2018. Current methods for mycotoxins analysis and innovative strategies for their reduction in cereals: an overview. *J Sci Food Agric*, 98(11), 4003–4013.

Singh, J. & Mehta, A. 2020. Rapid and sensitive detection of mycotoxins by advanced and emerging analytical methods: a review. *Food Sci Nutr*, 8(5), 2183–2204.

Smith, M. C., et al. 2016. Natural co-occurrence of mycotoxins in foods and feeds and their in vitro combined toxicological effects. *Toxins (Basel)*, 8(4), 94.

Solcan, C., et al. 2013. The hepatoprotective effect of sea buckthorn (Hippophae rhamnoides) berries on induced aflatoxin B1 poisoning in chickens 1. *Poult Sci*, 92(4), 966–974.

Song, S., et al. 2014. Multiplex lateral flow immunoassay for mycotoxin determination. *Anal Chem*, 86(10), 4995–5001.

Valenzano, S., et al. 2014. Determination of deoxynivalenol in wheat bran and whole-wheat flour by fluorescence polarization immunoassay. *Food Analytical Methods*, 7(4), 806–813.

Vardali, S., et al. 2023. Recent advances in mycotoxin determination in fish feed ingredients. *Molecules*, 28(6), 2519.

Vasapollo, G., et al. 2011. Molecularly imprinted polymers: present and future prospective. *Int J Mol Sci*, 12(9), 5908–5945.

Wang, H. & Liu, G. 2018. Advances in luminescent materials with aggregation-induced emission (AIE) properties for biomedical applications. *J Mater Chem B*, 6(24), 4029–4042.

Wang, Y., et al. 2024. Recent progress of molecularly imprinted technique for the detection of mycotoxins in food. *Foods*, 13(24), 4125.

Wang, Y., et al. 2009. Study of superparamagnetic nanoparticles as labels in the quantitative lateral flow immunoassay. *Mat Sci Eng C*, 29(3), 714–718.

Wang, Z., et al. 2021. A rapid and sensitive fluorescent microsphere-based lateral flow immunoassay for determination of aflatoxin B1 in distillers' grains. *Foods*, 10(9), 2109.

Wei, H., et al. 2023. Strategies to control mycotoxins and toxigenic fungi contamination by nano-semiconductor in food and agro-food: a review. *Crit Rev Food Sci Nutr*, 63(33), 12488–12512.

Xing, Y., et al. 2025. AFB1 consolidates HBV harm to induce liver injury and carcinogenic risk by inactivating FTCD-AS1-PXR-MASP1 axis. *Toxicology*, 511, 154057.

Xue, M., et al. 2025. Aspergillus mycotoxins: the major food contaminants. *Adv Sci (Weinh)*, 12(9), 2412757.

Xuhan, X., et al. 2018. Dual-terminal stemmed aptamer beacon for label-free detection of aflatoxin B1 in broad bean paste and peanut oil via aggregation-induced emission. *J Agric Food Chem,* 66(46), 12431-12438.

Yang, C. K., et al. 2019. Prevalence of mycotoxins in feed and feed ingredients between 2015 and 2017 in Taiwan. *Environ Sci Pollut Res Int*, 26(23), 23798–23806.

Zhang, K. & Banerjee, K. 2020. A review: sample preparation and chromatographic technologies for detection of aflatoxins in foods. *Toxins (Basel)*, 12(9), 539.

Zhang, J., et al. 2011. Simultaneous detection of deoxynivalenol and zearalenone by dual-label time-resolved fluorescence immunoassay. *J Sci Food Agric*, 91(2), 193–197.

Zhang, X., et al. 2017a. Fluorescence polarization immunoassay based on a new monoclonal antibody for the detection of the zearalenone class of mycotoxins in Maize. *J Agric Food Chem*, 65(10), 2240–2247.

Zhang, X., et al. 2017b. Multiplex lateral flow immunoassays based on amorphous carbon nanoparticles for detecting three fusarium mycotoxins in Maize. *J Agric Food Chem*, 65(36), 8063–8071.

Zhang, X., et al. 2018. Dual-wavelength fluorescence polarization immunoassay to increase information content per screen: Applications for simultaneous detection of total aflatoxins and family zearalenones in maize. *Food Control*, 87, 100–108.

Zhang, Y. Y., et al. 2023. Comparison of two commercial methods with a UHPLC-MS/MS method for the determination of multiple mycotoxins in cereals. *Food Chem*, 406, 135056.

Zhu, Y., et al. 2019. Label-free fluorescent aptasensing of mycotoxins via aggregation-induced emission dye. *Dyes Pigm*, 170, 107572.

Chapter 10

Improved batching and micro-scaling systems for more precise feed control

Oreofeoluwa Akintan, Michigan State University, USA; Kifle G. Gebremedhin, Cornell University, USA; and Daniel Uyeh, Michigan State University, USA

1 Introduction

Over the past century, the feed industry has undergone long-term restructuring driven by scale, consistency, and demand for cost efficiency. Early feed mills relied on disjointed production equipment with manual control to produce animal rations in bulk, often falling short of meeting nutritional and safety requirements. These standalone units – such as feed mixers used for producing total mixed rations (TMRs) – are still commonly used in small-scale or on-farm feed production, where the limited scale allows for sufficient oversight to maintain precision feed control (Umutoni and Ayantunde, 2024). However, as livestock production intensified and the scale of operations increased, manual and fragmented systems began to reveal critical limitations, particularly their inability to adapt to fluctuations in ingredient quality or to meet the diverse

http://dx.doi.org/10.19103/AS.2026.0172.03

nutritional needs of different animal species (Dadi et al., 2021; Duguma and Janssens, 2021).

Feed formulation has become increasingly complex in the modern era, influenced not only by factors such as variations in nutrient concentrations, particle size of raw materials, price volatility, environmental sustainability mandates, and evolving feed safety regulations. These factors have driven a shift away from manual, single-unit, batch-by-batch production toward partly/fully automated, data-driven systems, including the adoption of advanced batch control algorithms that manage the weighing, mixing, and monitoring of feed ingredients to ensure each batch strictly meets predefined nutritional and process standards (Cabrera et al., 2020; Control and Authority, 2017). Modern feed mills integrate advanced control systems, real-time monitoring, and intelligent decision-making frameworks that ensure improved nutrition, ingredient traceability, and safety compliance throughout manufacturing.

The batch control system enables comprehensive, real-time documentation and oversight of every stage in the feed manufacturing process – from raw ingredient sourcing to final feed production (Pomar et al., 2019). This documentation capability, which also constitutes a data collection process, is crucial for traceability, ensuring regulatory compliance, and enabling rapid identification and correction of safety issues that may arise during and after production. Critical components of these batch control systems include programmable logic controllers (PLCs) – specialized digital computers designed to manage industrial processes by executing control sequences and algorithms; human-machine interfaces (HMIs) – user-friendly control stations that enable operators to monitor and adjust production processes in real-time; feed processing machinery – mechanical equipment used to prepare, mix, and deliver animal feed, such as grinders, mixers, conveyors, and pelletizers, ensuring consistency and efficiency in feed production; and advanced feed formulation software, which automates ingredient handling, maintains strict process tolerances, and dynamically adjusts ingredient proportions during production. Parallel to batch control advancements, developments in micro-scaling technologies have enabled precision in ingredient measurement and nutrient delivery, particularly in smaller or more individualized feeding systems. Micro-scaling involves accurately measuring and controlling minute quantities of feed ingredients, allowing tailored nutritional delivery to individual animals or specific groups based on their unique requirements.

Advanced sensing technologies such as near-infrared spectroscopy (NIRS) – a rapid, non-destructive method for assessing nutrient composition – and hyperspectral imaging (HSI) – which captures detailed spatial and spectral data to analyze feed composition – enable rapid chemical and physical characterization of feed within a batch process (Clark et al., 1995; Jimaré et al., 2008; Uyeh et al., 2021a).

Integrating batch control and micro-scaling technologies within feed production represents a leap forward in the feed industry's capacity to deliver consistent, safe, and highly tailored nutritional solutions. Batch control and micro-scaling technologies collectively enhance animal nutrition practices' effectiveness, efficiency, and sustainability, addressing both economic and environmental pressures modern livestock systems face.

This chapter provides a technical overview of advanced batch control and micro-scaling technologies in animal feed production. It details modern batch control systems, highlighting key computational software and hardware components. Subsequent sections cover advancements in automation, INternet of Things (IoT) integration, artificial intelligence (AI), machine learning (ML), and monitoring systems. Finally, the chapter discusses challenges and future trends, emphasizing continuous process monitoring and intelligent automation in feed manufacturing.

2 Batch process in feed production

A batch refers to a defined quantity of material processed as a single, self-contained production run governed by a specific formulation and operational protocol. In contrast to continuous processes that maintain a steady-state flow, batch processes operate in discrete time-bound cycles, allowing for targeted intervention, traceability, and flexibility necessary for animal feed production (Li, 2013). As described by Paulk et al. (2022), batch processes in feed production enable precision in ingredient handling, formulation, and quality assurance, which is essential to optimizing feed efficiency and reducing cost. The batch process structure is especially suited for incorporating ingredient quality variability and implementing tailored nutritional strategies. Each batch typically proceeds through defined phases – charging of raw materials, transformation via operations like grinding, mixing, and pelleting, and finally, discharge and quality assessment. The structure inherently supports granular monitoring, traceability, and compliance with nutritional and safety standards (Control and Authority, 2017). Moreover, the episodic and dynamic nature of batch systems makes them ideal for integration with sensor-driven control architectures and advanced algorithms such as Model Predictive Control (MPC) and Iterative Learning Control (ILC), which refine performance over successive batches (Cabrera et al., 2020). The accompanying diagram (Fig. 1) illustrates a modern feed batch production workflow, highlighting key stages from ingredient storage to batch quality assessment, underscoring the importance of systematic sequencing in high-precision feed systems.

After being delivered to the plant or factory, the raw feed ingredients are stored in controlled storage tanks to initiate the batch process properly. Inspectors must assess the stored raw material feed to meet safety standards.

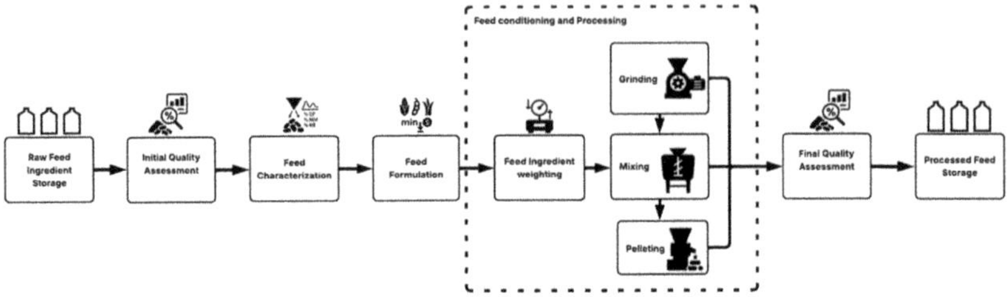

Figure 1 Illustration of the end-to-end feed production workflow, from ingredient storage to final product storage. The process includes initial quality assessment, feed characterization, and formulation, followed by feed ingredient weighing and core processing steps – grinding, mixing, and pelleting – encapsulated within a feed conditioning and processing unit. The final quality assessment ensures nutritional and physical standards before transferring to processed feed storage.

Feed characterization is an essential step that involves rapidly determining the chemical and nutritional composition of livestock feed materials, which can vary in composition; therefore, relying solely on nutrition composition tables is not adaptable to variations in feed composition. Nutritionists or feed formulation software use the results from feed characterization to formulate the feed. They typically apply a least-cost approach to maximize essential nutrients while minimizing cost. Once formulated, the feed mixture ingredients are weighed through automated load cells according to the predetermined mixing ratio to meet batch objectives. The grinding, mixing, and pelleting processes are carried out on the weighed feed material either sequentially or simultaneously (Table 1).

3 Integrated batch control and micro-scaling technologies in livestock feed systems

Figure 2 illustrates the subsystems involved in animal feed production, highlighting the interconnected roles of ingredient storage, sensors, actuators, and control logic layers supporting both batch control and micro-scaling technologies. Batch control systems manage feed preparation in discrete, repeatable units, regulating ingredient inclusion, mixing intensity, sequence timing, and delivery. Unlike traditional manual methods, modern systems employ automation with real-time monitoring and closed-loop feedback, significantly reducing variability while adhering strictly to nutritional and safety standards. Core computational components include PLCs, which utilize advanced control algorithms such as ILC and MPC for adaptive, batch-by-batch decision-making based on real-time sensor feedback and historical data (Li, 2013). After mixing, automated delivery mechanisms dispense feed accurately according to preset specifications, supported by robust traceability and data logging systems that document ingredient sources, sensor data, and operator inputs for quality assurance and compliance.

Table 1 Key process parameters in feed production – functions and impact on feed quality and safety

Parameters	Function	Measurement Impact on Feed Quality and Safety
Weight	The final feed must meet the desired nutritional formulation by adding each ingredient proportionally. Accurate weight control minimizes waste, reduces cost, and prevents under- or over-supplementation.	Accurate weight measurement prevents over- or under-formulation, reducing nutrient imbalances and enhancing animal health by meeting dietary requirements. Excessive or deficient nutrients can lead to metabolic disorders.
Humidity	Moisture is necessary to maintain optimal feed consistency and prevent microbial growth. It also affects shelf life, pelleting quality, and nutrient stability, impacting energy consumption during drying or pelleting.	Controlling humidity ensures the feed is not prone to spoilage or mold growth. Spoiled feed can lead to mycotoxin contamination, which poses health risks to livestock and affects nutritional intake.
Nutritional content	Verify that the feed delivers the intended protein, energy, fiber, and micronutrient levels. This ensures that the food meets animal performance needs and meets nutritional standards.	Assessing nutritional content supports the formulation of balanced rations, ensuring animals receive adequate nutrition and preventing deficiencies or excesses that can impact health and production. NIRS technologies allow for the efficient determination of this content.
Temperature	To prevent degradation of nutrients (like vitamins and enzymes), ensure safe processing (e.g. avoid overheating), and inhibit pathogen development. Also monitored to maintain pellet quality and system safety.	Maintaining appropriate temperature levels prevents spoilage and degradation of nutrients. High temperatures can promote the growth of pathogens in feed, leading to potential food safety risks for animals.
Flow rate	Controls the speed at which raw materials or final feed move through the system.	Maintaining a steady flow rate ensures uniform mixing, prevents clumping, blockages, or overflows, and synchronizes automated processes. Irregular flow can result in under-processed batches, leading to uneven nutrient distribution or increased risk of contamination.

Motor speed	To adjust the intensity and duration of mixing or grinding. Controlled motor speed improves homogenization, reduces energy usage, and ensures mechanical components operate within safe limits.	Adjusting motor speed affects the efficiency and quality of mixing. Inconsistent motor speed can disrupt the uniformity of feed, potentially leading to segregation or incomplete blending of ingredients, which can cause quality issues such as disfiguration of pellets and improved rationing of nutrients, reducing the impact of performing feed formulation.
NIRS	To provide real-time verification of feed quality (e.g. protein, fiber, moisture) without lab testing. It allows dynamic formulation adjustments and supports traceability, especially under variable ingredient quality.	NIRS enables assessments of feed nutritional quality, thus facilitating timely adjustments in formulations that enhance feed safety and compliance with health standards (Balehegn et al., 2021; Boote et al., 2021; Mendoza et al., 2023; Zahera et al., 2022). The technique reduces reliance on chemical analyses, lowering the risk of errors associated with such procedures.
Water activity (Aw)	Measures the availability of free water for microbial growth, not just total moisture.	Critical for predicting mold and bacterial growth potential. Low Aw (<0.6) inhibits microbial activity, improving shelf life and reducing spoilage risks.
Mycotoxin levels	Detects the presence of toxic fungal metabolites such as aflatoxins or fumonisins.	Directly correlates with feed safety. High mycotoxin levels can cause acute or chronic toxicity in animals; monitoring helps reject contaminated batches early.
pH	Indicates feed acidity or alkalinity, influencing microbial growth and nutrient availability.	Extreme pH levels can encourage spoilage organisms' growth or inhibit animal nutrient absorption. Maintaining an optimal pH helps preserve feed quality and supports healthy gut function.

Micro-scaling technologies complement batch control by enabling individualized nutrient delivery at spatial (individual animal), temporal (feeding cycle), and nutritional (ingredient micro-dosing) resolutions. Advanced sensing methods like NIRS and HSI provide real-time, on-site nutrient profiling (Clark et al., 1995; Jimaré et al., 2008; Rego et al., 2020). These data inform automated feeding systems utilizing RFID tags for animal identification and load cells to monitor feed consumption, exemplified by systems such as Grow Safe that optimize feed rations based on detailed consumption data. Micro-formulation algorithms recalibrate feed recipes daily or per feeding event, dynamically adjusting for ingredient quality and animal nutritional requirements (Sperger et al., 2020). Additional interventions include co-extrusion and microencapsulation technologies, delivering controlled-release nutrients that are particularly beneficial for vulnerable animals (Rojas and Stein, 2016; Tepe

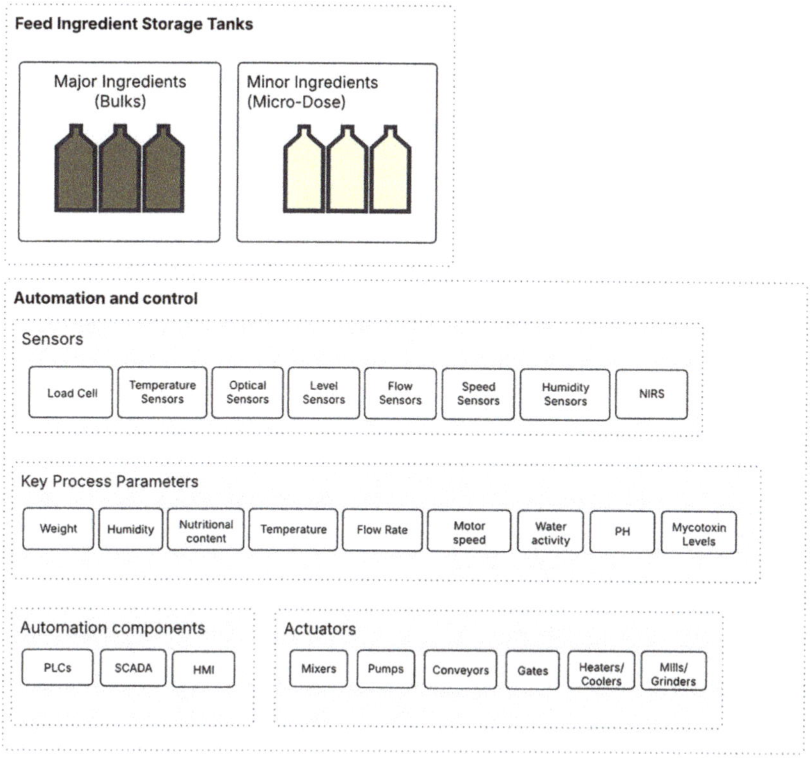

Figure 2 A modern automated feed production line's system architecture illustrates the separation of bulk and micro-dose ingredient storage tanks alongside integrated sensor networks, controlled variables, automation components (PLCs, supervisory control and data acquisition (SCADA), HMI), and actuators such as mixers, pumps, and grinders. The layout highlights the interplay between ingredient handling, real-time monitoring, and intelligent control for consistent and traceable feed formulation.

and Altaş, 2024). Emerging mobile evaluation units integrate NIR sensors, cloud analytics, and mixing capabilities to enable real-time TMR adjustments, facilitating localized, adaptive feed delivery.

At the foundational level, ingredient storage and supply subsystems – comprising primary tanks for bulk ingredients (e.g. corn, soybean meal) and minor tanks for micronutrients (vitamins, minerals) – automatically dose ingredients based on predefined feed formulations, minimizing manual variability (Sperger et al., 2020). The sensor and measurement subsystem ensures real-time batch consistency, employing load cells, temperature and humidity sensors, and optical technologies for nutrient analysis. Actuator subsystems perform critical physical tasks: mixers homogenize ingredients, conveyors transport materials, valves regulate material flow, milling units enhance digestibility, and pelleting machines produce uniform feed pellet

Table 2 Potentials for AI and ML applications across feed production stages

Feed Production Process	AI/ML Potentials
Raw feed ingredient storage	Predictive inventory monitoring using sensor data + ML forecasting.
Initial quality assessment	Computer vision for visual defect detection and spoilage identification.
Feed characterization	Spectral data analysis with ML regression models for nutrient profiling.
Feed formulation	Supervised learning models (random forests, GBM, artificial neural network (ANN)) for predictive formulation optimization.
Feed ingredient weighing	Reinforcement learning to adjust dosing based on previous batch success rates.
Grinding	Adaptive feedback loops based on energy usage or particle size prediction using ML.
Mixing	Dynamic MPC and reinforcement learning to adjust speed and time for uniformity (Kordabad et al., 2023).
Pelleting	Real-time temperature and pressure prediction using regression models to optimize pellet quality (You et al., 2022a).
Final quality assessment	Computer vision + convolutional neural networks (CNNs) to detect color/texture anomalies indicative of spoilage; automated classification.
Processed feed storage	Anomaly detection algorithms for environmental control (e.g. humidity, mold risk); the determinations of optimized storage condition.

sizes suitable for digestion. The integrated control and feedback subsystem, utilizing PLCs and HMIs and increasingly enhanced by AI, ML, and SCADA systems, dynamically adjusts operations in response to measured parameters like weight, moisture, and spectral data, driving continuous optimization (Cabrera et al., 2020; Dadi et al., 2021).

Together, batch and micro-scale technologies represent a comprehensive framework that bridges automation, intelligent sensing, and individualized nutritional strategies, significantly advancing efficiency, animal health, and sustainability in modern livestock feed systems (Table 2).

4 Technological advances: automation and Internet of Things integration

Automation and IoT integration have redefined the operational structure of batch control and micro-scaling technologies in livestock feed production by creating interconnected systems capable of executing consistent and traceable feed production processes. These technologies allow mechanical and digital components within a feed production facility to function in coordinated, closed-loop systems with minimal manual input. Modern batch control systems incorporate PLCs, HMIs, and real-time SCADA platforms. Engineers configure these components to regulate the sequence of operations, such as ingredient dosing, mixing, and discharge, with built-in fault detection and self-correction protocols. For example, the PLC executes a predetermined recipe by controlling feeders, motors, and actuators. At the same time, the HMI provides operators with a live interface to monitor process variables and override commands if needed.

Integrating IoT-enabled sensors has allowed feed systems to collect high-frequency data from different process points. These sensors measure parameters not limited to temperature, moisture content, weight, flow rate, and vibration, transmitting data wirelessly to centralized control units or cloud servers. In ingredient silos, for example, weight sensors track the exact quantity of raw materials dispensed, while moisture sensors positioned at conveyor belts or mixing chambers ensure that the feed maintains its targeted hydration levels. The data from these sensors support feedback mechanisms that allow dynamic adjustments mid-process. If a deviation is detected – for instance, an ingredient's moisture content falls outside acceptable thresholds – the control system can halt operations or reroute the batch for correction. Automation platforms are increasingly cloud-connected, using IoT protocols such as MQTT to integrate with enterprise resource planning systems, databases, and mobile dashboards (Mendes et al., 2018).

This networked environment enables remote supervision, predictive maintenance alerts, and cross-plant coordination. For example, cloud-based dashboards visualize trends in ingredient consumption, alert operators to anomalies, and generate automated compliance reports. These interfaces also synchronize with actuators in the physical plant to initiate commands or recalibrate hardware components based on the system's analysis. Micro-scaling applications benefit from IoT using smart metering units and RFID-based identification systems. In individual feeding stations, RFID tags detect animal identity and trigger actuators to dispense customized feed volumes. These stations have IoT modules that record individual consumption behavior and synchronize with cloud databases, creating a detailed feed history for each animal (Pomar et al., 2019). The system then uses real-time feedback from such stations to recalibrate feed plans for the next cycle, ensuring tighter control over nutrient delivery.

Additionally, automated calibration routines are now embedded within IoT-enabled systems to maintain precision without constant manual tuning. For example, if a load cell shows drift over time, the system can prompt automatic recalibration based on historical reference weights or initiate a maintenance alert. This self-checking mechanism preserves system accuracy and reduces downtime. The synchronized interaction between automated control logic and distributed IoT sensing networks results in continuous data exchange across all nodes of the feed production system. These interactions support real-time decision-making, operational traceability, and closed-loop optimization without external intervention.

The IoT-enabled closed-loop control system illustrated in Fig. 3 provides an automated and efficient feed batch control solution. Integrating advanced IoT technologies, the system comprises sensors, actuators, a gateway controller,

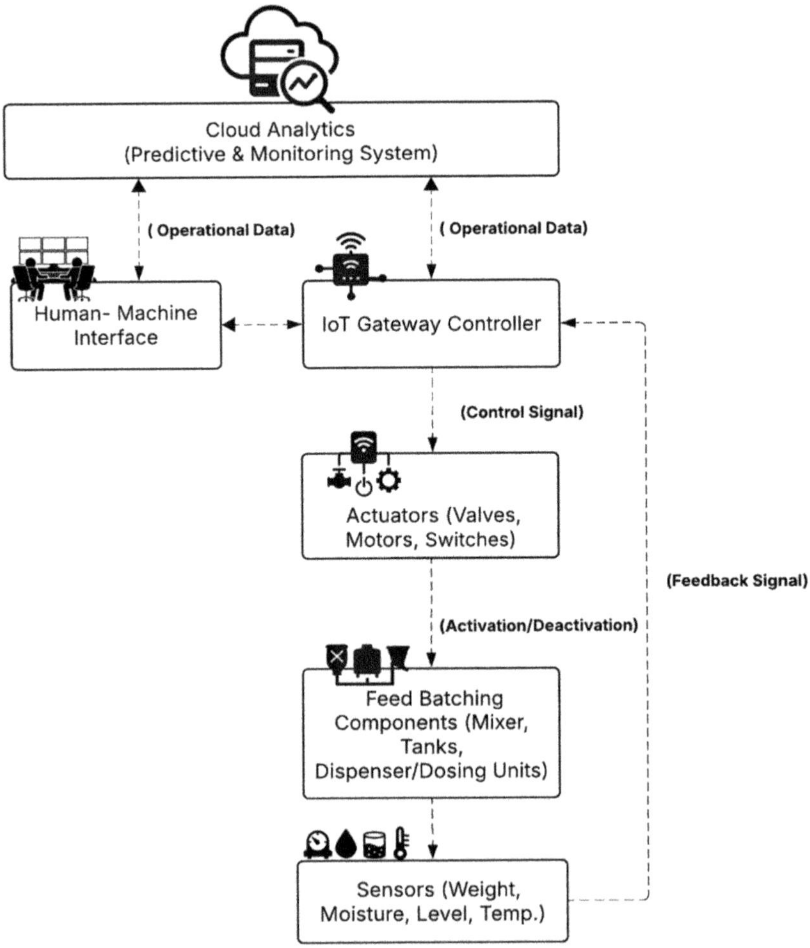

Figure 3 Diagram illustrating a feed batching control system powered by IoT and cloud analytics. Sensor data on weight, moisture, level, and temperature are collected from the feed-batching components and sent as measurements of the measurand signals. These are processed by actuators under the control of an IoT gateway, which receives operational feedback from both the sensors and the HMI. Control signals are issued to manage mixers, tanks, and dosing units. At the same time, cloud-based analytics systems monitor performance, generate predictions, and feed insights into the system.

cloud analytics, and an interactive user interface to streamline livestock feed processing and management. Initially, IoT sensors – comprising weight, moisture, level, and temperature sensors – continuously monitor real-time conditions within the feed batching system. These sensors collect critical data points such as feed quantity, moisture content, tank levels, and temperature, directly influencing feed quality and animal nutrition (Mota et al., 2023). The IoT gateway controller receives the sensor data, processes it at the edge, and securely forwards it to the cloud analytics platform (Gupta, 2022).

The cloud system analyzes the data and sends control signals back to the IoT gateway controller, which then actuates or deactivates system components accordingly. Specifically, IoT actuators, including motors, solenoid valves, and electronic gates (deactivate), are remotely activated or deactivated to maintain optimal batching conditions. Finally, the user interface provides a comprehensive overview of the process, displaying real-time analytics and enabling manual intervention if necessary. The closed-loop nature of the system ensures self-regulation through continuous feedback from sensors and cloud analytics, enhancing operational robustness and reliability (Senoo et al., 2024). The proposed IoT-enabled closed-loop system demonstrates significant advancements in automation capabilities within livestock management, aligning with contemporary precision.

5 Technological advances: Artificial Intelligence and machine learning-based opportunities

The advent of AI and ML models has advanced the prediction capabilities of batch control and micro-scaling technologies for feed production, allowing systems to move beyond rule-based automation (i.e. set thresholds) and function with adaptive intelligence. Unlike traditional control systems that rely on fixed logic or threshold-based decision rules, AI-driven systems utilize data from past and real-time operations to detect patterns and optimize decisions at both batch and micro scales. One of the potential applications of ML in batch control is predictive formulation. Here, supervised learning models trained on historical batch data, including sensor readings, ingredient quality profiles, ambient environmental conditions, and animal performance outcomes, can forecast the optimal mix of ingredients for upcoming batches. Researchers can theoretically use algorithms such as random forests, gradient boosting machines, and ANNs to model the nonlinear relationships between raw material variability and feed efficacy (Cabrera et al., 2020). These models can account for fluctuating nutrient concentrations in base ingredients (e.g. silage, soybean meal) and adjust inclusion rates to meet specific nutritional targets.

You et al. (2022a) demonstrated that ML regression models, particularly Support Vector Regression (SVR), can accurately predict pellet durability index (PDI) in commercial feed mills using a combination of feed formulation, environmental, and process data, achieving a mean absolute error of just 3.28% and explaining over 99% of the variance as random error. Complementing this, Ittiphalin et al. (2017) applied a hybrid AI approach combining local linear maps (LLM) and backpropagation neural networks (BPNN) to estimate the optimal ratio of mixer-added versus post-pelleted fat, significantly improving production efficiency and pellet quality by reducing reprocessing rates in a Thai feed mill. These studies illustrate how AI models can move feed production beyond static formulation rules to adaptive, data-driven systems that optimize process control and product quality. Pathumnakul et al. 2009) applied ANNs to predict the pelleting rate and dust levels using process parameters, nutrient profiles, and raw material inclusion, training three models. Model 3 achieved the best accuracy for production rate (MAPE = 9.34%), while Model 2 was most accurate for dust level prediction (MAPE = 15.51%). The approach enabled feed mills to assess formulation feasibility and production cost without test runs, enhancing efficiency and decision-making in feed production.

Engineers also deploy AI in adaptive control strategies, such as reinforcement learning and dynamic MPC, where the system learns to modify mixing durations, ingredient sequencing, and delivery timing based on real-time feedback. These methods allow systems to make micro-adjustments during a batch cycle to correct deviations and enhance uniformity. Reinforcement learning agents, for instance, can learn optimal mixing strategies by interacting with the feed processing environment and receiving feedback based on batch consistency or energy efficiency. In micro-scaling environments, such sensor networks can fine-tune nutrient delivery. Researchers may use k-means clustering to segment animals into nutritional groups while applying support vector machines (SVMs) or deep learning models to predict individual requirements. Computer Vision – another branch of AI – is increasingly utilized in feed quality assessment (You et al., 2022b).

CNNs can detect anomalies such as foreign particles, color deviations indicating spoilage, or improper mixing consistency by analyzing images captured from camera systems installed in feed mills or along the processing line. These visual inspection systems operate at high throughput and feed directly into control units, triggering alerts or corrective actions without human intervention (Chandrasekar et al., 2024). AI-powered data management systems further enhance traceability and compliance by automatically tagging and organizing batch records, flagging inconsistencies, and recommending corrective actions. Natural language processing (NLP) tools can even assist with generating regulatory reports or summarizing historical batch performance trends, reducing the administrative burden in large-scale operations. Collectively, the integration of AI and ML into batch control and micro-scaling systems potentially enables feed manufacturers to operate with a level of precision that was previously unattainable using conventional automation alone.

6 Technological advances: advanced monitoring and quality control systems

Quality control in feed production involves systematically coordinating ingredient handling, process monitoring, and final product assessment to ensure consistency and safety across batches. It begins with carefully sourcing and storing raw materials, followed by controlled handling to prevent contamination or degradation. During processing, quality control measures include visual inspection, physical testing (e.g. texture, color, and particle size), and chemical analysis to verify that the feed aligns with desired nutritional specifications. Operators conduct checks at various stages – such as during grinding, mixing, and pelleting – to detect irregularities like clumping, poor mixing uniformity, or excess fines. Feed mills routinely sample and analyze finished feed to confirm it meets formulation requirements before packaging and distribution. Quality control teams maintain detailed records to ensure full traceability of inputs, process conditions, and final outputs. By maintaining these standards, feed producers can ensure that livestock receive nutritionally adequate diets free from contaminants and consistent from batch to batch.

Advanced monitoring systems in feed production utilize optical sensing technologies such as NIRS and HSI to characterize the chemical composition of feedstuffs. These systems measure the reflectance or absorbance of light at specific wavelengths, corresponding to molecular vibrations of key constituents like moisture, protein, fiber, and fat. NIRS instruments, typically installed along conveyor lines or at intake points, scan feed ingredients as they move through the system (Fig. 4). The system compares spectral data against pre-calibrated models to quantify nutrient levels with high throughput and minimal sample preparation (Clark et al., 1995; Jimaré et al., 2008). HSI systems extend this capability by capturing spatial and spectral data, enabling pixel-wise analysis of feed surfaces. Such detailed imaging allows the detection of local inconsistencies or foreign materials across batches. These systems use cameras sensitive to hundreds of narrow spectral bands, typically ranging from the visible to short-wave infrared regions. Integrated into production lines, they continuously scan feed particles, generating spectral signatures analyzed through multivariate techniques or ML algorithms for classification and quantification (Gowen et al., 2012). Figure 4 illustrates using an NIRS spectrometer to assess the nutritional composition of feeds. Figure 5 illustrates an example configuration of an HSI system used to acquire spectral images from TMR feed samples (Uyeh et al., 2021b).

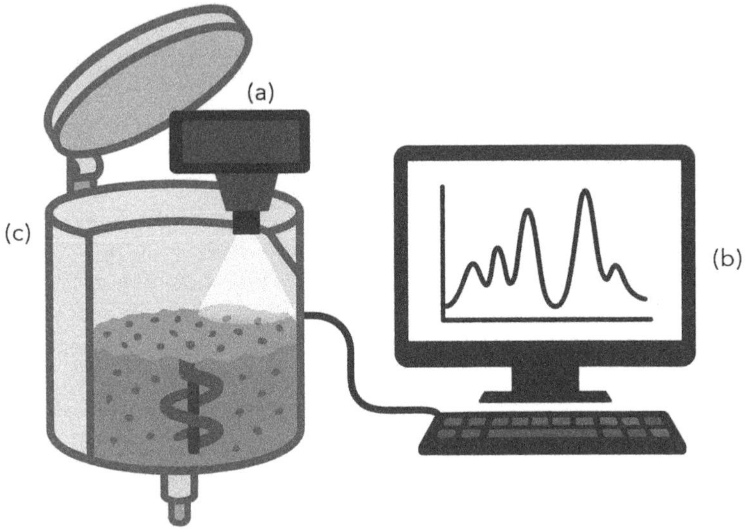

Figure 4 Illustration of an NIRS setup used to analyze the nutritional composition of feed mixtures in real time. The spectrometer captures spectral data from within the mixing chamber and transmits it to a connected computer for processing and prediction. (a) A spectrometer, (b) a computer workstation, (c) a storage tank.

NIRS and hyperspectral systems often integrate with software platforms that trigger alerts or adjustments when deviations occur. For instance, if the moisture content in ground corn exceeds a defined threshold, the system can halt processing or divert the batch. These monitoring solutions document batches meticulously, recording analytical outcomes alongside timestamped sensor metadata. Regular calibration routines using reference samples ensure consistent accuracy despite changing environmental conditions and feed matrix variations. Chemometric modeling techniques, such as Partial Least Squares Regression (PLSR) and Principal Component Analysis (PCA), further support these optical monitoring tools by helping interpret complex spectral data and reveal meaningful patterns. Figure 5 illustrates the use of HSI to characterize the distribution of crude protein in by-product samples. Consequently, monitoring transcends basic quality assurance, providing dynamic profiling of ingredient variability throughout the mixing, grinding, and pelleting stages.

7 Challenges and potential solutions

7.1 Challenges

Feed mills face several practical and financial constraints that hinder the integration of automated batch processing systems. Among the most common limitations are:

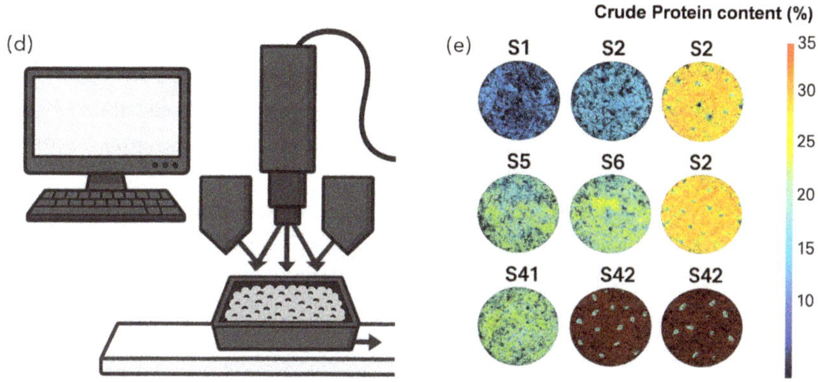

Figure 5 An example is the configuration of an HSI system designed to capture spatial and spectral information from TMR feed samples on a conveyor. The system enables detailed analysis of feed uniformity, composition, and potential contaminants through spectral imaging. (d) An HSI system and (e) an illustration of the spatial distribution of crude proteins in feed samples S1–S4.

1. High initial and maintenance costs: Advanced sensing technologies require a substantial upfront and maintenance cost for equipment, calibration, system integration, and personnel training.
2. Limited storage infrastructure: The number and volume of ingredient storage bins restrict the ability to procure and utilize local grains or by-products. For a new ingredient to be economically viable, the mill must be able to store and include it at levels that justify the costs of procurement, logistics, and nutrient matrix development. Adequate storage of raw materials requires controlled environmental conditions to prevent spoilage and nutrient degradation
3. Uncertain ROI: Mills find it challenging to justify the initial expenditure, especially in cost-sensitive markets, without clear visibility into potential savings and operational efficiencies.
4. Carbon Emissions and Sustainability Trade-Offs: While advanced technologies can improve feed efficiency and safety and reduce feed waste, their initial energy demands and manufacturing footprints may contribute to higher short-term CO_2 emissions.

7.2 Potential solutions

A structured cost-benefit analysis should be conducted before implementing automated batching systems to overcome these barriers. This analysis should:

1. Assess capital and operational expenditures in relation to anticipated gains such as reduced nutrient variability, minimized feed loss,

enhanced batching accuracy, and decreased dependency on external testing services.

2. Estimate the economic impact of improved ingredient utilization, particularly for underutilized or locally sourced feedstocks and by-products often excluded due to inconsistent quality, limited storage options, or rigid formulation constraints.
3. Emphasize the broader benefits of real-time process characterization – including, but not limited to, NIRS, such as improved responsiveness to ingredient variability, enhanced formulation agility, and more uniform feed output, all of which contribute to increased productivity and reduced environmental impact.

By quantifying the return on investment, feed producers can make informed decisions that account for financial realities and operational bottlenecks, paving the way for broader adoption of precision technologies in the feed industry.

8 Future trends in research

Livestock feed manufacturing has steadily moved toward tighter process control through batch management and micro-scaling systems. These setups now handle ingredient dosing, careful mixing, and continuous monitoring with minimal manual involvement. However, the next phase signals a shift – not in tools alone, but in how knowledge from raw materials is folded directly into the act of making feed. Technologies like NIRS and HSI, once used after production or in isolated labs, are on course to be embedded within the very mechanics of feed assembly (Uyeh et al., 2021a). This would allow nutritional readings to shape decisions as ingredients flow, letting machines correct proportions and timing on the fly based on actual values, not assumptions. Learning models that improve through trial allow feed mills to move beyond simply following instructions. They begin to develop operational instincts. These systems detect subtle shifts in feedstock properties, such as moisture or fiber variation, and adjust their internal parameters to preserve nutritional quality.

In the future, distributed sensors and data points scattered throughout the facility – and even across the delivery chain – will create a fabric of input signals. This fine-grained data will alert the system, adjusting to the grain's quality or an unusual humidity level before it becomes problematic. With local processing units (edge devices) and compact sensor clusters, these capabilities may take root directly on farms, allowing for quick tailoring of feed blends based on what is growing in nearby fields and the current needs of the animals. Ledger-based tracking systems record every nutritional reading and feed adjustment throughout the journey, from ingredient intake to delivery at the barn. These systems create a transparent trail that captures what was used, not just what was intended. As a result, modern feed production systems listen, respond, and adapt in near-real time – not because someone instructed them to, but because they are equipped to observe and act. In this environment, characterization serves as a measurement tool and a subtle force guiding the production process from within. Looking ahead, reinforcement learning offers a powerful pathway to strengthen batch control systems further. Unlike traditional models, reinforcement learning allows the system to learn optimal responses through continuous interaction with real-world data. The control system refines its decision-making over time by receiving feedback from its actions, rewarding successful adjustments, and penalizing poor outcomes. When applied to feed production, this approach could enable machines to learn how to respond to unpredictable ingredient quality, shifting environmental conditions, and new dietary targets without explicit reprogramming. As more data become available through sensors and real-time characterization tools, reinforcement learning agents can develop nuanced control strategies that are self-correcting, anticipatory, and tailored to the evolving needs of the livestock production environment.

9 Conclusion

The following conclusions can be drawn from this study:

1. **Precision through batch control and micro-scaling.** Batch control and micro-scaling systems have improved feed consistency, safety, and nutrient accuracy. Combining automation with responsive sensing and targeted delivery enables formulation strategies that align with the specific needs of individual animals or groups – ultimately supporting operational efficiency, animal health, and sustainable feeding practices.
2. **Beyond automation: intelligent decision-making.** Integrating data-driven tools, including adaptive models and algorithmic control, allows feed production to move beyond rigid, rule-based automation. Systems now learn from input variation, anticipate maintenance issues, and refine feed mixes on the fly – ushering in a new tier of precision nutrition and system autonomy.
3. **Real-time nutrient visibility with optical sensing.** Technologies such as NIRS and HSI provide real-time visibility into the chemical makeup of feed ingredients. These sensing methods enable immediate feedback and adjustments during processing, minimizing nutrient variability and ensuring formulation accuracy at production.
4. **Practical constraints and data-driven solutions.** Despite these technological advancements, challenges persist, particularly regarding the high upfront costs of rapid characterization systems and

infrastructure limitations, such as restricted ingredient storage capacity. To navigate these constraints, conducting detailed cost-benefit analyses is crucial. These evaluations can reveal the potential long-term savings and performance gains that justify initial investments, especially when systems are optimized to handle local grains or underutilized by-products efficiently.

10 Where to look for further information

1. Regulatory Guidelines: EFSA, FDA, GMP+ Feed Safety Assurance Scheme.
2. Technological Innovations: Research from leading agricultural engineering institutions.
3. Industry Applications: Case studies from commercial feed manufacturers.

11 References

Balehegn, M., et al., 2021. Near-infrared reflectance spectroscopy for forage nutritive value analysis in Sub-Saharan African Countries. *Agron J* 114, 100–114. https://doi.org/10.1002/agj2.20801.

Boote, K.J., et al., 2021. Fodder development in sub-saharan africa: an introduction. *Agron J* 114, 1–7. https://doi.org/10.1002/agj2.20924.

Cabrera, V.E., et al., 2020. Symposium review: real-time continuous decision making using big data on dairy farms. *J Dairy Sci* 103, 3856–3866. https://doi.org/10.3168/jds.2019-17145.

Chandrasekar, A., et al., 2024. Image-based modeling and control for batch processes. *J Process Control* 143, 103314. https://doi.org/https://doi.org/10.1016/j.jprocont.2024.103314.

Clark, D.H., et al., 1995. Near infrared reflectance spectroscopy estimation of 13 C discrimination in forages. *J Range Manag* 48, 132. https://doi.org/10.2307/4002799.

Dadi, V., et al., 2021. Agri-food 4.0 and innovations: revamping the supply chain operations. *Prod Eng Arch* 27, 75–89. https://doi.org/10.30657/pea.2021.27.10.

Duguma, B. and Janssens, G., 2021. Assessment of livestock feed resources and coping strategies with dry season feed scarcity in mixed crop–livestock farming systems around the gilgel gibe catchment, Southwest Ethiopia. *Sustainability* 13, 10713. https://doi.org/10.3390/su131910713.

European Centre for Disease Prevention and Control (ECDC), European Food Safety Authority (EFSA), & European Medicines Agency (EMA)., 2017. ECDC/EFSA/EMA second joint report on the integrated analysis of the consumption of antimicrobial agents and occurrence of antimicrobial resistance in bacteria from humans and food-producing animals. *Efsa J* 15. https://doi.org/10.2903/j.efsa.2017.4872.

Gowen, A., et al., 2012. Terahertz time domain spectroscopy and imaging: emerging techniques for food process monitoring and quality control. *Trends Food Sci Technol* 25, 40–46. https://doi.org/10.1016/j.tifs.2011.12.006.

Gupta, S.K., 2022. Application of predictive analytics in agriculture. *Technoarete Trans Intell Data Min Knowl Discov* 2. https://doi.org/10.36647/ttidmkd/02.04.a001.

Ittiphalin, M., et al., 2017. An artificial intelligence model to estimate the fat addition ratio for the mixing process in the animal feed industry. *J Intell Manuf* 28, 219–228. https://doi.org/10.1007/s10845-014-0972-x.

Jimaré, M.T., et al., 2008. Process analytical chemistry: applications of near infrared spectrometry in environmental and food analysis: an overview. *Appl Spectrosc Rev* 43, 452–484. https://doi.org/10.1080/05704920802031382.

Kordabad, A.B., et al., 2023. Reinforcement learning for MPC: fundamentals and current challenges. *IFAC-PapersOnLine* 56(2), 5773–5780. https://doi.org/10.1016/j.ifacol.2023.10.548.

Li, G., 2013. Batch-to-batch iterative learning control for end-point qualities based on kernel principal component regression model. *J Comput* (Taipei) 8. https://doi.org/10.4304/jcp.8.12.3184-3190.

Mendes, C.R., et al., 2018. Application of big data and the internet of things in industry 4.0. *Eur J Eng Technol Res* 3, 20–24. https://doi.org/10.24018/ejers.2018.3.11.967.

Mendoza, P.T.D., et al., 2023. NIR spectral imaging for animal feed quality and safety. *Appl Eng Agric* 39, 553–564. https://doi.org/10.13031/aea.15051.

Mota, G. de A., et al., 2023. Smart Sensors and Internet of Things (IoT) for sustainable environmental and agricultural management. *Revista Caderno Pedagógico* 20, 2692–2714. https://doi.org/10.54033/cadpedv20n7-014.

Pathumnakul, S., et al., 2009. A neural network approach to the selection of feed mix in the feed industry. *Comput Electron Agric* 68, 18–24. https://doi.org/10.1016/j.compag.2009.04.001.

Paulk, C.B., et al., 2022. Feed processing technology and quality of feed. In: L.I. Chiba (ed.), *Sustainable Swine Nutrition*. Wiley, pp. 429–444. https://doi.org/https://doi.org/10.1002/9781119583998.ch16.

Pomar, J., et al., 2019. Precision livestock feeding, principle and practice. In: 397–418. https://doi.org/10.3920/978-90-8686-884-1_18.

Rego, G.M., et al., 2020. A Portable IoT NIR spectroscopic system to analyze the quality of dairy farm forage. *Comput Electron Agric* 175, 105578. https://doi.org/10.1016/j.compag.2020.105578.

Rojas, O.J. and Stein, H.H., 2016. Use of feed technology to improve the nutritional value of feed ingredients. *Anim Prod Sci* 56, 1312. https://doi.org/10.1071/an15354.

Senoo, E.E.K., et al., 2024. IoT solutions with artificial intelligence technologies for precision agriculture: definitions, applications, challenges, and opportunities. *Electronics* (Basel) 13, 1894. https://doi.org/10.3390/electronics13101894.

Sperger, J., et al., 2020. The future of precision health is data-driven decision support. *Stat Anal Data Min Asa Data Sci J* 13, 537–543. https://doi.org/10.1002/sam.11475.

Tepe, A. and Altaş, T., 2024. Technological processes applied to laboratory animal feeds and new feeding approaches. *Düzce Tıp Fakültesi Dergisi* 26, 24–29. https://doi.org/10.18678/dtfd.1504013.

Umutoni, C. and Ayantunde, A.A., 2024. Awareness and use of improved livestock feed technologies among agro-pastoral households in the West African Sahel. *Trop Anim Health Prod* 56(6), 194. https://doi.org/10.21203/rs.3.rs-4017463/v1.

Uyeh, D.D., et al., 2021a. Animal feed formulation: rapid and non-destructive measurement of components from waste by-products. *Anim Feed Sci Technol* 274, 114848. https://doi.org/https://doi.org/10.1016/j.anifeedsci.2021.114848.

Uyeh, D.D., et al., 2021b. Rapid and non-destructive monitoring of moisture content in livestock feed using a global hyperspectral model. *Animals* 11, 1299.

You, J., et al., 2022a. Using machine learning regression models to predict the pellet quality of pelleted feeds. *Anim Feed Sci Technol* 293, 115443. https://doi.org/https://doi.org/10.1016/j.anifeedsci.2022.115443.

You, J., et al., 2022b. Using machine learning regression models to predict the pellet quality of pelleted feeds. *Anim Feed Sci Technol* 293. https://doi.org/10.1016/j.anifeedsci.2022.115443.

Zahera, R., et al., 2022. The Use of Near-Infrared Reflectance Spectroscopy (NIRS) to predict dairy fibre feeds in vitro digestibility. *IOP Conf Ser Earth Environ Sci* 951, 012100. https://doi.org/10.1088/1755-1315/951/1/012100.

Index

www.ingramcontent.com/pod-product-compliance
Ingram Content Group UK Ltd.
Pitfield, Milton Keynes, MK11 3LW, UK
UKHW021826100926
8556IPUK00003B/183

9 781835 453117